Classic Geology in Europe 15

The Southern Pennines

CLASSIC GEOLOGY IN EUROPE

1 *Italian Volcanoes* Chris Kilburn & Bill McGuire
ISBN: 9781903544044 (2001)
2 *Auvergne* Peter Cattermole
ISBN: 9781903544051 (2001) *Out of Print*
3 *Iceland* Thor Thordarson & Ármann Höskuldsson
ISBN: 9781780460925 (Third edition 2022)
4. *Canary Islands*
ISBN: 9781903544075 (2002) *Out of Print*
5. *The north of Ireland* Paul Lyle
ISBN: 9781903544082 (2003) *Out of Print*
6. *Leinster* Chris Stillman & George Sevastopulo
ISBN: 9781903544136 (2005) *Out of Print*
7. *Cyprus* Stephen Edwards *et al.*
ISBN: 9781903544150 (2005)
9. *The Northwest Highlands of Scotland* Con Gillen
ISBN: 9781780460406 (2019)
10. *The Hebrides* Con Gillen
ISBN: 9781780460413 (forthcoming 2023)
11. *The Gulf of Corinth* Mike Leeder *et al.*
ISBN: 9781903544235 (2009) *Out of Print*
12. *Almeria* Adrian Harvey & Anne Mather
ISBN: 9781780460376 (2015)
13. *The Southern Pennines* John Collinson & Brian Roy Rosen
ISBN: 9781780461007 (2024)

Classic Geology in Europe 13

The Southern Pennines

John Collinson; D.Phil., FGS
formerly Keele University and University of Bergen

Brian Roy Rosen; PhD, FIBiol, FGS, FLS
Scientific Associate, Natural History Museum, London

LIVERPOOL UNIVERSITY PRESS

First published 2024 by
Liverpool University Press
4 Cambridge Street
Liverpool
L69 7ZU

British Library Cataloguing-in-Publication data
A British Library CIP record is available

ISBN 978-1-78046-100-7 (paperback)
ISBN 978-1-78046-676-7 (epub)
ISBN 978-1-78046-678-1 (PDF)

Typeset by Carnegie Book Production, Lancaster
Printed in the Czech Republic via Akcent Media Limited

Note on Illustrations

1.5
4.3
4.4–4.5
5.1
5.3
10.3
11.5–11.7
11.9
11.11
11.17–11.22
11.25–11.29
11.31–11.35
11.37–11.38
11.40–11.44
11.46–11.55

Contents

Dedicated to the memory of
Bill Groves and Harold Reading

Preface

The Southern Pennines is an area of high country stretching from Nidderdale in the north to Ashbourne in the south. Its great scenic diversity ranges from the limestone areas of the Yorkshire Dales and the White Peak to the high moorlands of the Dark Peak. On this higher ground, dissected by deep valleys, human activity has created a patchwork of different landscapes, a diversity determined, in no small measure, by the underlying geology. It is an area in which important geological ideas have been developed going back to one of the earliest known geological accounts and maps, making it an area of Classic Geology.

The geology involves a fascinating story of the development and infilling, over tens of millions of years, of a large sedimentary basin complex located close to the equator some 330 million years ago. It shows an evolving geography from tropical seas with highly biodiverse shallow-water carbonate platforms and enigmatic reef-like structures (mud mounds) to a deep basin that was progressively infilled by huge deltas. The elimination of deep water led to a vast plain of rivers and swamps in which the Coal Measures were laid down. The northern limit of the area coincides with the rapid thinning of the Millstone Grit north of the Craven Faults and the southern limit is roughly the limit of Carboniferous outcrop.

Some existing field guides to parts of the area are rather dated both in geological thinking and regarding access to localities. This Guide is an up-to-date account of current ideas and the status and significance of key localities. Some quarries mentioned in earlier descriptions are now infilled or otherwise inaccessible, and we hope that the Guide will help stimulate much-needed geoconservation measures.

Unlike other guides in this series, this one describes an area where human activity has had major impacts on the environment, and we highlight examples where geology has been an underlying factor. This is part of a wider consideration about the way geology has influenced an area's geography, history and natural history, and follows the precedent of numerous older publications. Furthermore, current developments in the discipline of geology are moving away from its traditional focus on exploration and exploitation of resources like fossil fuels, which have led to adverse anthropogenic changes. Accordingly, we include remarks on the area's industrial past, notably mining, textile mills, transport and large-scale quarrying and the roles of agriculture and tourism. The mass trespass of Kinder Scout in 1932 led to the opening up of the moorlands to leisure pursuits. The southern sector of the Pennine Way crosses the area, and both Derbyshire and the Yorkshire Dales are classic areas for cave exploration. We recognize the need to make our often-complex subject accessible to a wider readership and we suggest different ways of learning about the geology through 'geo-biking' and armchair fieldwork using Google Earth and Google Street View and indicate localities suitable for those with limited mobility.

The localities described are grouped in six chapters. The tectonic, stratigraphic and sedimentological basis for environmental reconstruction is explained in introductory chapters.

We have been friends and colleagues since we met as undergraduates, and we subsequently immersed ourselves in the geology (and ecology) of numerous places around the globe. Therefore, we bring to this Guide different experiences. JC's early research was on the Kinderscout Grit of North Derbyshire, and he continues to develop ideas about the Millstone Grit, particularly through research students. BRR's background has largely been outside the area apart from teaching. His geology degree sparked an interest in modern carbonate environments, leading him into marine ecology and research on living coral reefs and climate change, and to palaeocological and biogeographical studies of numerous ancient reefs, coral facies and faunas. He brings these perspectives to the limestones of the Derbyshire Platform, emphasizing sedimentary processes and palaeoenvironments.

Acknowledgements

The book has benefited greatly from discussions, often in the field, with Ole Martinsen, Colin Jones, Peter Gutteridge, Ian Chisholm, Hugh Torrens, Duncan McLean, Vanessa Banks, Alessandro Carniti, Peter Jones, Patrick Cossey and the late Trevor Ford. Gilbert Kelling, Colin Jones and Ole Martinsen read some early text. Alexandra O'Rorke and Michelle O'Grady, of the now defunct British Geological Survey office and shop at the Natural History Museum (NHM), London, made BGS resources available and Hellen Pethers and colleagues at NHM Library and Archives helped continually with library resources.

Very special big thanks go to BRR's wife, Jill Darrell (Curator of Fossil Corals at NHM) who provided constant field assistance and discussion throughout the five years preparing this Guide, much of it during her own spare time. Her insights contributed to many ideas in the Guide, while she also helped with observations of her own. In the NHM, she used her official time to help with specimens, access to NHM collections in her charge, and interaction with librarians, an especially important task during two years of Covid-related restrictions when BRR was not allowed to enter the workplace.

The following generously made available material used in the preparation of illustrations, which have been extensively re-drafted: Colin Jones; Peter Gutteridge, Ole Martinsen, Ian Kane, John Baines, Steve Okolo, Charlie Bristow, Colin Waters and Ian Chisholm. The British Geological Survey and the Yorkshire Geological Society gave permission to use material from their publications. Nigel Mountney and Roman Soltan kindly provided photographs. Figures 1.1; 8.3; 10.1 & 12.11 are based on Google Earth images. All other illustrations are the work of the authors.

Individual sources are noted in figure captions and full bibliographic details are in the Bibliography (pp. 339–351).

For help in clarifying access to localities we thank Emma Armstrong (Waddington Fell Quarry), Chris Harris (Mouselow Quarry), Evelyn Southwell (Darby Delph) and Richard Bean (Once-a-Week Quarry).

Enormous thanks to Anthony Kinahan, of Dunedin Academic Press for his encouragement, advice and patience, to Anne Morton for her careful and patient copy editing and to Patrick Brereton of Liverpool University Press and Carnegie Book Production for seeing the book through the press.

This Guide is, to some extent, a spin-off from a succession of 'geo-picnics' that we enjoyed in the company of Jill Darrell and of our great friend, the late Bill Groves (1942–2021), an outstanding and enthusiastic teacher of geology, especially in the field.

This book is dedicated to Bill's memory and to that of Harold Reading (1924–2019) who taught us as an inspiring lecturer and tutor, and who fostered our enthusiasm for ancient environments and fieldwork.

Chapter 1

Introduction

The Pennines of northern England form a spine of high ground stretching from the Scottish border to the Midlands lowlands. This book is about the southern part of this distinctive area which, for many years, has offered a welcome escape from the towns and cities of the flanking industrial areas. Much of the higher ground is extensive peat moorland given over to sheep, shooting, walking, rock climbing and cycling. The uplands are dissected by deep valleys, the dales, formed by major rivers discharging to the east in Yorkshire, to the south in Derbyshire and to the west and south along the western flanks. In limestone-dominated areas, agriculture prevails, whilst where the hills are formed of sandstones and shales, industry is more important, much of it historically related to various textile industries and their need for water power and abundant soft water. Stone of various types is the main building material, contrasting with the predominance of brick in the towns and cities of the flanking coalfield lowlands. In particular, the area is dominated by its drystone walls, which give a distinct character to the landscape.

Except for the Aire Gap, which allows an easy low-level route, the Pennines have historically posed challenges to east–west travel. Whilst the Leeds–Liverpool Canal exploits the Aire Gap, the other two canals that cross the hills are major engineering feats. The Rochdale Canal climbs to over 180 m between Todmorden and Littleborough, whilst the Huddersfield Narrow Canal crosses the hills through the Standedge Tunnel, nearly 5 km in length. South of Airedale, three railway routes cross the Pennines and all depend on major tunnels. Prior to the building of the M62, high-level road routes often presented hazardous conditions in the winter, with the Snake Pass, Woodhead and most roads into Buxton frequently closed. Milder winters have made these closures less common, and the construction of the M62 in the late

1960s greatly improved trans-Pennine travel. Its construction created extensive cuttings through the Millstone Grit, and these added significantly to geological understanding.

Much of the scenery is striking, and often spectacular, and has attracted tourists from the seventeenth century onwards, some to take the waters at Buxton, at Matlock or, in the case of Charles Darwin, at Ilkley. Others visited to indulge a romantic appreciation of the picturesque, and the area has attracted and inspired many artists and writers. Thomas Hobbes and Charles Cotton eulogized the 'The Wonders of the Peake', features that proved decidedly underwhelming to Daniel Defoe. J.M.W. Turner was a frequent visitor to Wharfedale, and James Ward's painting of Gordale Scar is powerfully atmospheric, as are paintings of Dovedale by Wright of Derby. The Brontë sisters drew inspiration from the bleak country around their home in Haworth and Ted Hughes from the equally striking landscape of Calderdale, whilst the work of the current Poet Laureate, Simon Armitage, is deeply influenced by the high moorlands. The television series *Last of the Summer Wine* and *Happy Valley* introduced the rugged Pennine landscape to a wide audience. Today, tourists visit the area to enjoy the varied scenery, towns and villages and to pursue a wide variety of activities and interests such as rock climbing, caving, industrial history, heritage railways and architecture. The Pennine Way, Britain's first long-distance footpath, traverses the northern part of the area covered by this Guide.

This diverse landscape falls partly within two National Parks (Peak District and Yorkshire Dales) and two Areas of Natural Landscape (Nidderdale and Forest of Bowland). In addition, this Guide covers other examples of striking and important upland and industrial landscapes, each of which is underpinned by its geology. This book aims to explain these larger-scale relationships and to illustrate and explain the geology at a more detailed level. In particular, it emphasizes the environments in which the sedimentary rocks that make up the area were deposited and how these environments relate to the evolving palaeogeography over a time interval of some 45 million years.

The Carboniferous rocks that make up the area comprise the limestones of the Derbyshire White Peak and the Craven District of Yorkshire, and the Millstone Grit of the Dark Peak, the Lancaster Fells and the Pennine Moors. Each area shows its own distinctive landscape, influenced by both the rock types and by the ways in which they have been altered, deformed and eroded over millions of years. After deposition, these sedimentary rocks were subjected to large-scale folding and faulting, and this deformation, along with subsequent uplift and erosion largely determines the present-day distribution of the various stratigraphic units at outcrop, and the close relationship between the geology and the present-day topography (Fig. 1.1). This Guide is based on our own research, on supervising research students and running courses and field trips, and on the extensive geological literature that the area has stimulated. The book is aimed primarily at those who lead or take part in geological field trips and courses from universities, schools, industry, and geological societies, but we hope that it will also interest those with a general curiosity about the world around them, perhaps visiting as walkers, cyclists, climbers, cavers or tourists. An extensive glossary should help the less experienced reader to better follow the Guide. In addition, we hope that students of industrial and agricultural history and of natural and cultural landscapes will find a basis for understanding the wider context of their studies.

1.1 Boundaries of area

The area covered by this Guide is determined mainly by the underlying geology, in particular the area of outcrop of the Carboniferous limestones and the overlying Millstone Grit. Some of the lowest parts of the overlying Coal Measures locally fall within the area but, for the most part, the base of the Coal Measures is the effective boundary. As a result, the area has a somewhat irregular shape that extends from Nidderdale and the limestones of the Craven area in the north, to the Staffordshire Moorlands and the White Peak in the south. In the west, it includes parts of the Lancaster Fells, the Blackburn and Rossendale areas, and the hills to the east of Manchester. In the east it is bounded by the Yorkshire and Derbyshire coalfields. Within this overall area, there are significant outcrops of Coal Measure strata, in synclines such as the Burnley Coalfield.

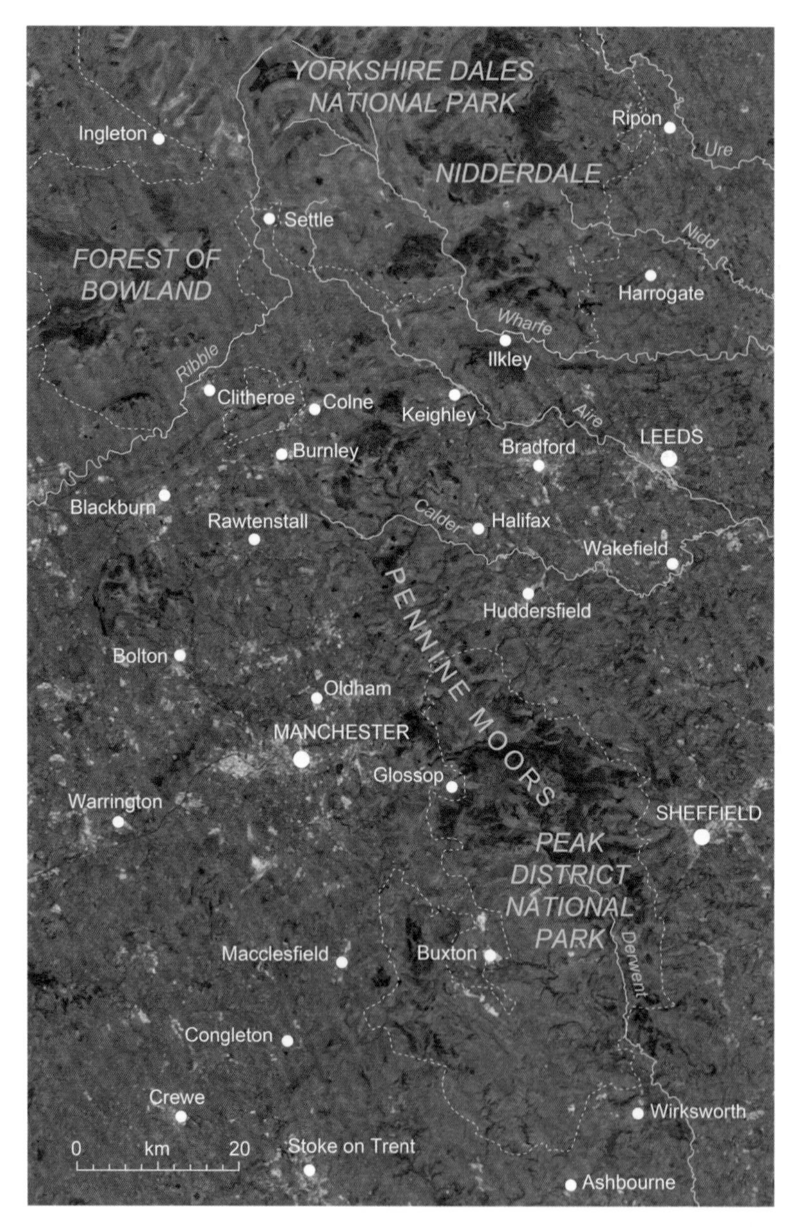

Figure 1.1 Satellite image of the Southern Pennines. The main Pennine hills are picked out by brown moorland. Larger cities are in lower-lying areas at the fringes of the coalfields. Google Earth image.

1.2 Stratigraphic scope

The focus of this Guide is the lower part of the Carboniferous succession, up to the base of the Westphalian Coal Measures (Fig. 1.2). This interval comprises two distinct elements. The older, limestone-dominated succession (Tournaisian-Viséan or Dinantian) includes the shallow-water limestones of the Askrigg Block and of the Derbyshire Massif and the deeper-water limestones and mudstones of the Bowland Basin. The younger (Namurian) succession, dominated by sandstones and finer-grained clastic sediments, constitutes the 'Millstone Grit'.

On the Askrigg Block, to the north of the North Craven Fault, Viséan (late Dinantian) shallow-water limestones lie unconformably on Lower Palaeozoic basement, which crops out in inliers between Ingleton and Malham, with classic examples of angular unconformity exposed at Ingleton, Crummackdale and Horton-in-Ribblesdale.

In Derbyshire, the Carboniferous limestones at outcrop are all Viséan in age and comprise a thick succession of shallow-water sediments deposited on a platform that was surrounded by deeper water. Microbial carbonate mud mounds, sometimes referred to as 'reefs', occur along the margin of the platform and in adjacent off-platform areas. Sub-Carboniferous basement rocks are known only from boreholes. In basin areas, deep-water limestones, some of earlier, Tournaisian, age occur locally at outcrop.

The Namurian succession within the main basin areas includes deep-water mudstones, turbidite sandstones and overlying deltaic and fluvial successions that make up the 'Millstone Grit', an early quarryman's term for the coarse-grained, pebbly sandstones that characterize many rugged 'edges' and tors. On the southern part of the Askrigg Block there is a thinner, shallow-water Millstone Grit succession that is separated from the shallow-water Viséan limestones by an interval of 'Yoredale' facies.

The Westphalian Coal Measures that conformably overlie the Millstone Grit do not feature strongly in this Guide. The Coal Measures outcrop extensively in flanking areas where they have been extensively mined both at outcrop and in deep mines, an industry that powered the Industrial Revolution, but which has now largely died out. The coalfield areas generally have lower-lying and less rugged topography than areas

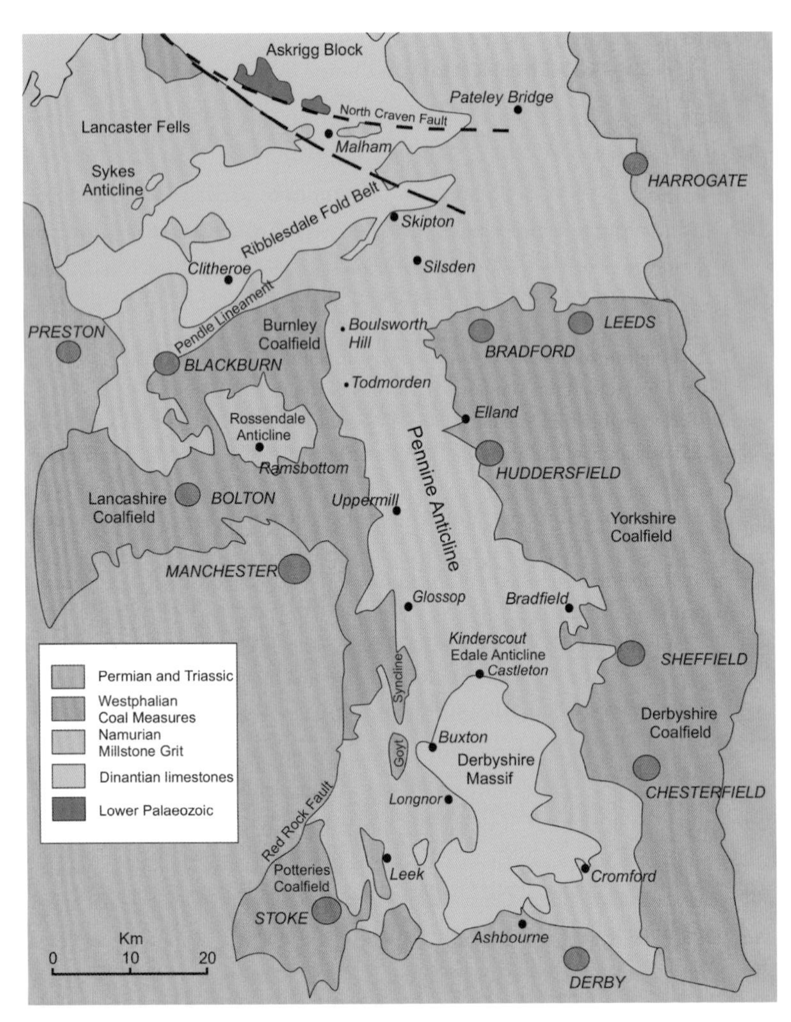

Figure 1.2 Simplified geological map, showing the outcrops of the main subdivisions of the Carboniferous, the unconformably underlying Lower Palaeozoic and the unconformably overlying Permo-Triassic. The main tectonic structures are mostly the result of Variscan compression and these largely determine outcrop patterns (much simplified from BGS 1:624 000 mapping).

of Millstone Grit outcrop as the Coal Measure succession has fewer extensive sandstones to form features in the landscape.

Younger sediments, deposited after the major phase of folding, faulting and large-scale erosion (Variscan Orogeny) are mainly confined to the Triassic sediments of the Cheshire Plain. These fall outside the scope of this book, although small outliers of Triassic sandstones in Staffordshire, within palaeovalleys eroded on the Variscan Unconformity, occur within the general area. The youngest sediments are very local deposits of Miocene age that occur in small solution pits in the limestones of Derbyshire.

Chapter 3 provides a fuller discussion of the stratigraphy.

1.3 History of research and wider significance

Earlier geologists

Coinciding with the lead mining industry, the early nineteenth century saw the Pennine area play important roles in the development of fundamental geological thinking. In Derbyshire, some of the earliest contributions came from the works of **John Whitehurst** (1778) who documented the strata of Derbyshire as an appendix to an early attempt at understanding the larger geological picture. The term 'Millstone Grit' first appeared in print in a predominantly mineralogical description by **John Mawe** (1802). Another early contributor was **White Watson**, a stone merchant and supplier of mineral and fossil specimens. He and Mawe both illustrated the stratigraphy and geological structure of the county with inlaid stone panels, made from samples of the actual rock units being illustrated. These showed structural profiles in different parts of the county. Watson wrote a supporting pamphlet in 1788, and a more substantial publication in 1811. However, Watson's and Mawe's panels were largely derived from profiles produced by the extraordinary polymath **John Farey** (Fig. 1.3). Farey was, amongst other things, a geologist and a mining and agricultural surveyor who applied the stratigraphic principles of his mentor, William Smith, to produce highly sophisticated geological maps and profiles of the Ashover area for his patron, Sir Joseph Banks. These were sadly not published for nearly 200 years. Farey also published in 1811, a wide-ranging report on the geology and agricultural potential of Derbyshire, a comprehensive report that included a thorough account of the Carboniferous succession.

Early geological investigations in Yorkshire were concentrated on the Jurassic and Cretaceous successions exposed at the coast, and the first systematic account of Carboniferous rocks of the Pennines was the second volume of **John Phillips's** *Geology of Yorkshire*, published

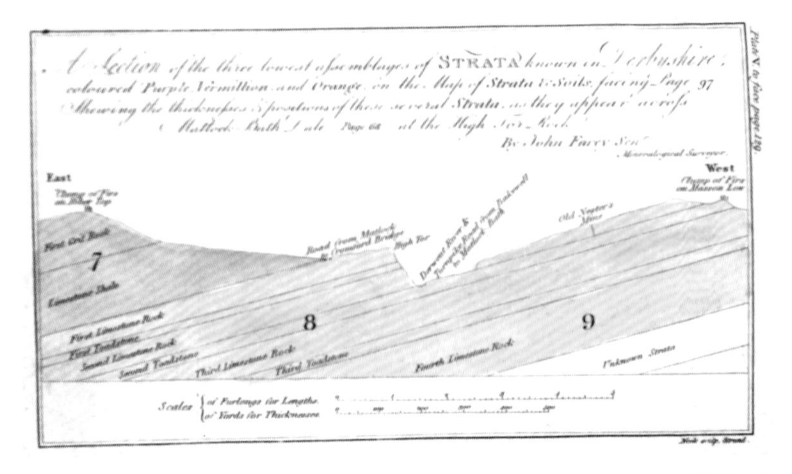

Figure 1.3 An east–west cross-section of the inferred geology in the region of Matlock Bath, Derbyshire. The section shows both the rock types and the thicknesses of the major units. This is probably one of the earliest published geological cross-sections (Farey, 1815).

in 1836. This focused on the northern part of our area, concentrating on the Dinantian limestone and Yoredale succession of the Askrigg Block. Phillips clearly recognized the cyclic nature of the Yoredale succession on the Alston and Askrigg blocks, an important insight that anticipated later sedimentological thinking. However, attempts to extend the 'Yoredale' interval further south towards Derbyshire led to considerable confusion, which was not fully resolved until the early twentieth century, when the pioneering work on goniatite biostratigraphy by **R.G.S. Hudson** and **W.S. Bisat** established a robust stratigraphic framework. At about the same time, **P.F. Kendall** and **H.E. Root** published generalized interpretations of the Millstone Grit and the Coal Measures in terms of their environments of deposition, drawing on publications in the *Proceedings of the Yorkshire Geological Society,* which remains a major source of information to the present day, as does the *Mercian Geologist,* a publication of the East Midlands Geological Society.

Later developments

Studies in the Pennine area have contributed to more widely significant developments in stratigraphy and sedimentology. The Dinantian limestone areas of both Yorkshire and Derbyshire have provided material for better understanding of the development of carbonate platforms and microbial mud mounds as well as of the effects of sea-level changes in generating intraformational karst, whilst studies of the Namurian have led to major advances in both stratigraphy and sedimentology.

The stratigraphic framework of the Dinantian was mainly worked out in areas distant from the Pennines, for example the Bristol area, the Mendips and southern Belgium. The application of this framework to the Pennine area, led by **W.H.C. Ramsbottom** and **N.J. Riley**, has permitted a closer understanding of sea-level changes and facies distribution. Detailed sedimentological and palaeoecological studies have led to greater understanding of both basins and carbonate platforms and, in particular, the development and distribution of microbial mud mounds. The term 'mud mound' is preferred in this Guide, but these features have been variously termed 'reefs', 'reef knolls', 'Waulsortian reefs' and 'microbial build-ups'. Earliest investigations by **R.H. Tiddeman** on the Cracoe 'reef' belt were followed by a succession of studies by **W.W. Black**, **D.J.C. Mundy** and, most recently, by **C.N. Waters** and co-workers. The so-called 'Waulsortian reefs' of the Bowland Basin have been described by **A. Lees** and **J. Miller**, **A.E. Adams** and **P. Kabrna**. **P.H. Bridges**, **G.M. Walkden**, **C. Oakman** and **P. Gutteridge** contributed to the understanding of the platform limestones of the Derbyshire Massif, where they have described facies variations and the cyclic organization of the sediments, including the presence of intra-formational karstic surfaces. They also refined interpretations of the mud mounds that fringe the Derbyshire carbonate platform. These form an extensive 'fringing reef' that has been described in a series of important papers by **D. Parkinson**, **E.B. Wolfenden**, **F.M. Broadhurst**, **I.M. Simpson** and **T.D. Ford**., Ford also made many important contributions to understanding the mineralization, cave development and mining history of the area. The interaction of tectonics and sedimentation has been well documented by **R. Gawthorpe** and **R.F. Grayson**.

These advances, and those for the Namurian, were largely made possible by the maps and memoirs produced by geologists of the British Geological Survey (BGS). Papers by **A.G. Lee**, **M.R. Leeder**, **A.H. McMahon** and **R.P. Steele** in the conference volume edited by **B.M. Besly** and **G. Kelling** are important in documenting the tectonic framework of the area, as is the BGS subsurface memoir by **G.A. Kirby *et al*.** on the Craven (Bowland) Basin. This drew heavily on reflection seismic data acquired during onshore exploration for oil and gas.

Within the Namurian succession, early basin-wide correlation was fraught with problems until the 1920s when **W.S. Bisat** began to establish a robust biostratigraphic framework based on the occurrence of goniatites. Prior to that, correlations within the Millstone Grit had been largely based on the major sandstone units, and these were often misleading. Detailed study of goniatites, which evolved rapidly through Namurian times and which occur in thin 'marine bands', allowed the development of a high-resolution chronostratigraphic framework that was independent of sandstones. This provided a template for understanding the depositional history of the Millstone Grit and helped to establish a global correlation scheme for the Namurian, with goniatite-bearing marine bands now correlated across and between continents.

Henry Clifton Sorby, a remarkable polymath who is widely regarded as the father of sedimentology, lived near Sheffield, where he was active through the second half of the nineteenth century. He not only pioneered the study of rocks in thin section and the hydrodynamics of sand bedforms, but he was the first to realize that cross-bedding and cross-lamination in sandstones are indicators of palaeoflow direction. Cross-bedding, he suggested, also provides a basis for quantifying burial compaction in sandstones.

Arthur Gilligan's pioneering work on heavy minerals, in the early years of the twentieth century, helped to identify the source area of the Millstone Grit sandstones as metamorphic and granitic basement lying far to the north. This analysis has been extensively developed and refined by **J.I. Chisholm, A.C. Morton** and **C.R. Hallsworth** in recent times. After Gilligan's study, most publications dealt with local stratigraphic issues, with little of sedimentological relevance, until the early 1960s when **J.R.L. Allen** interpreted the Mam Tor Sandstones as turbidites. **B.K. Holdsworth** came to the same interpretation for sandstones in the Staffordshire Basin and recognized that turbidites could occur in non-geosynclinal settings. Prior to that, turbidites were entangled with the concept of 'flysch', an Alpine term that was widely recognized as a key stage in the development of 'geosynclines', then the dominant component in orogenic models prior to the advent of plate-tectonic concepts in the late 1960s.

The refined stratigraphic framework provided by marine bands allowed more robust and detailed correlations of Namurian strata, leading to more precise interpretation of basin-filling history. **W.H.C. Ramsbottom** and **H.G. Reading** demonstrated bathymetric controls on sandstone distribution, and progressive and sequential basin filling. The recognition of turbidites within the Namurian allowed **R.G. Walker** to develop early ideas about ancient submarine fans. The relationships between turbidites and associated deltaic intervals suggested to **J.D. Collinson** that sediment by-passing was important, and his recognition of giant cross-bedding at several stratigraphic levels promoted discussions by **P.J. McCabe**, **G. Hampson** and **C.M. Jones** about the effects of sea-level changes and water depth on depositional patterns. **C.S. Bristow** synthesized the regionally extensive Rough Rock channel complexes and, more recently, **I. Kane** documented the interaction of turbidity current deposition and local syn-depositional tectonics. **O.J. Martinsen**, **M.J. Brettle**, **P.B. Wignall**, **J.R. Maynard**, **C.N. Waters**, **A.C. Benfield** and **A.M. O'Beirne** developed ideas in sequence stratigraphy, including the nature of flooding events, forced regressions and the role of fluvial incision.

The relationships between geology and industry, particularly mining and quarrying, have been extensively documented. The extraordinarily versatile **Arthur Raistrick** pulled together geology, industrial archaeology, lead mining history and landscape in the Askrigg Block limestone area in a succession of popular books. In Derbyshire, the history of lead mining has been recorded by, amongst others, **T.D. Ford** and **J.H. Rieuwerts**, whilst the history of quarrying and the uses of building stones have been extensively documented by **I.A. Thomas**, both in print and through the establishment of the National Stone Centre at Wirksworth.

1.4 Human influences on geology and landscape

The Pennines not only provide an interesting narrative of geological development over the past 400 million years, but also record the influence of human (anthropogenic) activity in more recent times.

The earliest refashioning of the physical landscape of the Pennines was probably the establishment of agricultural terraces, such as those in Wharfedale and dating back some two thousand years. However, the main irreversible physical changes relate to quarrying and mining which have been major features of Pennines life since well before the Industrial Revolution of the nineteenth century. Caves in limestone areas provided shelter for early human habitation, and local stone was later used for building enclosures and dwellings, as shown by the ubiquitous drystone walls. Early quarries for building stone exploited very local sources. However, economic development from the seventeenth century onwards led to some building stones being quarried as a high-value product, worth exporting further afield, a practice that continues to the present day. This type of quarrying reached its peak in the nineteenth century, providing building and paving stones for the developing towns and cities. These activities declined in the twentieth century as costs, particularly of manpower, increased. Other products, particularly concrete for flags and kerbs, became more economically attractive and, more recently, cheap imports of flagstones from China and India have further eroded local production. Sandstone quarries now mainly produce crushed stone for aggregates and pre-cast concrete, although a few still produce dimension stone, often marketed as 'Yorkshire Stone'. A few quarries exploited more specialized products such as high-silica sandstones, known as 'ganisters', which were used as refractories in furnaces. Other quarries, mainly in the Coal Measures, continue to exploit shales for brick-making and other ceramic uses.

Limestones have been quarried from early times for burning for lime, as shown by the abandoned lime kilns. They were later extensively quarried as a flux in iron smelting, for cement making, as a chemical feedstock, and for acid scrubbers in coal-fired power stations. The high chemical purity of some of the limestones means that they are now extensively exported as powders used for glass making and the pharmaceutical industries. Limestone is also quarried extensively for aggregate and cement making. It is ironic that cement production should be a feature of areas celebrated for their drystone walls.

In addition to high-volume quarrying, limestone has been quarried and sometimes mined for specialized decorative and artistic uses, with several quarries producing polished 'marbles' for fireplaces, tabletops etc., notably the Ashford Black Marble. The Hopton Wood Stone of Derbyshire has been described as 'England's premier decorative stone' and was used by several of the twentieth century's great sculptors including Barbara Hepworth, Jacob Epstein and Henry Moore and for some 120 000 headstones by the Commonwealth War Graves Commission.

Some quarries were opened specifically for local projects such as the building of reservoir dams. Such reservoirs not only provide water for the populated areas in the valley bottoms and further afield but also help stabilize the discharge of rivers downstream, particularly during dry periods. Reservoirs become settling tanks for sediment derived from their catchments, which in the Pennines is often rich in eroded peat. Without the dam, this sediment would have found its way further downstream to be deposited on floodplains and at the coast. Furthermore, major road building has involved large-scale remoulding of the landscape. Construction of the M62 in the late 1960s created huge cuttings, with large volumes of rock being shifted sideways to build the Scammonden Dam that holds Deanhead Reservoir.

Quarries, once abandoned, become potential sites for fresh deposition as landfill sites, which can unfortunately bury potentially important geological exposures, a fact that may eventually impact on some of the localities described in this Guide. It is to be hoped that these losses can be minimized by sympathetic discussion between operators and geologically informed planners. In limestone quarries, the permeability of the rocks will usually deter tipping of other than quarry waste because of the risk of groundwater contamination, and some quarries are now preserved as sites for climbing and nature conservation.

There is a long history of mining in the Pennines. Whilst the main coal resources lie in the Westphalian Coal Measures, which occur in the low-lying coalfield areas bordering the Pennines, thin coal seams were exploited in a minor way within the Millstone Grit succession. Coal mining that fuelled the Industrial Revolution took place in the coalfields flanking the Pennines, but the canals and railways that were built to transport the coal led to cuttings and tunnels that are irreversible changes in the Pennine landscape, they fortuitously enhanced the geological understanding by creating extensive exposures.

In the limestone areas, mining of metal ores was important from at least Roman times. The main mineral was the lead ore, galena, although it has been suggested that its initial mining was driven by its subsidiary content of silver. In areas of North Yorkshire, such as the Greenhow area and across the Derbyshire Limestone Massif, thousands of miners and mineral processors left their mark on the landscape. Many miners worked small, shallow concessions, the traces of which can still be seen in disturbed land where trenches and lines of bell pits follow the trends of underlying veins or rakes. In other areas, deeper and more extensive mines were sunk, and these are still seen as surface headworks and pumping houses (Fig. 1.4). Deep mines also led to the excavation of extensive drainage tunnels ('soughs') that allowed mine waters to drain to nearby valleys. These have had the effect of permanently lowering the water table over wide areas, creating problems for some surface streams and for wells. Some soughs continue to discharge to this day, and thereby help to maintain river discharge in periods of drought. Along with galena, associated minerals such as barite and fluorite were extracted and commonly discarded as waste, although from the late eighteenth century barite was exploited as a minor component of Wedgwood's jasperware body. More recently, some spoil tips have been reworked to extract these minerals for various industrial processes.

Limestone is host to important copper mineralization at Ecton, in Staffordshire, on the western margin of the limestone platform. Mining for copper at Ecton began in the Bronze Age and continued into the later part of the nineteenth century. At one time it accounted for a high proportion of the total copper production of England.

Potentially reversible changes to the Pennine landscape are mainly related to changes in vegetation, some directly attributable to human activity, others less so, and probably reflecting climate fluctuations in post-glacial times. The earliest widespread changes were the transformation of the uplands from birch and pine forest to the peat-covered moors that we see today. Tree remains recovered from below the peat range in age from around 7600 to 4000 years BP, suggesting that deforestation occurred episodically over a long period. Much was probably due to climatic fluctuations, but the onset of grazing and burning associated with upland settlement in the Bronze Age was also

Figure 1.4 The headworks of Magpie Mine near Monyash, Derbyshire, historically preserved surface features of a deep lead mine.

a factor. This movement to higher ground followed settlement of lower lying areas where the natural woodlands were cleared for crops and grazing from around 6000 years BP. These changes were consolidated by the building of agricultural terraces and walls as well as more permanent, stone-built settlements. Increased burning of coal in nearby industrial areas since the eighteenth century inflicted acid rain on some parts of the Pennines, with serious consequences for woodland and moorland vegetation. Sphagnum moss was widely almost eliminated through acidification but is now recovering where peat moors have been allowed or encouraged to retain water.

The development of extensive conifer plantations by the Forestry Commission, water companies and private landowners is driven mainly by commercial considerations, with timber as the eventual crop. Such afforestation typically produces monocultures of closely spaced trees with very restricted biodiversity, and can hardly be considered reforestation. It does, however, help to store precipitation and buffer run-off, thereby reducing the risks of flooding. On many moorlands, devoted to shooting, restoration of earlier vegetation is not a priority, rather the opposite, with draining of bogs and burning of heather being the norm. The 'improved' drainage may have sometimes sometimes contributed to downstream flooding in valleys through accelerating

Figure 1.5 A turbine of the Carsington Wind Farm at Harboro' Rocks at the southern edge of the Derbyshire Massif. In the distance to the southeast, the soon-to-be-closed Ratcliffe-on-Soar coal-fired power station is highlighted by the sun, the whole view capturing transition to pollution free, renewable energy.

run-off during storms. Attempts at authentic reforestation occur as part of rewilding projects on the fringes of moorland, where fencing to eliminate grazing sheep has allowed the slow re-establishment of mixed woodland.

The most recent changes to the Pennine landscape relate to the mitigation of climate change. Mill chimneys, those icons of the Industrial Revolution, as captured by L.S. Lowry and his many followers, have gradually been demolished, to be replaced on the higher ground by wind turbines, the new anthropogenic icons of the landscape (Fig. 1.5).

Chapter 2

Tectonics and Structure

2.1 Tectonic controls on basin development and sedimentation

The evolution of the Carboniferous sedimentary basins of northern England, including those that accommodated the Pennine successions, sits in time between two major mountain-building episodes (Fig. 2.1). The first was the Caledonian orogeny, which ended in early Devonian times (*c*.400 my ago). This widespread and complex episode, involving closure of major ocean basins and continental collision, produced vast mountain ranges in Scotland, Scandinavia, East Greenland and eastern North America. Erosion of these mountains provided most of the clastic detritus that filled the Carboniferous basins. The Caledonian orogeny also created widespread tectonic structures, particularly major faults, which were reactivated when sedimentary basins began to develop in the late Devonian as a result of crustal stretching or extension. Furthermore, large granite intrusions, emplaced in the early Devonian, were important in determining rates of subsidence and, hence, the nature and positions of blocks and basins when crustal extension began.

The second important mountain-building episode was the late Variscan orogeny, the main effects of which are concentrated further south, from Germany through northern France and including southwest England and South Wales. The Variscan episode involved widespread compressional tectonics that led to major uplift. The resulting mountains provided a source of clastic sediment in latest Carboniferous times (*c*.300 my ago), particularly in the southern North Sea, southwest England and South Wales. The Variscan episode also led to the folding, faulting and uplift of the Pennine area, creating major structures such as the Pennine Anticline, the Pendle Monocline and the synclines that preserve Coal Measures on the flanks of the Pennines. Some of

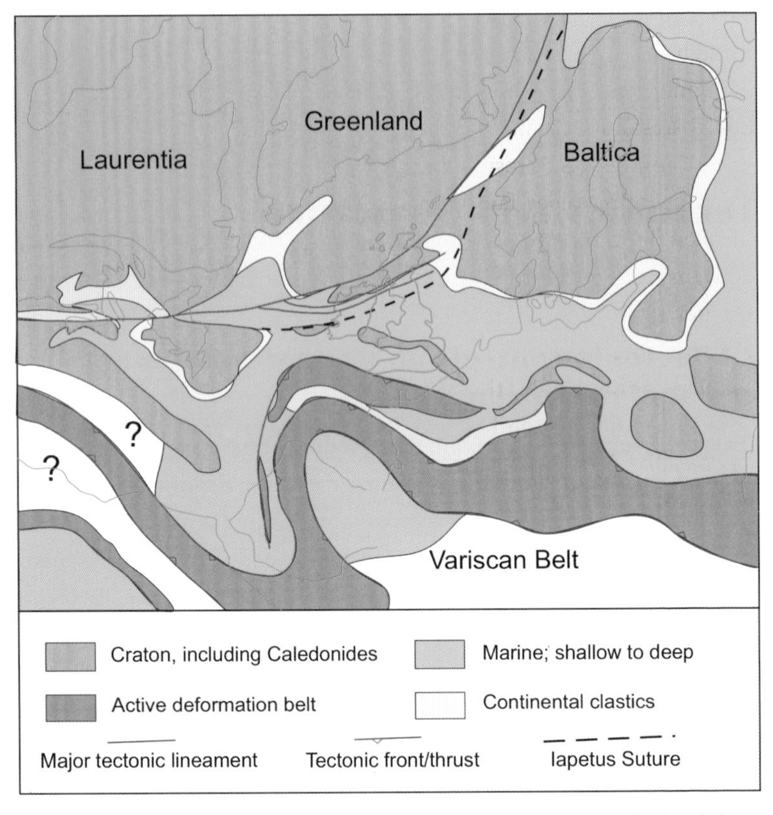

Figure 2.1 Palaeogeographic map showing the position of the British Isles in relation to the large-scale distribution of continents and major tectonic zones, during the early Carboniferous Variscan tectonic episode. The northward migration of Gondwana, lying to the south, was a major driving force, creating compressional stresses in the south but more extensional stresses further north (after Glennie, 2005; based on Zeigler, 1987).

the structures that developed under the Variscan compressive regime resulted from reactivation (often inversion) of structures that had controlled the development of basins earlier in the Carboniferous. These structures, in turn, had been localized on older Caledonian structures. Tectonically, therefore, the evolution of the Pennine Basin Complex involved multiple reactivations of old, persistent structures.

Between the Caledonian and Variscan orogenies there was a phase of crustal extension. This led to the development of a series of basins that formed a belt extending from western Ireland to the southern North Sea. In the Pennine area, the main phase of extension took place mainly in early Carboniferous times, beginning in the late Devonian, continuing throughout the Dinantian and largely dying out in the early Namurian. The dominant extensional stress was oriented north–south, driven by crustal stresses transmitted from the main Variscan belt to the south. In the Pennine area it resulted in the brittle fragmentation of the crust, often exploiting older fault lines. The resulting fault-bounded blocks subsided because of crustal thinning and became progressively tilted as extension continued. This led to up-tilted, slowly subsiding footwalls and more rapidly subsiding hanging-wall areas.

To the north of the Pennine Basin Complex, which is the focus of much of this Guide, the Askrigg and Alston fault blocks subsided more slowly due to the buoyant effect of underlying early Devonian granites. The Askrigg Block tilted down to the north, so that a thin late Viséan succession is found near the southern margin compared with the thicker and more stratigraphically extensive succession in the Stainmore Trough to the north. The western margin of the Askrigg Block is the Dent Fault, a north–south normal fault that throws down to the west. The North Craven Fault marks the southern limit of the Askrigg Block and delimits the northern boundary of the Pennine Basin Complex. Immediately south of that fault, a series of synthetic normal faults splay out to the southeast. The South Craven Fault extends to link up with Morley–Campsall Fault that separates the Gainsborough Trough from the Askern–Spittal High in Lincolnshire. This fault picks up the older 'Charnian' northwest–southeast structural trend that characterizes the basement in the East Midlands.

The southern boundary of the basin complex against the Midland Landmass (Anglo-Brabant Massif or St. George's Land in some literature) is less well defined, although its boundary with the Widmerpool Gulf, the southernmost sub-basin, is a fault. This landmass subsided more slowly than the basins to the north and was sufficiently elevated to have been a minor source of clastic detritus from time to time.

Between the Askrigg Block and the Midland Landmass, during the late Devonian to mid-Carboniferous extension, the crust fragmented into blocks of the order of 20–30 km in dimension, bounded by faults that largely reflect earlier Caledonian structural trends. In the west, these follow a northeast–southwest trend. The Pendle Fault, which can be projected southwards into older structures in North Wales, defines the south-eastern margin of the Bowland Basin. Further to the northwest, the Bowland Fault forms the margin between the Bowland Basin and the Bowland High. South of the Pendle Fault, other faults with a similar trend define the Central Lancashire High, the Rossendale Basin and the Heywood High. To the east, the basin configuration is dominated by east–west trending faults that demarcate the Huddersfield Basin, the Holme High and the Alport Basin, the last of which extends south to the northern margin of the Derbyshire Massif. These basins are known mainly from seismic data, supported by sparse boreholes. However, some of the associated normal faults are seen at outcrop, in some cases with reversed displacement following later Variscan compression.

Across the Derbyshire Massif, during Carboniferous extension, shallow-water limestones were deposited on three adjacent blocks that showed different degrees of tilting. The Bakewell and Cronkston–Bonsall faults, which bound these blocks, cross the massif with a northwest–southeast 'Charnian' trend. As well as delimiting different tilting regimes, these faults were intermittently sufficiently active to create local sub-basins on the platform. Faults also controlled the position of the northern and western margins of the massif, although the edge of the limestone platform does not everywhere coincide with an underlying fault.

Whilst normal faulting and tilting dominated the tectonic response to extensional stresses, associated shearing gave rise to folding in the Bowland Basin where the Ribblesdale Fold Belt had a complex history. Details of the structural development are difficult to establish because of poor exposure due, in part, to burial by extensive Quaternary deposits. However, restricted evidence, such as local intra-Dinantian unconformities, suggests that some folds began to develop transpressively during the extensional phase and then developed further during Variscan compression.

Crustal extension continued throughout the Dinantian but gradually died away in early Namurian times. However, subsidence continued because of cooling of the thinned crust. During this so-called 'sag' stage, subsidence was more widespread and less localized by faults. Across the Pennine Basin Complex, the thickness variations observed in Late Namurian and Westphalian strata suggest that maximum thermal subsidence occurred in south Lancashire. This style of subsidence

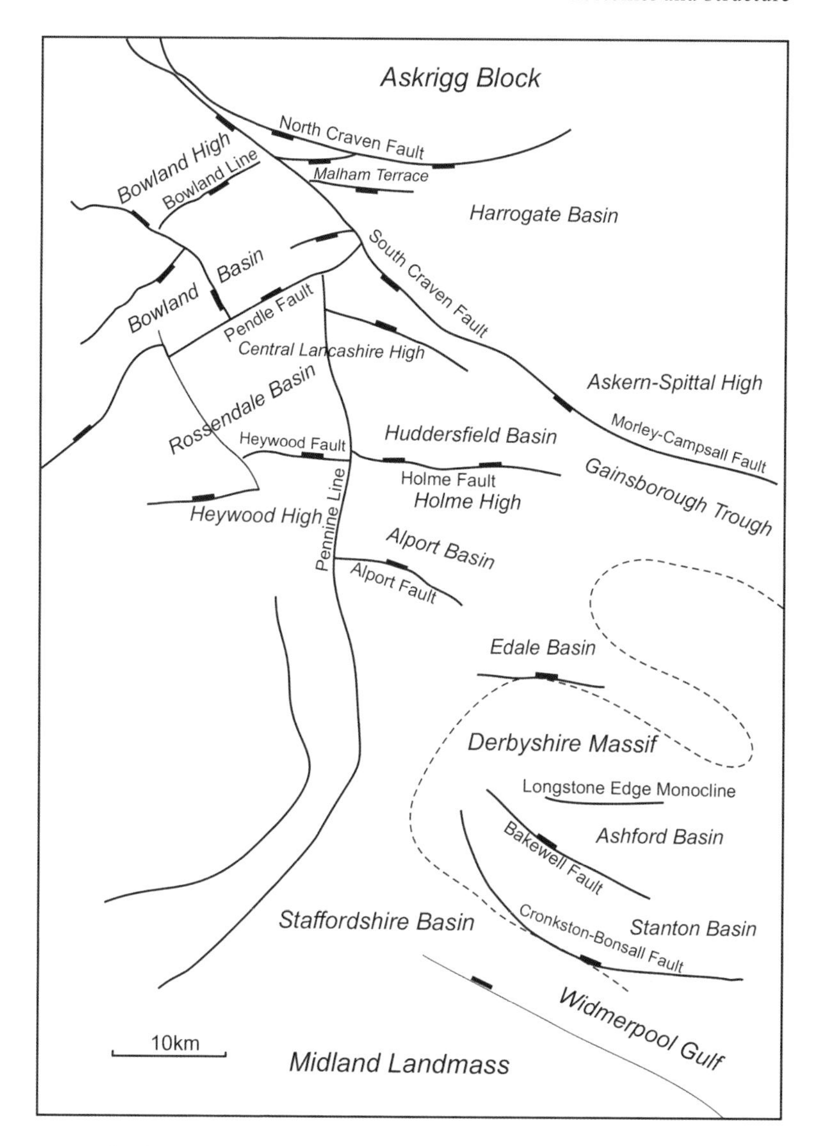

Figure 2.2 Map showing the major extensional structures which were active during the Dinantian, and which strongly influenced sedimentation into the Namurian (based in part on Kirby *et al.*, 2000).

persisted into the Westphalian when the major Coal Measure successions were deposited. It was ended by the Variscan compression episode, which started in late Westphalian times.

2.2 Variscan deformation and present-day structural configuration

Compressional stress in late Carboniferous to early Permian times resulted from increased activity in the Variscan orogenic belt to the south. This led to the inversion of several earlier extensional faults, to the development of major folds, and to widespread tectonic uplift. This resulted in deep erosion during Permian and Triassic times, when up to around 3 km of Carboniferous sediments (mainly Coal Measures) were removed from the Pennine area, and the present outcrop pattern was largely established. Further movements in Tertiary times, due to Alpine events, led to some further uplift and erosion, but there was no major reactivation of faults at that time. The compressional structures are discussed below, broadly from north to south, with reference to Figure 1.2.

Askrigg Block: Craven and Dent Faults

The northernmost structural element of the area is the Askrigg Block. This is bounded on its southern side by the North Craven Fault, which experienced only minor Variscan inversion. Its dominantly normal, down-to-the-south displacement is shown by the presence of Namurian sediments on its southern side, as at Fancarl Crag. Towards the west, the fault has clear topographic expression at Giggleswick Scar. The Middle and South Craven Faults also seem to have experienced little inversion, although anticlines between the faults, as at the Skyreholme and Greenhow Anticlines, record some local compression.

In contrast, the Dent Fault, which defines the western margin of the Askrigg Block, underwent significant inversion, to the extent that Lower Palaeozoic basement to the west of the fault is now displaced higher than the Viséan limestones of the block. Given that regional Variscan compression was dominantly oriented north–south, there was probably a significant transpressive element to the Variscan deformation along the fault.

The Askrigg Block itself remained rigid throughout the Variscan deformation and underwent little internal deformation. However, the dominant joint sets within the Dinantian limestones, which are commonly picked out by the karstic weathering as grykes, result from the Variscan stress field.

Ribblesdale Fold Belt and Pendle Lineament

The major rigid blocks that developed during crustal extension remained relatively stable during Variscan compression, with strain concentrated mostly on their bounding faults, which typically show evidence of inversion. However, the Bowland Basin, between the Bowland High and the Central Lancashire High, underwent significant plastic deformation of its dominantly fine-grained sediments. There was further growth of northeast–southwest trending folds, several of which were initiated during an earlier extensional phase. These comprise the Ribblesdale Fold Belt, an area some 20 km wide and 50 km long at outcrop, and probably extending further to the southwest under Permo-Triassic cover. Within this belt there are some fifteen named anticlines that are elongate domes with broadly similar trends. Locally intense deformation is evident from some very steep dips (e.g. Skipton Anticline). The Sykes and Skipton Anticlines are associated with two of the localities described in this Guide (Figs 1.2 & 7.1). The significant uplift associated with this folding resulted in some of the earliest Carboniferous strata being exposed in the centre of this zone.

This deformed zone is bounded to the southeast by the Pendle Monocline (Pendle Lineament) which resulted from major inversion of the basin-bounding normal fault. The major monocline dips to the southeast as the north-western limb of the Burnley Coalfield syncline.

Rossendale Anticline

This broad, domal structure occurs at the southern end of the Burnley Coalfield syncline. It splits the outcrop of the Coal Measures that occur in synclines to east and west. The eastern syncline involves quite a tight fold between the Rossendale and Pennine Anticlines, whilst the syncline to the west is asymmetric with its steep, north-western limb being the Pendle Monocline. The Rossendale Anticline probably reflects inversion of the earlier Rossendale Basin, which was controlled during extension by the Heywood Fault to the south. This fault was inverted during Variscan compression.

At its core, the Rossendale Anticline exposes the Namurian succession down to the mid-Marsdenian. The outcrop pattern is fragmented by a network of faults trending northwest–southeast.

Pennine Anticline

The Pennine Anticline (or Monocline) is the dominant large-scale structure of the Central Pennines. It trends north–south from its junction with the Pendle Lineament in the north to its merging with the Derbyshire Massif in the south. It is asymmetric with a steeper western limb, particularly opposite the Rossendale Anticline. In the earlier extensional phase, the role of the Pennine Lineament, the precursor to the Pennine Anticline, is not clear. It was probably a normal fault that was later inverted, but details of its behaviour are uncertain. In the Variscan deformation it grew as a major uplift whose subsequent erosion now exposes sediments down to the Kinderscoutian. In places, it is affected by quite intense faulting, particularly around Todmorden. The folding that extends around the western side of the Derbyshire Massif can be considered as an extension and complication of the Pennine Anticline deformation. Whilst the structure is dominantly Variscan, it underwent further uplift in Cretaceous and Pliocene times (see section 2.3).

Towards the southern end of the main anticline, the structural pattern is characterized by several east–west folds that abut against the northern edge of the Derbyshire Massif. The Alport Dome and the Edale Anticline and their intervening synclines extend for some 10 km north of the massif margin.

Derbyshire Massif

The Derbyshire Massif, consisting of mainly rather massive Viséan limestones, appears to have behaved largely as a rigid unit during Variscan compression, with more ductile deformation concentrated around its flanks. This resulted in local intense folding of thinly bedded off-platform limestones (Fig. 12.3 & 12.4) and this deformation passes out westwards into broader folds in the Staffordshire Basin (Fig. 2.3).

Under Variscan compression some local folds such as the Stanton Syncline were perhaps accentuated but, for the most part, rather uniform dips prevailed, with an overall gentle dip to the east. Mineral veins, which typically follow major joints, trend both northwest–southeast and northeast–southwest (Fig. 5.2). Blocks of Bowland Shale preserved in Miocene pocket deposits in the southern part of the massif suggest that post-Variscan erosion across the massif did not everywhere remove post-Dinantian Carboniferous sediments.

North–south folds of Staffordshire

West of the Derbyshire Massif, and extending northwards towards Manchester, is a series of folds with broadly north–south axes. These result from deformation of the more ductile sediments that make up the succession from deep-water Viséan limestones and mudstones up to the Coal Measures. Anticlinal folds are mainly elongate domes, plunging north and south. The main anticlines are the Mixon–Morridge Anticline within the Staffordshire Basin, which brings Viséan sediments to outcrop, the Gun Hill Anticline which involves lower Namurian strata, and the Todbrook and Kettleshulme anticlines, further north, which have Namurian outcrops in their axial zones.

Major synclines are the Potteries and Goyt Synclines, both of which have Coal Measures in their axial areas. These folds have clear topographical expression as the thick sandstones in the upper part of the Namurian succession give prominent ridges that can be traced across the landscape The Rough Rock forms the conspicuous ridge round the northern end of the Potteries Syncline, including Mow Cop and Congleton Cloud. The Roaches and Chatsworth Grits and the Rough Rock all give topographic features that help to define the Goyt Syncline (Fig. 2.3).

The western margin of this folded belt is the Red Rock Fault, which marks the eastern edge of the Cheshire Basin, which is filled with

Figure 2.3 View from Morridge of the southern end of the Goyt Syncline, showing the closure of the escarpments of the Roaches and Chatsworth Grits (Marsdenian). Orange arrows indicate tectonic dip.

sediments of Permian to early Jurassic age with Triassic sediments making up the majority. Pleistocene, mainly glacial, erosion of the soft Triassic sediments has led to a very flat topography that contrasts with the rugged topography of the folded Carboniferous.

2.3 Post-Variscan tectonics

The extensive erosion of uplifted areas during and following Variscan deformation led to sediments of a wide range of Carboniferous ages being exposed at the Variscan Unconformity, which appears to have had considerable topographic relief. The thickest Permo-Triassic successions, above the unconformity, occur in flanking areas such as the Cheshire Basin. The extent to which Permo-Triassic sediments were deposited across the Pennine area is difficult to judge, as there is no direct evidence.

During Cretaceous times, uplift and eastward tilting began, associated with subsidence in the southern North Sea area. This probably led to further erosion of Carboniferous sediments and to the removal of any remaining Permo-Triassic cover. Later reworking of Triassic sediments is suggested by the mid-Miocene Brassington Formation sediments preserved in karstic dissolution collapse pockets near the southern edge of the Derbyshire Massif. These are thought to be remnants of a sheet of mainly fluvial sediments, which were probably derived from Triassic sediments that cropped out to the south. The depositional surface of the Miocene sediments would have had very low relief and was probably close to sea level. The karstic processes that led to the dissolution of the pockets were probably initiated by uplift in the Pliocene. On the Derbyshire Massif this uplift has been estimated at around 350 m on geomorphological grounds, and this was part of a more widespread uplift phase that helped define today's southern Pennine topography.

Chapter 3

Stratigraphy

3.1 Pre-Carboniferous

Knowledge of the pre-Carboniferous geology of the Pennine area is extremely patchy. Pre-Carboniferous strata are only seen at outcrop in inliers below the Base-Carboniferous Unconformity near the southern edge of the Askrigg Block between Ingleton and Malham Tarn. The oldest unit is the 'Ingletonian' of the Ingleton inlier, which is of Arenig (early Ordovician) age. The folded Lower Palaeozoic sediments of the Horton and Austwick inliers are significantly younger.

Over the rest of the area, the sub-Carboniferous geology is known only from geophysics and from a few deep boreholes. The Lower Palaeozoic sediments of the Askrigg Block are intruded by the Wensleydale Granite, an early Devonian batholith which is known from gravity surveys and from a borehole. Within the Pennine basins, no boreholes penetrate the Base-Carboniferous Unconformity with certainty, although the Boulsworth Hill borehole, east of Burnley (Fig. 1.2), penetrated red beds of indeterminate but possibly late Devonian age. On the Derbyshire Massif, the Eyam and Woo Dale Boreholes (Fig. 4.3) both reached pre-Carboniferous rocks. The Eyam Borehole penetrated Ordovician (Llanvirnian) mudstones below 1800 m of Dinantian sediments, whilst the Woo Dale Borehole encountered lavas and pyroclastic sediments of probable Devonian age below 274 m of Dinantian limestones.

3.2 Carboniferous

The stratigraphic subdivision of the Carboniferous of the British Isles has a long and complex history. A full account is beyond the scope of this book, but full discussions can be found in Waters *et al.* (2011). The earliest schemes of subdivision were lithostratigraphic, and essentially local in extent. Biostratigraphic approaches, based on 'Marker Beds' containing various fossil types, led to more widespread correlations and to regionally applicable schemes. Recent truly international schemes are based in part on stratotypes (type sections) from the Pennine area. The current European stages and their component sub-stages are shown in Figure 3.1 along with their relationships to international and North American units. For the most part, discussion in this Guide will be at the level of stages or sub-stages, with finer resolution occasionally used, especially in the Namurian. Much of the literature on the Lower Carboniferous uses the term 'Dinantian' for the combined Tournaisian and Viséan, and this usage is followed here in referring to the overall succession.

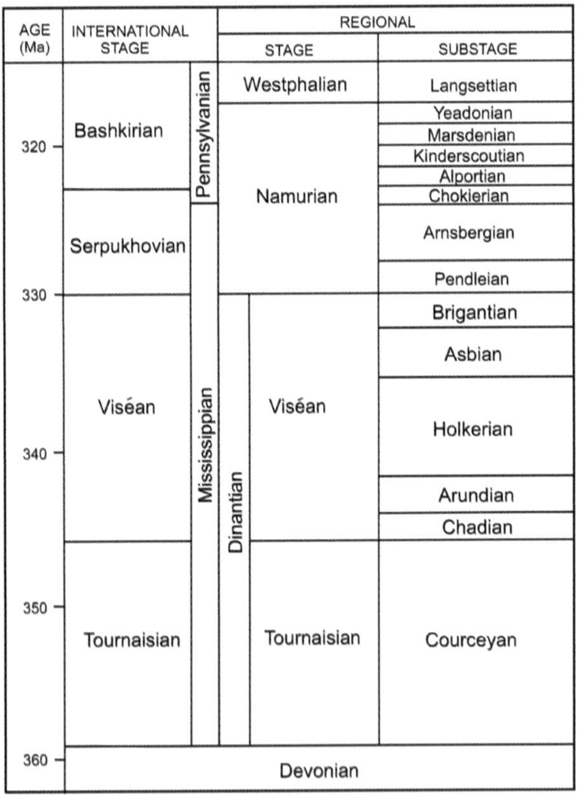

AGE (Ma)	INTERNATIONAL STAGE		REGIONAL		
				STAGE	SUBSTAGE
	Bashkirian	Pennsylvanian		Westphalian	Langsettian
				Namurian	Yeadonian
					Marsdenian
320					Kinderscoutian
					Alportian
	Serpukhovian	Mississippian			Chokierian
					Arnsbergian
					Pendleian
330	Viséan		Dinantian	Viséan	Brigantian
					Asbian
					Holkerian
340					Arundian
					Chadian
350	Tournaisian			Tournaisian	Courceyan
360	Devonian				

Figure 3.1 The stages and substages of the Carboniferous and their relationship to geological time. This book mainly uses regional sub-stages. The Tournaisian and Viséan together have traditionally been referred to as the 'Dinantian' and, where appropriate, this usage is retained.

Tournaisian and Viséan (Dinantian)

The Tournaisian and Viséan stages are recognized on the occurrence and distribution of several different types of fossil, some macrofossils such as corals and brachiopods, some microfossils such as foraminifers and conodonts. Stratotypes of the European stages are defined in northern England, except for the Courceyan (Ireland) and the Arundian (S. Wales). The Courceyan stage equates with the Tournaisian, whilst the Viséan is divided into five constituent stages. Within this chronostratigraphic framework, local successions are characterized by lithostratigraphic names.

On the southern part of the Askrigg Block, the Viséan succession rests unconformably on the pre-Carboniferous basement. It has two major components, a lower **Great Scar Limestone Group** (Holkerian–Asbian) and an upper **Yoredale Group** (Brigantian). The lowest formation of the Great Scar Limestone that is well exposed is the **Kilnsey Formation**, comprising dark-coloured limestones. This is overlain by the **Malham Formation**, which is generally paler in colour. Some formation names reflect facies differences; for example, the **Cracoe Limestone Formation** is characterized by the occurrence of large microbial mud mounds ('reef

STAGE	ASKRIGG BLOCK CRAVEN FAULT BELT	CRAVEN BASIN		DERBYSHIRE MASSIF		
BRIGANTIAN		Craven Group	Bowland Shale Formation *(Lower Bowland Shale)*	Peak Limestone Group		Eyam Limestone Formation Monsal Dale Limestone Formation
ASBIAN	Malham Formation (Cracoe Lst Fmn)		Pendleside Limestone		Hopedale Limestone Formation	Bee Low Limestone Formation (Ecton Lst Fmn)
HOLKERIAN	Kilnsey Formation		Hodderense Limestone			Woo Dale Limestone Formation
ARUNDIAN			Hodder Mudstone Formation *(Worston Shale)*			
CHADIAN						(Milldale Lst Fmn)
COURCEYAN		Bowland High Group	Clitheroe Limestone Formation Chatburn Limestone Formation			

Figure 3.2 The lithostratigraphic nomenclature of the Dinantian formations across the South Pennines. Names in italics were commonly used in the past, and occur in much relevant literature. The names in purple refer to off-platform facies equivalents which have more local formation names (based on Waters *et al.*, 2011).

knolls') and is a lateral equivalent of both the Kilnsey and Malham Formations. This nomenclature is summarized in Figure 3.2.

In the Bowland Basin, the Lower Carboniferous succession has been sub-divided in a variety of overlapping and, in some cases, contradictory schemes. This partly results from the sediments' cropping out in an area with thick Quaternary cover, where most exposures are widely spaced, isolated anticlines. This Guide tries to follow the latest scheme, but readers of earlier (pre-2011) literature must be prepared for some confusion.

Tournaisian (Courceyan) sediments, which are assigned to the **Bowland High Group**, are of shallow-water origin. The Viséan succession is assigned to the **Craven Group** of generally deeper-water origin. Its constituent formations are set out in Figure 3.2. In the older literature, the commonly mentioned 'Worston Shale' is used at both group and formation level to include all strata between the top of the Bowland High Group (i.e. top **Chatburn Limestone Formation**) and the base of the Bowland Shale Group (note that Bowland Shale is now a formation). In the present-day system, the Craven Group is a rather variable series of limestones, often muddy and including limestone turbidites and large mud mounds, the 'Waulsortian reefs', which occur in the lowest unit, the **Clitheroe Limestone Formation**. The overlying **Hodder Mudstone Formation**, **Hodderense Limestone Formation** and **Pendleside Limestone Formation** are capped by the **Lower Bowland Shale Formation** (Brigantian). This drapes the Bowland and Central Lancashire Highs (known only in the sub-surface) as well as the reef knoll morphology of the Cracoe Limestone Formation at the northern margin of the basin. Some folds of the Ribblesdale Fold Belt were growing during deposition, and so the succession is punctuated by local intra-Dinantian unconformities. The detailed stratigraphy is accordingly complex.

Across and around the Derbyshire Massif, the Dinantian succession is assigned to the **Peak Limestone Group**, which is mainly characterized by shallow-water platform limestones over the massif but includes off-platform facies that comprise both resedimented (turbidite) limestones and large deeper-water mud mounds. The overlying Craven Group is characterized by deeper-water facies and continues into the deep-water succession of the early Namurian Bowland Shale. The age of the boundary between the Peak Limestone Group and the dominantly mudstone Craven Group varies from place to place, the earliest being Holkerian in the Staffordshire Basin. In most places the lithological change occurs within the Brigantian. The Peak Limestone Group on the platform area has been divided into four units, **Woo Dale Limestone Formation**, **Bee Low Limestone Formation**, **Monsal Dale Limestone Formation** and **Eyam Limestone Formation**. These range in age from Holkerian to Brigantian. Older limestones and dolomites ranging back probably to the Tournaisian are known from the Eyam and Woo Dale boreholes.

The Woo Dale Limestone Formation is locally exposed in inliers and the uppermost 130 m (Holkerian) is seen at outcrop. It attains greater thickness in boreholes where it passes down into the Courceyan and is extensively dolomitized.

In the Bee Low Limestone Formation (Asbian), which comprises mainly rather pale limestones, there is a clear differentiation of facies between shallow-water platform limestones and surrounding platform-margin limestones. The platform succession is locally punctuated by intraformational palaeo-karstic surfaces that typically define cyclothems. These surfaces provide a basis for local correlation. In the northern part of the block, the Bee Low Limestone Formation includes a unit of basic volcanic rocks, the **Lower Miller's Dale Lava**. Where present, this has been used to separate the **Chee Tor Rock** below from the **Miller's Dale Beds** above, although these terms are now redundant.

The Monsal Dale Limestone Formation is of early Brigantian age and outcrops widely, mainly in the eastern part of the Derbyshire Massif. Locally, the **Upper Miller's Dale Lava** occurs within the Monsal Dale Formation when the limestones below it were earlier consigned to the **Station Quarry Beds**, again a redundant term. The Monsal Dale Limestones are largely calcarenites, with several distinctive fossil-rich bands, and are commonly bioturbated. Compared with the Bee Low Limestone Formation, the limestones are commonly more thinly bedded and darker in colour, partly due to cherts and mud-rich partings. The top of the formation is an unconformity with considerable local relief. In places, the upper surface of the Monsal Dale Formation shows the development of mud mounds, which were subaerially exposed during formation of this unconformity.

The uppermost formation of the Peak Limestone Group is the Eyam Limestone Formation of late Brigantian age. This unit records submergence after a period of uplift and erosion, although the transgression appears not to have reached the western side of the platform. The Eyam Formation limestones are dominantly rather dark, thinly bedded and of shallow-water origin but record the start of a deepening trend that led eventually to the deposition of the Craven Group mudstones.

In off-platform settings, around the southern and western sides of the Derbyshire platform, thick intervals of calcareous mudstones and resedimented limestones and a series of large deep-water ('Waulsortian') mud mounds, similar to those at Clitheroe, are assigned to the **Milldale Limestone Formation** and the **Hopedale Limestone Formation.** The **Ecton Limestone Formation** is a local equivalent of the upper part of the Hopedale Limestone Formation. The ages of these formations are somewhat loosely established, but the Milldale Limestones are generally regarded as Chadian whilst the Hopedale Limestones seem to extend from early Arundian up into the Asbian.

Namurian

The substages of the Namurian are defined by thin, well-defined 'marine bands' that punctuate the dominantly siliciclastic succession. The marine bands, which are typically a few centimetres thick, have fully marine faunas, thin-shelled pectenoid bivalves, *Lingula* and, most importantly, thick-shelled goniatites. The marine bands are associated with flooding events, deeper water and fully marine salinity in basins that were probably brackish in intervening periods of lower sea level throughout all but the earliest Namurian. The boundaries of the seven Namurian substages, five of whose stratotypes were designated in the Pennines (the other two are in Belgium) are defined at marine bands containing diagnostic goniatites. Whilst only eight marine bands are needed to define these substages (coinciding with goniatite zones), there are some seventy known marine bands in the Namurian. This means that a higher resolution biostratigraphic framework of zones is possible based on intermediate marine bands because goniatites evolved very rapidly throughout the Namurian and are distinctive in each band. As goniatites occur almost exclusively within marine bands, these units can be regarded as timelines related

SUB-STAGES	INDEX	ZONES & KEY MARINE BANDS
YEADONIAN	G1	*Ca. cumbiense* (G1b) *Ca. cancellatum* (G1a)
MARSDENIAN	R2	*B. superbilinguis* (R2c) *B. bilinguis* (R2b) *B. gracilis* (R2a)
KINDERSCOUTIAN	R1	*R. reticulatum* (R1c) *R. eoreticulatum* (R1b) *H. magistrorum* (R1a)
ALPORTIAN	H2	*V. eostriolatus* (H2c) *H. undulatum* (H2b) *Hd. proteum* (H2a)
CHOKIERIAN	H1	*H. beyrichianum* (H1b) *I. subglobosum* (H1a)
ARNSBERGIAN	E2	*N. stellarum* (E2c) *C. edalensis* (E2b) *C. cowlingense* (E2a)
PENDLEIAN	E1	*C. malhamsense* (E1c) *C. brandoni* (E1b) *C. leion* (E1a)

Figure 3.3 The stages of the Namurian and their relationships to the main goniatite zones. The marine bands or their codes are commonly used to identify cyclothems, which take their name from the marine band at the base of the cyclothem, e.g. the 'R_{1c} cyclothem' would refer to the succession overlying the *R. reticulatum* Marine Band.

to short-lived periods of full marine salinity. They therefore define a very high-resolution chronostratigraphic framework with time intervals between marine bands, of the order of 100 000 years.

The sub-zones and some of the more important marine bands are shown in Figure 3.3. The index designations are commonly used to identify zones (e.g. R_{1c} for the latest zone of the Kinderscoutian) and smaller subzones.

In addition to a biostratigraphy based on goniatites, there is also a scheme based on palynomorphs (spores and pollen). These floral elements provide a much coarser subdivision of the Namurian, but they have the advantage of being more widely dispersed than goniatites in mostly fine-grained sediments. They are, therefore, useful in areas of poor exposure or in un-cored boreholes where macro-fauna cannot be identified.

The lithostratigraphy of the Namurian is historically complex with numerous local names, particularly for the sandstones, which were extensively quarried. Prior to a secure chronostratigraphic framework being available, knowledge of the relationships of the various sandstones had been, at best, approximate and, at worst, highly misleading. The main sandstones were correlated locally, but more widespread correlation was largely based on counting down from the base of the Coal Measures. The result was that the Rough Rock, the highest Namurian sandstone, was correlated with reasonable confidence, but progressively older sandstones were increasingly mis-correlated. The Yoredale succession of the Askrigg Block sits between the Great Scar Limestone and the base of the local Millstone Grit. Accordingly, the interval of mudstones and interbedded sandstones between the limestones of the Derbyshire Massif and the base of the Millstone Grit in north Derbyshire were also termed 'Yoredales'. The basal unit of Millstone Grit in north Yorkshire is the Grassington Grit, of Pendleian age, whilst the basal Millstone Grit unit in north Derbyshire is the Lower Kinderscout Grit, of late Kinderscoutian age, and in Staffordshire it is the Roaches Grit of mid-Marsdenian age. The scope for confusion is obvious.

The allocation of sediments into the well-established chronostratigraphic framework resolved these earlier problems and led to some rationalization of the lithostratigraphic terminology at a formal level, although the original names of most individual sandstones and associated cyclothems are still widely used. In the revised scheme (Fig. 3.4A), which is found only in the most recent publications, the Namurian succession is divided into a lower **Craven Group** overlain by the **Millstone Grit Group**. The Craven Group is dominated by mudstones whilst the base of the Millstone Grit Group is drawn at the incoming of sandstones, irrespective of composition or provenance.

The Craven Group comprises only one formation, the **Bowland Shale Formation**, a term that has replaced 'Edale Shales', long used in the southern part of the basin. Its chronostratigraphic age varies widely from sub-basin to sub-basin. In the Bowland Basin and the Staffordshire Basin it is confined to the Pendleian, but in Derbyshire it extends up into the Marsdenian. The Millstone Grit Group comprises a sandstone-bearing interval that varies in facies and in age from place to place. It includes turbidite and deltaic successions, including the cyclothems that make

			N. Yorkshire N.W. Lancashire	Central-East Lancashire	W. Yorkshire Derbyshire	Staffordshire
Langsettian	G2		Lower Coal Measures			
Yeadonian	G1	G1a	Rossendale Formation			
		G1b				
Marsdenian	R2	R2c	Marsden Formation			
		R2b				
		R2a				
Kinderscoutian	R1	R1c	Hebden Formation			
		R1b				
		R1a				
Alportian	H2	H2c				
		H2b				
		H2a	Samlesbury Formation			Morridge Formation
Chokierian	H1	H1b			Bowland Shale Formation	
		H1a				
Arnsbergian	E2	E2c				
		E2b	Silsden Formation			
		E2a				
Pendleian	E1	E1c	Pendleton Formation			
		E1b	Bowland Shale Formation			
		E1a				

			N. Yorkshire N.W. Lancashire	Central-East Lancashire	W. Yorkshire Derbyshire	Staffordshire
Langsettian	G2			Woodhead Hill Rock	Crawshaw Sst	Woodhead Hill Rock
Yeadonian	G1	G1a	Rough Rock	Rough Rock Upper Haslingden Flags	Rough Rock Rough Rock Flags	Rough Rock
		G1b		Lower Haslingden Flags		
Marsdenian	R2	R2c		Brooksbottoms Grit	Huddersfield White Rock	Chatsworth Grit
		R2b		Fletcher Bank Grit	Midgley Grit /Ashover Grit	Roaches Grit
		R2a			Scotland Flags/Heyden Rock	
Kinderscoutian	R1	R1c	Addingham Edge Grit	Kinderscout Grit	Kinderscout Grit Mam Tor Sst	Longnor Sandstone
		R1b	Parsonage Sst	Todmorden Grit		
		R1a				
Alportian	H2	H2c				
		H2b				
		H2a				
Chokierian	H1	H1b	Sabden Shale		Bowland Shale (Edale Shales)	Morridge Formation Shales
		H1a				
Arnsbergian	E2	E2c	Middleton Grit			
		E2b	Nessfield Sst			
		E2a				Minn Sandstone
Pendleian	E1	E1c	Pendle Grit			
		E1b				
		E1a	Bowland Shale			

Figure 3.4 A) The recommended lithostratigraphic names for the Namurian in the Southern Pennines. The names reflect the large-scale facies changes across the basin (after Waters *et al.*, 2011). **B**) Lithostratigraphic names applied historically to Namurian sediments, emphasizing the names of major sandstones. These names are generally used in this book, as they allow greater precision and equate closely with most BGS maps and memoirs.

up most of the upper part of the Namurian, and is capped by the Rough Rock, the uppermost Namurian sandstone. Up to seven formations are recognized within the Millstone Grit Group. The three uppermost, the **Hebden**, **Marsden** and **Rossendale Formations** are widespread, and represent mainly the cyclothemic succession that typifies the Millstone Grit. The lower part of the Group comprises formations of more limited extent. In the Bowland Basin, the **Pendleton**, **Silsden** and **Samlesbury Formations** are geographically restricted and comprise a mixture of deep- and shallow-water sediments. In the Staffordshire Basin the **Morridge Formation** spans much of Namurian time and comprises mainly deep-water sediments up to the mid-Marsdenian. It is equivalent to the Bowland Shale Formation further north and differs mainly by having a significant component of thin fine-grained, quartzitic sandstones.

The various formations that make up the Millstone Grit Group include both sandstones and finer-grained sediments, in many cases organized in coarsening-upwards cyclothems that commonly coincide with goniatite sub-zones. Typically, there are several such cyclothems in any formation, although there has been no formal attempt to identify the individual cyclothems or their components as 'members'. Instead, it is convenient to refer to cyclothems by the goniatite that characterizes the basal marine band (e.g. R_{2c1}, *B. superbilinguis* cyclothem) or to the named sandstone that caps the cyclothem (e.g. Chatsworth Grit cyclothem). The problem with the latter approach is that the sandstones can vary in name (e.g. the Chatsworth Grit equivalent in the north is the Huddersfield White Rock) and so using the goniatite index (e.g. R_{2c1}) is the most secure means of communicating correlation. The various names of sandstones are, however, still valid and, whilst some rationalization has occurred, lateral name changes are common, particularly between older BGS map sheets (Fig. 3.4B).

Westphalian

Within the region covered by this Guide, the Westphalian is represented by **Coal Measures** sediments that outcrop mainly on the flanks of the Pennines and lie outside the boundaries of the study area. The Lower Westphalian succession includes several major and widespread sandstones that form prominent features in the landscape around the margins of the various coalfields. By contrast the Middle Westphalian, which is the main coal-bearing interval, lacks laterally extensive sandstones. This gives rise to lower ground with few natural exposures and is, in many places, heavily populated due to coal mining and its associated industries.

The Westphalian is subdivided into four stages, of which only the earliest, the **Langsettian**, is mentioned further in this Guide. In earlier literature, the stages are commonly referred to Westphalian A (**Langsettian**), B (**Duckmantian**), C (**Bolsovian**) and D. The three lowest stages are defined from sections on the eastern flanks of the Pennines and their bases are defined at goniatite-bearing marine bands. The highest stage (Westphalian D) is now termed the **Asturian** and is defined in southern Spain. Whilst goniatite-bearing marine bands serve to define stage boundaries, they are less abundant compared with the Namurian, and other biostratigraphic criteria are important. Palynomorphs (spores and pollen) and macroflora are typically used, whilst non-marine bivalves also have a role.

The lithostratigraphy of the Coal Measures is simple at the coarsest level, with the **Pennine Coal Measure Group** being subdivided into the **Pennine Lower**, **Middle** and **Upper Coal Measures Formations**. The first of these coincides with the Langsettian. Several major sandstones occur, most obviously in the Langsettian. Amongst the more important are the Crawshaw Sandstone, the Elland Flags, the Greenmoor Rock and the Grenoside Sandstone.

3.3 Permo-Triassic

Variscan compression, folding and uplift was followed by a period of deep erosion during Permian and Triassic times, during which up to several kilometres of Carboniferous sediments were removed. All this eroded material was transported away from the area and there are no Permian sediments preserved. However, some indirect evidence of this phase is found in the local reddening of Carboniferous sandstones, for example the Pendle Grit at Waddington Fells Quarry. This suggests that the Permo-Triassic land surface may not have been far above the top of the present-day outcrop.

Triassic sediments in the area are mainly confined to the Cheshire Basin. There, the unconformity with the Carboniferous succession is mainly deeply buried, and Carboniferous rocks are not encountered to the west until North Wales, near Wrexham. However, at the edge of the Pennine area, local outliers of Triassic sediments, up to 200 m thick, are present around Leek, where red pebbly sandstones are equated with the **Sherwood Sandstone Group**, a thick and extensive interval that occurs widely across the Midlands and Cheshire.

3.4 Miocene

Miocene deposits are preserved only locally in pocket deposits on the Derbyshire limestone massif, typically associated with areas of dolomitization. The pockets are remnants of a wider sheet of sands and clays, deposited in fluvial settings and probably derived from reworked Triassic sediments. The sediments are assigned to the **Brassington Formation,** the uppermost part of which contains pollen and spores indicating a mid to late Miocene age (*c.*10 Ma). These sediments are preserved in karstic dissolution pockets. During dissolution, the sediments were deformed as they collapsed into the developing voids, probably following the phase of Pliocene uplift.

3.5 Quaternary

The Cheshire Plain and the Craven Lowlands south of Skipton are covered by thick deposits of Quaternary glacial and pro-glacial sediments. In the Craven Lowlands, the hummocky nature of the landscape records glacially moulded drumlins, whilst in the Cheshire Plain there are major outwash complexes that record glacial standstills. The higher Pennine areas lack significant Quaternary sediments. On the fringes of upland areas, glacial deposits such as boulder clay extend up some lower hillsides and occur in some lower-lying valleys, for example around Leek. The extent to which higher areas were glaciated varies considerably. In the Craven Uplands, along the southern margin of the Askrigg Block, glaciation was widespread, and major topographic features such as Malham Cove and Kilnsey Crag are moulded by glacial and sub-glacial processes. The glaciation of this area is also a major factor in the development of widespread limestone pavements.

Further south, the Pennine Moors and the Derbyshire Massif display less evidence of significant glaciation during the Late Pleistocene. This

Figure 3.5 Land-slipped masses at Mam Nick on the southern side of Edale where the Mam Tor Sandstone has slipped over the Bowland (Edale) Shale mudstones.

may explain the lack of limestone pavements in Derbyshire, compared with the Craven area. These southern areas were, however, subjected to periglacial processes whose products commonly mask the underlying geology, part of the 'Drift' of some BGS maps. These are mainly solifluction deposits ('Head' in some maps) on the steeper hillsides, where loose material was moved downslope by freeze-thaw processes.

The landslips that are present on many steep hillsides are of post-glacial origin (Fig. 3.5). Some, like Mam Tor (Figs 10.4 & 10.8) and along the Snake Pass are still active, whilst others, such as those on the western flanks of Kinder Scout, Alport Castles and the southern side of Longdendale, are static.

3.6 Anthropocene

In March 2024, after over 15 years of deliberation, the official bodies concerned with stratigraphic names rejected formal adoption of the 'Anthropocene' as a discrete geologic period by a wide voting margin, primarily due to a dispute over the proposed start time.

Consequently, in this Guide, we use the term 'anthropogenic' where describing human influence that has led to significant permanent or long-term changes to the landscape. At the coast, these might be changing rates of erosion or deposition. On land, changes are mainly associated with removal and redistribution of rock and building products through quarrying and, locally, by the building of dams.

Chapter 4

Sedimentology, Provenance and Palaeogeography

4.1 Palaeogeographic context

To fully understand the depositional history of the Carboniferous successions, it is important to appreciate the wider context of the Pennine basins. Chapter 2 pointed out that the Pennine area was located within an area of continental crust that lay close to the contemporaneous equator. To the north lay a widespread belt of late Silurian–early Devonian continental collision, the Caledonides, which was a mountainous area comparable to the present-day Himalayas. The Atlantic Ocean, of course, was a long way in the future. To the south lay a zone of active Variscan tectonism that extended through northern continental Europe (Fig. 2.1). The Caledonide mountains provided catchment areas for large river systems that drained source areas for much of the clastic detritus that found its way into the Pennine basins (Fig. 4.1). Other catchment areas supplied material from time to time. The Variscan mountains shed sediment northwards in late Carboniferous times but none was carried beyond the Midland Landmass to reach the Pennine basins.

As well as tectonic activity, the other main control on Carboniferous sedimentation from mid-Viséan times onwards was a pattern of large-scale oscillations in sea level due to waxing and waning of glaciers in Gondwana, around the contemporaneous South Pole. These eustatic changes of sea level were important in controlling the water depth in depositional areas, helping to drive periods of transgression and regression and, throughout the Namurian, controlling the salinity of basin waters. The interaction of eustatic sea-level changes and on-going subsidence controlled *relative* sea level and the generation or destruction of accommodation space. These cyclic changes, which occurred over a range of amplitudes and periodicities, led to flooding surfaces, to fluvial

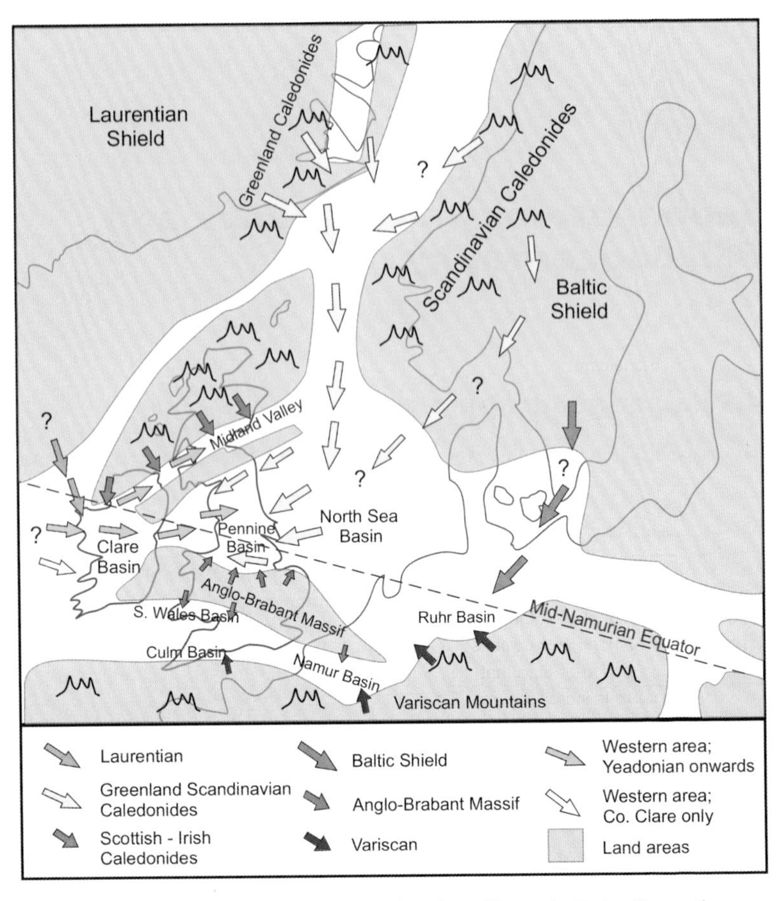

Figure 4.1 Large-scale palaeogeography of northern Europe in Carboniferous times, showing the main routes by which clastic sediments were supplied to the various basins (based on Jones, 2022).

incision and to the accumulation of sediment in repetitive cyclothems (Figs 4.2 & 4.3).

The large-scale palaeogeographic setting has important implications for understanding Carboniferous sedimentation. First, the large distances from an open ocean severely limited access to the basins for tides, other than weak ones generated entirely within the basins. Some Dinantian limestones show cross-bedding that suggests effective tidal currents, but in Namurian sediments there is no convincing evidence for tidal activity. This bears out the results of computer modelling, which predict minimal tidal activity in these basins at that time.

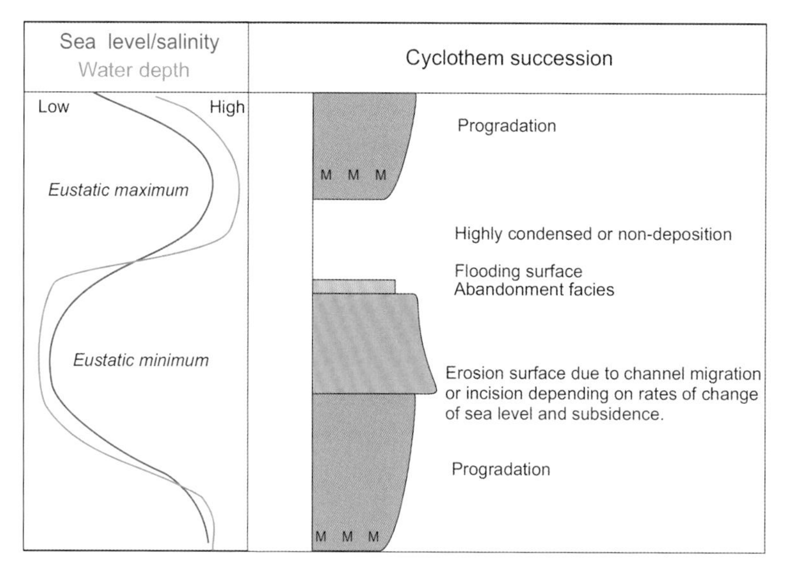

Figure 4.2 A schematic profile showing the relationship between the elements of a typical progradational cyclothem and the changing water depth, sea level and salinity as a result of fluctuations in eustatic sea level and an on-going steady subsidence. Note particularly the period of time that may be represented by very little or no sediment when water depth is increasing most rapidly.

Second, whilst normal salinity prevailed during the Dinantian, distance from the open ocean seems to have had consequences for the salinity of water in the basins during the Namurian. The intra-cratonic setting meant that basins were a long way from open ocean, and the inflow of fully marine sea water must have depended on long-distance connecting straits. It is difficult to be sure exactly where the open ocean and the connecting straits lay and how they behaved at different times throughout the Carboniferous. There was some communication to the east via the Donetsk Basin, but this may not have been the only connection. The dimensions of connecting straits would have determined the volumes of water that could be exchanged between basins and ocean. Large fluvial sandstones are present in the wider Carboniferous basin complex from early Viséan times onwards. The rivers that delivered such volumes of sand must also have delivered huge volumes of fresh water, which can only have reduced basin-water salinity in constricted basins, rather like the present-day Baltic Sea, where salinity is lowered by the large rivers that enter it, especially from Sweden. A similar situation occurs with Lake Maracaibo in northern Venezuela. Both examples can help to envisage conditions in north European Carboniferous basins.

Fossil evidence suggests that, throughout the Dinantian, the basin waters were fully marine. During Tournaisian times, they were locally hypersaline. However, fossil evidence suggests that, from early Namurian times, basin waters were not always fully marine. The occurrence of discrete, fossiliferous marine bands isolated between thicker intervals of basin mudstones that lack macrofauna suggests that fully marine conditions were punctuated by long periods when reduced salinity prevailed. During intervals of high sea level, water exchange with the ocean seems to have been sufficient for fully marine waters to invade the basins, whilst during periods of lowered sea level, high river discharge diluted some or all of the water column. Whether this dilution was confined to the upper part of a stratified water column or involved a total flushing out of marine water is uncertain. Some geochemical evidence suggests that desalination was confined to the upper part of the water column, which would have eliminated pelagic marine faunas there, whilst the lower part of the stratified water column may have been sufficiently anoxic to eliminate organic activity there. The distribution of trace fossils suggests that desalination became more prevalent through the Namurian.

The early Namurian (Pendleian) records the diminution of the widespread precipitation of limestones from warm tropical waters and the onset of deposition of dominantly siliciclastic sediment. These changes probably reflect a major change in the configuration of ocean-connecting straits, as a similar change occurs across South Wales, Ireland and northern continental Europe, as well as eastern North America.

The Milankovitch climatic cycles that controlled the glacial regime in Gondwana, and hence sea-level changes worldwide, would also have driven palaeoclimatic changes beyond the southern hemisphere. Palaeoclimate, particularly patterns of precipitation, would have undoubtedly influenced the delivery of clastic sediment. However, it is not clear how the medium-term changes that drove the expansion and decay of glaciers in the southern hemisphere were expressed in the near-equatorial setting of present-day Europe, particularly as they would also have influenced the extent and types of vegetation. It is likely that vegetation changed in response to the same climatic cycles that drove changes in sea level. The interplay of topography, changing patterns of vegetation cover and precipitation in the sediment source area would have determined the nature and volume of the sediment delivered. However, for each of these factors we have only indirect evidence, and there is uncertainty about the feedback relationships between them. The only direct evidence of vegetation comes from coals from reworked plant material within the sedimentary successions.

A wet tropical climate would have favoured deep chemical weathering, particularly in low-relief lowland areas. However, in the main upland sediment source areas, it seems that weathering was less extreme, since feldspar is a prominent component in northerly derived sandstones, suggesting high topographic relief and rapid erosion with limited chemical weathering. The main source area of clastic sediment, lying to the north or northeast in Scandinavia and Greenland, was a major mountain range or ranges created during the Caledonian orogeny in late Silurian and early Devonian times (Fig. 4.1). Such a mountain range would have interacted with changing equatorial weather systems, perhaps giving a monsoonal regime at times.

4.2 Dinantian sedimentology

Throughout the Dinantian, the supply of clastic sediment to the Central Pennine Basins was very restricted, with only small volumes of sand and clay finding their way to the northern and southern margins. However, further north, from North Yorkshire to Northumberland, clastic sediment forms a significant component of Yoredale cyclothems, the products of deltas which extended as far south as the Askrigg Block. The fluvial Fell Sandstone (Holkerian-Arundian) is largely confined to the Northumberland Trough. At the southern boundary of the Pennine Basin complex, limited amounts of sand were delivered from small catchments on the Midlands Landmass. The absence of significant clastic sediment meant that warm tropical seas prevailed across the Southern Pennine area, giving ideal conditions for precipitation of calcite and aragonite, by algae and other organisms that accumulated over a range of water depths. Shallow-water carbonate precipitation broadly kept pace with subsidence on the more slowly subsiding parts of tilting fault blocks, and these therefore remained shallow, developing into carbonate platforms. By Holkerian times, the Central Pennine Basin Complex had evolved a highly differentiated bathymetry with relief up to several hundred metres. Platforms experienced variable levels of wave and tidal energy. At times, platform margins were fringed by belts of microbial mud mounds or by belts of high-energy shoals. Beyond these margins, the sea floor fell off into deeper water, probably a few hundred metres deep, where basins accumulated mudstones and interbedded limestones, some of which are turbidites along with large, deep-water 'Waulsortian' mud mounds.

The limestone platforms are exposed at outcrop on the Askrigg Block and on the Derbyshire Massif. Between these areas, carbonate platforms are known in the sub-surface, detected by seismic surveys and by boreholes. The Bowland High, the Central Lancashire High and the Holme High all accumulated shallow-water carbonates, some of which were resedimented into adjacent deep water.

The facies and environmental interpretation of limestones depend in large measure on their constituent grains and on their textural fabrics. Some of these features are visible in the field, best aided by a hand lens, but their detailed description and interpretation really depend on microscopic petrography. Some of the interpretations presented here depend on published petrographic work, which is not discussed in great detail.

Limestone depositional facies are described below in relation to major sub-environments.

Platform limestones

These comprise the majority of limestone on both the Askrigg Block and on the Derbyshire Massif. Colours vary from palest grey to almost black, depending on content of clay, organic matter and the level of reducing conditions. Pale limestones tend to be more massive and thickly bedded, whilst darker limestones are often more thinly bedded, commonly associated with muddy interbeds and chert nodules. Platform limestones are commonly rather uniform and parallel-bedded, at least at the local scale (e.g. Fig. 11.38). Many are fine-grained micrites and wackestones with variable components of coarser grains, including faecal pellets, and whole and fragmented fossils, which occur in life position or as broken and reworked fragments. They may be widely scattered in a finer matrix or be concentrated into discrete beds or lenses.

Micritic limestones were largely precipitated as lime muds, mainly by algae and bacteria. In some micritic limestones, fine lamination is well preserved, suggesting a lack of burrowing organisms, perhaps due to hypersaline or anoxic conditions.

Coarser-grained limestones, calcarenites, usually have bioclastic debris and pellets as the main components. Calcarenites vary in texture from less well sorted with a micritic matrix to well sorted with sparry calcite cement. There is a relative scarcity of ooids in Pennine Dinantian limestones compared with equivalents south of the Midlands Landmass. Parallel and ripple lamination and occasional cross-bedding provide evidence of episodes of high-energy currents or waves but scarcity of oolites suggests that persistent tidal currents were less important. Less predictable events, such as storms, probably accounted for much of the high-energy sediment transport. Large-scale cross-bedding near the fringes of the Derbyshire platform is attributed to shoals that migrated towards the margin and down the upper slope. These may have been ebb-tidal deltas or have been driven by storm-related currents (Figs 4.5 & 10.2). However, detailed interpretation of the hydrodynamic regime in carbonates is commonly limited where depositional lamination is disturbed or destroyed by burrowing.

Asbian and Brigantian platform successions display cyclic repetitions of facies, cyclothems, at the scale of several metres thickness, within which compositions, textures and fossil content can vary in different ways (Fig. 4.3). However, such changes may be difficult to convincingly demonstrate in the field and often require careful petrography. The cyclothems are thought to record the build-out of nearshore flats, following a rise in sea level. Some cyclothems have a laminated crust in their upper part along with disturbance by rootlets, indicating sub-aerial emergence. Such surfaces show irregular scalloping or mammilation and are commonly overlain by thin clay layers locally termed 'clay wayboards' These clays include volcanic tuffs, typically potassium bentonites (Fig. 4.4). Clay wayboards define the main bedding, and their weathering can produce steps in the present-day topography. The surfaces record subaerial emergence, cementation as beach rock and, in some cases, colonization by plants to give a palaeosol, followed by karstic dissolution. Palaeokarstic relief in Derbyshire is mostly quite small, a few tens of centimetres, although there are isolated larger solution pits on the platform. Elsewhere, for example in Anglesey, relief of up to several metres occurs in platform limestones of similar age. Gordon Walkden, the pioneer of modern sedimentological studies in Derbyshire, recognized and documented

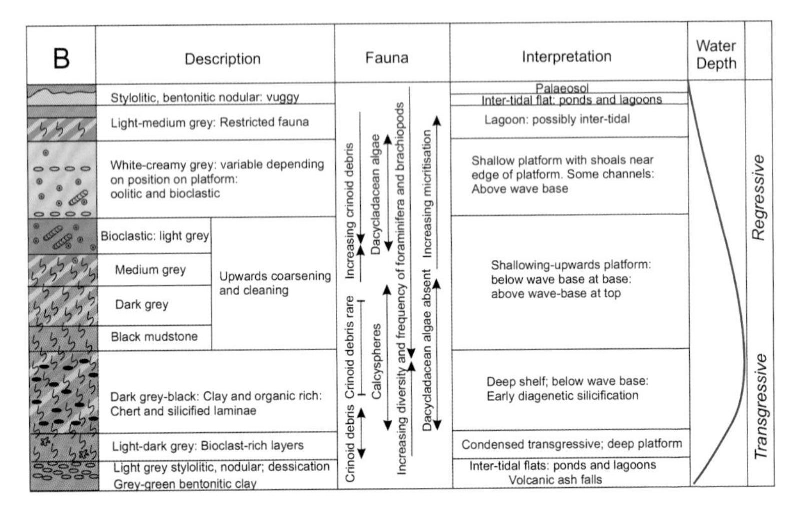

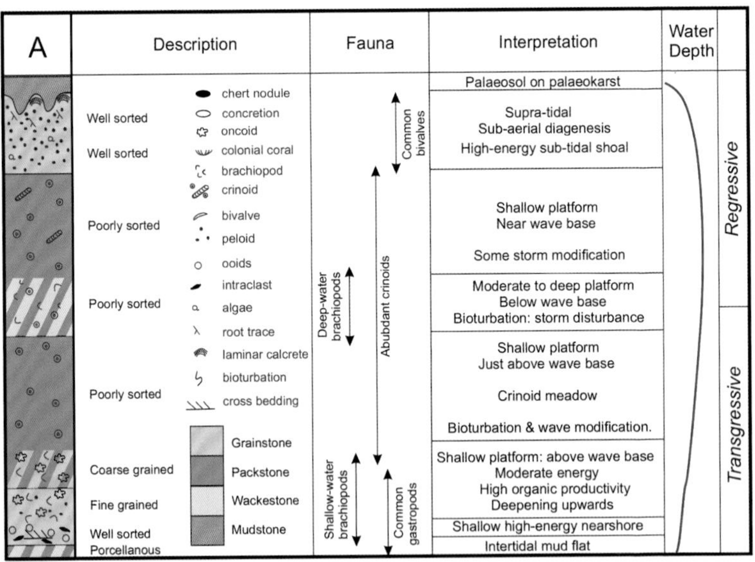

Figure 4.3 Idealized logs of the maximum development of a cyclothem in platform limestones of (**A**) the Asbian Bee Low Limestone and (**B**) the Brigantian Monsal Dale Limestone, based on the southern part of the Derbyshire Platform (based on Walkden & Oakman, 1982).

these palaeokarstic intervals and radically changed understanding of the platform. Coincidentally, an American geologist, Ed Purdy, carried out experimental studies on karstic erosion of limestone blocks and compared his results with the Quaternary history of modern reefs. This also changed perceptions of modern reefs and carbonate platforms, showing that subaerial karstic erosion during Pleistocene low stands could create forms of negative relief (like channels) and positive relief (like tower karst). These results provide analogues for palaeokarstic erosion on the Derbyshire Platform.

On the Derbyshire Platform, the character of the cyclothems changes with age. In the Bee Low Limestones (Asbian), clean limestones appear to have been deposited shortly after transgression (Fig. 4.3A) whereas in Monsal Dale Limestones (Early Brigantian) the lower parts of cyclothems have more abundant and widespread clays and organic matter, which persist upwards for a significant thickness (Fig. 4.3B). It has been suggested that this indicates decreasingly efficient carbonate productivity on the platform.

Figure 4.4 A palaeokarstic surface with erosional relief on top of the Bee Low Limestones. It is overlain by a clay horizon and irregularly bedded thin limestones at the base of the Monsal Dale Limestones. The surface represents a widespread unconformity. Redhill Quarry (see locality description for full context).

Shallow-marine limestones dominate the platform areas, but organic-rich and commonly laminated muddy limestones occur in local intra-platform basins along with redeposited limestones and sediments disturbed by slumping.

Platform margins and basin floors

As deep-water basins developed and persisted through the late Dinantian, the transition zones between carbonate platforms and deep basins took on a range of forms. Some, such as those that characterize the Asbian–Brigantian on the north and west sides of the Derbyshire Massif, are steep and are associated with a narrow but quite continuous 'fringing reef' belt of Asbian age. The margins are typically associated with underlying normal faults (e.g. Fig. 10.4). Elsewhere, and in pre-Asbian successions generally, more gradual deepening occurred, creating depositional ramps down which sediment was transported from the platforms and on which further carbonates were precipitated *in situ*. In the Bowland Basin, broad ramps extended from the Askrigg Block and from the Bowland and Central Lancashire Highs. Similar ramps were probably present around buried highs such as the Holme High. On the higher parts of ramps, shallow-water processes such as wave action and storm-related currents were active along with the generation of new biogenic carbonate. Lower on the ramps, slumps and sediment gravity flows, particularly turbidity currents, carried sediment down slope. Distally from the ramps, the deepest basin areas probably had very low gradients and hemipelagic muds dominated, locally interbedded with carbonate turbidites.

Less steeply inclined platform margins occur at outcrop around the southern and eastern flanks of the Derbyshire Massif, where platform sediments pass rapidly out into gently dipping ramp deposits. This margin, at times, included mud mounds and a belt of carbonate shoals (Fig. 4.5). Immediately basinward of the margin, the gradient of the upper ramp was locally sufficient to cause slumping.

Deep basinal sediments of early Carboniferous age are only seen at outcrop in areas of massive Variscan tectonic inversion. In the area covered by this Guide, such outcrops are confined to the Bowland Basin and to the south and west of the Derbyshire Platform. In the Bowland Basin, the Craven Group deep-water facies comprise limestone

Figure 4.5 Large-scale cross-bedding in the Monsal Dale Limestones (Brigantian) at Dale Quarry, Wirksworth produced by possible tidal shoals, close to the platform margin and formed by currents directed off the platform.

turbidites interbedded with dark, bituminous mudstones in thick, rather monotonous successions. The Pendleside Limestone (Asbian) is a thick development of such turbidites. Ongoing faulting and growth of folds during the Viséan within the Bowland Basin influenced basin-floor topography. This controlled the distribution of large ('Waulsortian') mud mounds and made the distribution of turbidites complex and unpredictable. Similar turbidites occur in the Milldale (Chadian) and Hopedale Limestones southwest of the Derbyshire Platform, where they are also associated with large Waulsortian mud mounds. Inferred limestone turbidites were also encountered in the Edale and Alport boreholes north of the Derbyshire Massif.

Microbial carbonate mud mounds

Within the generally well-bedded limestones of both the deeper-water ramps and the shallow-water limestone platforms are carbonate mud mounds of inferred microbial origin, ranging in age from Chadian to Brigantian. Where exposed at outcrop, more readily eroded, overlying muddy beds may be totally or partially removed so that the mud mounds are exhumed in the present-day landscape as conspicuous topographic features (Fig. 8.2). Rounded hills range in size from large features, over 100 m high and up to 1 km in width, to small features a few metres high and tens of metres wide. Some have steep flanks with gradients of the order of 30°, whilst others have low relief.

Chadian examples within the Bowland Basin (Fig. 8.2) occur in the Clitheroe Formation, a unit that was deposited during a period of deepening from shallow-water Chatburn Limestone below to deeper-water Hodder Mudstone above. Chadian examples in the Staffordshire

Basin occur as an extensive belt of within the Milldale Limestone Formation close to the southwest corner of the Derbyshire Massif between Lower Dovedale and the Manifold Valley. Mounds of this type are commonly described as 'Waulsortian', after similar features of comparable age in Belgium.

Asbian and Brigantian mud mounds, on the other hand, occur in on-platform and platform-margin settings in both Derbyshire and Yorkshire.

Under conditions that allowed vertical aggradation, mud mounds produce thick, near-vertical and narrow bodies of structureless limestone that sit close to a platform margin and have been termed 'fringing reefs'. They lie between flat-lying platform limestones and breccias and other beds that dip steeply towards the adjacent basin. Such a fringing reef belt characterizes much of the northern and west margins of the Derbyshire Platform from the Castleton area to Upper Dovedale. With less vertical aggradation, clusters of smaller mounds occur close to the platform edge, such as those at the top of the Monsal Dale Limestones at the National Stone Centre.

These features, which in this Guide are usually termed 'mud mounds', have variously been called 'reefs', 'reef knolls', 'apron reefs', 'fringing reefs', 'Waulsortian reefs' and 'bioherms', diverse terms that could suggest that the features are highly diverse. However, apart from size and shape, they have much in common. Most differences relate to depositional context, including inferred water depth. The use of the term 'reef' is potentially misleading as it is commonly associated with shallow water and a dependency on a robust organic framework. There is no evidence of framework building organisms in Dinantian examples.

Settings

Mud mounds occur in a variety of depositional settings. Smaller examples, some with rather tabular forms, are confined to the shallow-water Derbyshire platform. Off-platform mounds ('Waulsortian reefs'), which can occur in large and in multi-mound complexes, are associated with ramp or basin-floor settings, as in the Courceyan of the Bowland Basin (Fig. 8.2) and the Chadian of eastern Staffordshire, quite close to the southwestern margin of the Derbyshire platform. Microfacies analysis of these mounds suggests that they developed in water depths ranging from around 130 m to several hundreds of metres.

Mud mounds near the edges of platform areas that have been characterized as 'fringing' or 'apron reefs', although they probably grew under a significant depth of water and only emerged during periods of lowered sea level. Some examples occur as discrete, isolated mounds in a particular depth-controlled zone, as in the 'Cracoe Reef Knolls' at the southern fringe of the Askrigg Block (Fig. 4.6).

The 'Cracoe Reef Knoll' belt lies roughly along the Middle Craven Fault in strata ranging in age from Holkerian to Brigantian. The oldest, smaller 'mud mounds', of Holkerian and early Asbian age, developed on a ramp sloping from the Askrigg Block platform to the Bowland Basin. The largest mud mounds, the 'reef knolls', lie slightly to the north of the smaller features, just south of the Middle Craven Fault. They constitute a discontinuous 'apron reef' that separates the platform to the north from a ramp into the deeper-water Bowland Basin to the south. They developed in late Asbian times on a hanging-wall step when the fault was active and rapid subsidence created the necessary accommodation space. Some flank areas have boulder beds of limestone blocks, which probably resulted from Brigantian inversion on the Middle Craven Fault, allowing a period of erosion before burial by Bowland Shale mudstones.

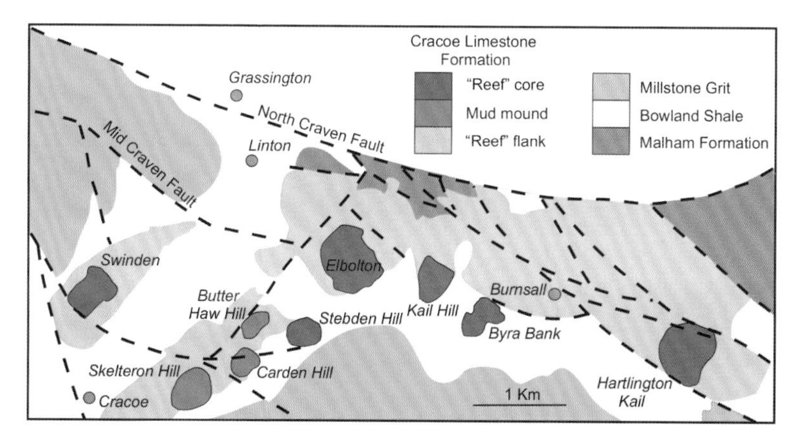

Figure 4.6 Map of the Craven Fault Zone showing the distribution of mud mounds ('reef knolls') and their associated facies in the Cracoe area. The Swinden mound is largely quarried away (based on Waters *et al.*, 2017).

They also occur as more continuous fringes at sharply defined platform margins around the northern and western flanks of the Derbyshire Platform where mud-mound deposits, of Asbian age, aggraded vertically for many tens of metres, although only a few tens of metres wide. They are laterally continuous and their position is close to the steep platform edge. Around the eastern and southern margins of the Derbyshire Massif, the setting of Brigantian mounds was at the top of a more gently sloping ramp where the fringing belt was less sharply defined. Their position shifted in response to changes in relative sea level in Brigantian times and they are preserved as relict features on top of the Monsal Dale Limestones. They were subaerially exposed during the intra-Brigantian fall in sea level and were modified by karstic processes before being buried by the late Brigantian Eyam Limestones when sea level rose.

Facies relationships

Mud mounds are most simply characterized as comprising two facies, a core and flanking deposits. Cores mainly consist of massive, structureless mud-grade limestones, micrite or wackestone, some of which may be pelleted. In some places, there is some reported stromatolitic lamination

Figure 4.7 An example of facies in the core of a mud mound. Structureless carbonate mud has small scour pockets containing fossils, mainly brachiopods. Mound core facies are quite variable. Monsal Dale Limestone, Coal Hills Quarry (Stop 6) National Stone Centre, near Wirksworth.

towards the top of mud mounds, but this does not account for the main volumes of the mounds. Scattered fossils, both whole and fragmented, are commonly abundant, typically brachiopods, crinoids and bryozoans, sometimes concentrated in scoured lenses, as well as microfossils such as foraminifera. In some examples, there is an increase in the occurrence of

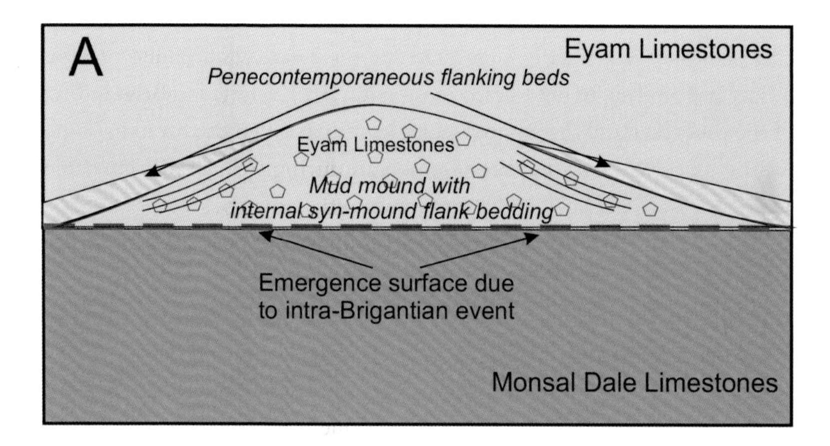

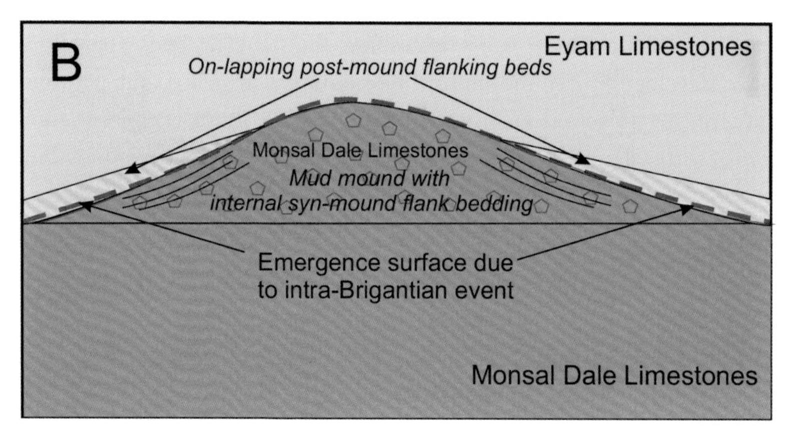

Figure 4.8 Alternative models for the stratigraphic relationships between smaller mud mounds and surrounding sediments in platform and platform margin settings of the Derbyshire Platform. **A**) This model, suggested by Ford (1977) and by BGS mapping (Aitkenhead *et al.*, 1985) has the mounds at the base of the Eyam Limestones with their flanking beds penecontemporaneous with the mounds or immediately following their 'death'. **B**) The more recent and current model, due to Adams (1980) and Gutteridge (1991b), has the mounds at the top of the Monsal Dale Limestones, preserved as relict features during emergence at the intra-Brigantian lowstand of sea level. Deposition of the flanking beds is then widely separated in time from the mounds. The red line indicates the inferred surface of subaerial emergence in each case.

frame-building organisms such as bryzoans and corals towards the top of the mounds, but in most examples there are no obvious framework-building or sediment-binding organisms. Calcite-filled stromatactis cavities are present in some cases, suggesting early cementation and/or positions of soft-bodied organisms. The sediments give little indication of the origin of the lime mud or of the sediment-binding processes that allowed the mounds to grow. Microbial activity (probably some combination of algae and bacteria) is generally suggested for the near-synchronous precipitation and binding of mud-grade carbonate, but a detailed understanding of the processes and the environmental conditions seems still to be some way off. In contrast, the flanks and tops of mud mounds have more variable sediments both between and around mounds.

The recognition of flank beds is not straightforward. Some are contemporaneous with the growth of the mound whilst others relate to later deposition around a relict mound following a later period of flooding. Internal flank beds are variably bedded and show depositional dips inclined away from the core, sub-parallel with the mound surface. In some examples, the core and flank sediments interfinger, suggesting episodic growth while the mound was growing. In other cases, probably deeper-water examples, the flanking beds appear to have developed after the mound had stopped growing, but without intervening subaerial emergence. In early Brigantian examples from Derbyshire, flanking beds that drape the cores of some mounds are separated from the cores by palaeosols and palaeokarstic features, indicating a period of subaerial emergence and a disconnect between the two phases of deposition (Fig. 4.8). This is well seen at Ricklow and at the National Stone Centre.

Some flank sediments, which closely follow mound growth without any emergence, are richly fossiliferous. Crinoid debris is particularly abundant, suggesting that the surfaces of some mud mounds provided a substrate for the growth of crinoids once mound growth had stopped, but there is no evidence that crinoids had any role in trapping or binding sediments. Contrasting fossil populations on different flanks of individual mounds may indicate varied exposure to prevailing currents and waves. However, there is no clear evidence that relict mounds influenced the distribution of facies in the draping sediments when the two units are separated by an emergence surface.

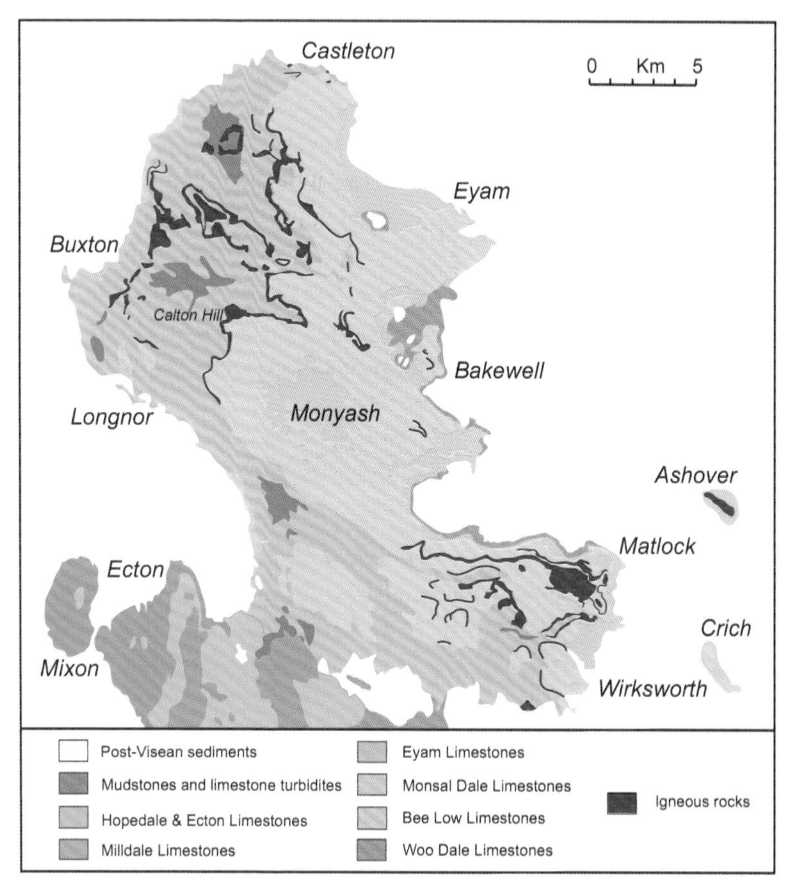

Figure 4.9 The distribution of volcanic rocks across the Derbyshire Massif. Most are lavas that follow the stratigraphy, but there are local intrusive centres. These have not been differentiated on the map (based on Ford, 2002).

The Asbian mound systems that make up the 'fringing reef' belt around the northern and western margins of the Derbyshire platform seem to have little obvious flank beds on the platform side, but the basinward side coincides with the steep platform margin and it often has coarse breccias and beds dipping into the basin. The timing of the deposition of this re-sedimented material can be conjectural.

Interbedded volcanics

Carboniferous volcanic rocks are largely confined to the Derbyshire Massif, where they occur within platform limestones close to the

Asbian–Brigantian boundary in several areas (Fig. 4.9). They are mainly basaltic lavas, often deeply weathered at outcrop so that most detailed knowledge of the volcanic intervals comes from boreholes, although worthwhile sections occur in some quarries and railway cuttings. Those seen at fresher outcrop show evidence of subaqueous extrusion in the form of hyaloclastites and pillow forms. Volcanic intervals are up to many metres thick and may comprise multiple flows, in some cases separated by beds of volcanic ash (tuff). Several volcanic centres have been inferred. A centre near Bonsall was responsible for the lavas around Matlock, whilst those near Tunstead and at Calton Hill probably supplied the Miller's Dale Lavas.

In addition to the lavas, volcanic material also occurs as thin beds of tuff, commonly only a few centimetres thick, which are interbedded within the platform limestones. 'Clay wayboards' that overlie many palaeokarstic surfaces are commonly volcanic, often potassium bentonites. It is not clear whether these air-borne tuffs are derived from the local vents or from more distant sources. Thin layers of pale bentonitic clay are also present in the early Namurian basin mudstones (Morridge Formation) of the Staffordshire Moorlands. The location of the volcanoes responsible has not been identified, but there was volcanic activity in the Midland Valley of Scotland at this time.

Dolomitization

Limestones in the Derbyshire Massif are locally dolomitized, in some places significantly so. Dolomitized limestones are typically rather porous, with a sugary texture, and are often brown in hue compared to the unaltered grey limestones, indicating the addition of some iron during diagenesis. Dolomites in Derbyshire occur in two distinct ways, at different times and involving contrasting settings and processes.

The *first* type, of limited extent, is the laminated dolomite mudstone associated with the development of an intra-platform basin on the Derbyshire Platform in the Brigantian. These record hypersaline conditions during a low stand of sea level under arid conditions.

The *second* type of dolomitization involved limestones reacting diagenetically with magnesium-rich brines. In the southern part of the Derbyshire Platform, the Bee Low Limestones are extensively dolomitized in two areas. One is a zone around 2 km wide and about

12 km long trending northwest from Matlock. The other zone is around 3 km wide and 6 km long between Wirksworth and Brassington, the best exposure being at Harboro' Rocks. The dolomitized limestone in these zones varies in thickness, up to 200 m, and its lower interface with unaltered limestone has considerable relief, suggesting downward percolation of dolomitizing fluids. This dolomitization cuts across the stratigraphy and pre-dates lead-fluorite mineralization.

This type of dolomitization results from alteration by brines expelled from the thick Dinantian and early Namurian mudstone succession of the adjacent Widmerpool Gulf. Continued burial up to late Westphalian times caused these muds to undergo compaction, to be heated and to expel hot, mineral-laden brines which migrated into the adjacent limestone platform. The fluids were probably channelled along major fractures as well as moving through limestone units with higher permeabilities. The same general model applies to the later lead/zinc mineralization (see Fig. 5.4).

Chert

Chert occurs quite abundantly as nodules, in the upper part of the Derbyshire platform limestone succession, particularly in the Monsal Dale and Eyam Limestones. This form of microcrystalline quartz occurs both as isolated nodules, up to tens of centimetres in size, and as strings of nodules that broadly follow bedding (Fig. 11.29). They result from the diagenetic precipitation of silica, whose origin is uncertain. Some may have been derived from siliceous skeletons of marine organisms such as sponges and radiolaria, whilst some may derive from alteration of clay minerals in impure limestones. Cherts were once extensively quarried as a raw material for the ceramic industry of Staffordshire, where they were calcined and milled as a component in stoneware bodies.

Dinantian palaeogeography and tectonics of the Derbyshire Massif

The Dinantian history of the Derbyshire Massif is a record of carbonate deposition under a complex regime of fluctuating sea level and variable subsidence and associated tilting, but with possible episodes of local uplift. The massif is crossed by two major northwest–southeast faults that fragment the massif into three distinct blocks, and these responded somewhat independently during extension, The faults also experienced local normal displacements that led to intra-platform basins and depocentres (Fig. 4.10).

The early history of the area is not entirely clear. There was a phase of deepening following the initial base-Dinantian transgression, and this led to a ramp and basin (proto-Widmerpool Gulf), lying to the south of the eventual shallow-water platform. In the deeper-water setting, turbidites and 'Waulsortian' mud mounds of the Milldale Limestones (Chadian) were deposited. The relationship between these basinal areas and the evolving platform is not entirely clear and more work is needed.

However, the well-defined carbonate platform was fully established by late Holkerian times, and shallow-water carbonate production continued on the platform throughout the Asbian with deposition of the Bee Low Limestones. Bioclastic debris from variably fragmented crinoids, brachiopods and corals, and mud-grade carbonate from likely algal sources are the main components. Asbian carbonate mud mounds occurred as a discontinuous fringe close to the platform margin. The mud mounds developed somewhat differently depending on the local tectonic setting and its influence on accommodation space. Close to fault-controlled steep margins, on the northern and western flanks of the Derbyshire platform, mud mounds stacked vertically to give thick, narrow zones, often termed a 'fringing reef', whilst in the south and east, a more gently sloping margin caused the mud mounds to be less constrained in position.

The flat-lying Bee Low Limestones of the platform interior are generally thickly bedded and show significant lateral facies changes. Calcarenites with rounded peloids and oolitically coated bioclasts, close to the marginal mud-mound zone, pass into micritic limestones, suggesting lower energy conditions on the platform. All these limestones are fossiliferous with a wide variety of fauna. The platform interior behind the mud mounds was not an extensive tranquil lagoon as is commonly associated with present-day atolls fringed by coral–algal reefs. Rather, the platform was a mosaic of areas with different biotas and levels of wave and tidal activity.

Deposition took place under a regime of fluctuating eustatic sea level, giving stacked shallowing-upwards cyclothems, separated by surfaces with palaeosols or palaeokarstic dissolution indicating prolonged emergence during low stands (Figs 4.3A & 4.4). Rising sea level gave rise to flooding and the creation of accommodation space for the next cyclothem. However, there is evidence that not all Asbian cyclothems are widespread and related to eustatic changes. In Asbian cyclothems, carbonate deposition was re-established soon after flooding, suggesting that carbonate productivity on the platform was high.

The end of the Asbian was characterized by widespread tectonic activity as well as a fall in sea level that created a significant unconformity at the Bee Low/Monsal Dale (Asbian–Brigantian) boundary. The top of the Bee Low limestones commonly shows erosional relief, much of it the result of karstic process during a prolonged phase of emergence, but some resulting from the erosion of channels with slumped infills on the more gently sloping southeast margin.

Deposition of the Monsal Dale Limestones in the early Brigantian began with the transgression of the platform area from the east, probably when the underlying fault block was tilting to the east. Along the eastern side of the platform, the transgression shifted the marginal break of slope back towards the platform. Mud mounds were extensively re-established at the displaced margin, whilst shoals of carbonate sand formed as tides and storm currents moved sediment off the platform towards the margin. On the platform, flat-lying Monsal Dale Limestones were deposited as stacked, somewhat thinner cyclothems, and the constituent limestones are more thinly bedded with clay interbeds, suggesting diminished carbonate productivity. Lower on the marginal ramp, thin bedded limestones with shaly interbeds resulted from gravity-driven flows.

Tilting led to westwards thinning of the Monsal Dale Limestones, to the extent that the north-western areas of the platform apparently experienced little or no sediment accumulation. However, local fault-controlled depo-centres led to the accumulation of significant thicknesses of limestone even when none accumulated on the surrounding platform. Small outliers have up to 100 m of Monsal Dale Limestones, as seen at Hindlow Quarry, which sits in the hanging wall of the Cronkston/Bonsall Fault. It seems likely that carbonate sediments produced on the platform were exported, either washed over the platform margin or transported to areas where tilting or faulting created more accommodation space.

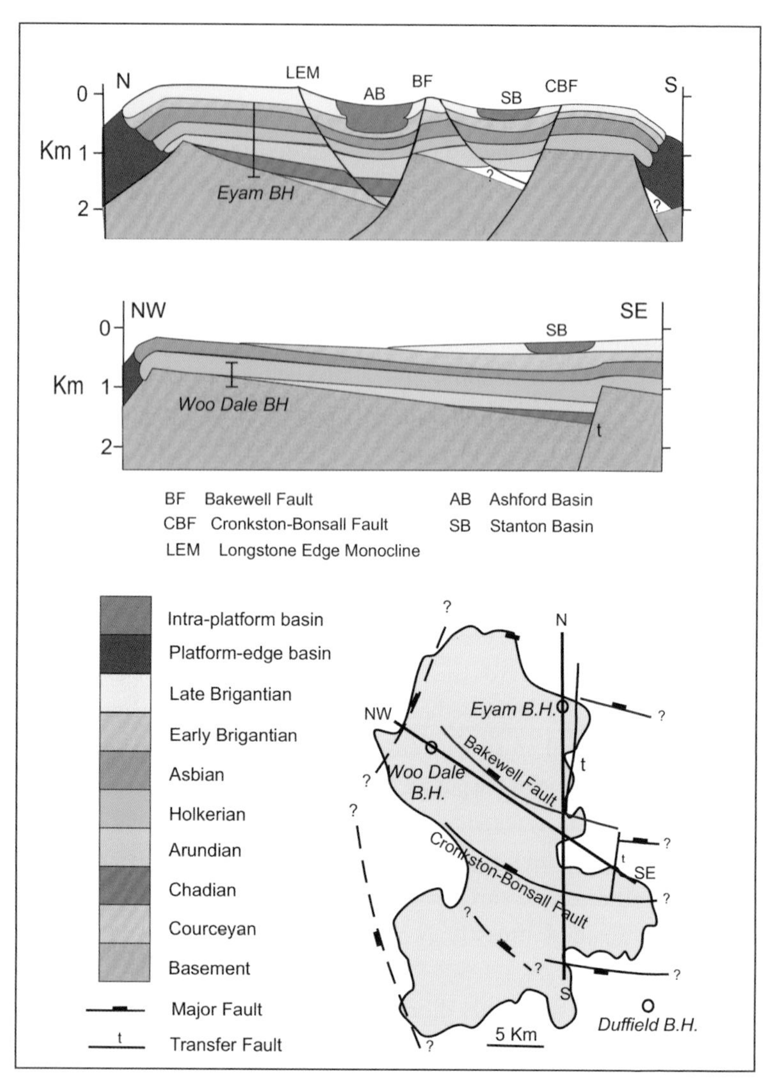

Figure 4.10 Schematic profiles across the Derbyshire Massif, to illustrate the influence on sedimentation of syn-depositional tilting and movement on extensional faults. The Brigantian intervals both show thinning to the northwest due to active tilting. The mud mounds that fringe the platform in many areas are not illustrated. The intra-platform Stanton Basin formed in the late Brigantian and persisted into the early Namurian. The platform margin sediments vary in thickness, character and age from place to place, and their illustration is schematic (adapted from Gutteridge, 1987).

Movement on major cross-cutting faults at this time also led to the development of local, intra-platform basins like the Ashford Basin (Fig. 4.10), giving important thickness and facies variations within the Monsal Dale Limestones. Deeper water and reducing conditions in these basins are inferred from the dark colour of the Rosewood Marble due to organic matter. Episodic resedimentation of adjacent shallow-water platform carbonates by turbidity currents was probably initiated by storms stirring sediment into suspension. A lacustrine phase developed in the basin towards the end of the Monsal Dale limestone deposition, giving rise to the Headstone Laminite, when brackish conditions are suggested by low faunal diversity and fringing calcretes. Extrusion of basalt lavas in the early Brigantian generated local topography which influenced thicknesses and facies of subsequent limestones.

Deposition of the Monsal Dale Limestones was ended by a large fall in eustatic sea level, giving widespread emergence of the platform top, with mud mounds being preserved as discrete features up to a few tens of metres high on the top surface, which is characterized by karstic dissolution and palaeosols. This emergence surface was draped and buried by Eyam Limestones following the Late Brigantian transgression. Recent understanding (herein termed the 'Adams–Gutteridge model', Fig. 4.8) has led to the mud mounds now being assigned to the top of the Monsal Dale Limestones rather than to the base of the Eyam Limestones, as previously thought.

Intra-Brigantian tectonic activity, along with a fall in sea level, may help explain features of north and west margins of the platform. The complex landscape in Upper Dovedale around Parkhouse and Chrome hills (Figs 12.5 & 12.6) involves large fringing mud mounds of Bee Low Limestones. Erosion of draping Namurian mudstones has exhumed a pre-Namurian topography, but this is not purely that of an 'exhumed reef', as is sometimes suggested. The margin is significantly faulted, and the faults, which do not extend up into mudstones, are clearly pre-Namurian. Furthermore, the faulting appears to be dominantly extensional and not related to later Variscan compression. Tectonic tilting and uplift during the Brigantian could have led to fault displacement along the western side of the platform, probably reactivating structures that localized the platform margin in Holkerian times. Upper Dovedale lies at the intersection of the platform margin and the Cronkston–Bonsall Fault Zone and this might have complicated fragmentation of the margin. The extent to which the limestone masses that make up Chrome and Parkhouse hills are simple fault-bounded blocks or reflect mound morphology, is difficult to judge, as karstic dissolution during late Brigantian emergence might have added further complications. Further detailed stratigraphic and structural analysis is needed.

Around some steep platform margins, coarse breccias occur where large blocks fell and accumulated, prior to burial by Bowland Shales. In the breccias at Treak Cliff, the blocks are of Bee Low Limestone, the same as the *in-situ* limestones on which they rest. It is uncertain when these breccias developed and whether that was in a sub-aqueous or subaerial setting. There is a general view that they are predominantly sub-aqueous, but if significant tectonic uplift coincided with a major fall in sea level, then the steep margin may have been partially subaerial during the Brigantian. Possibly relevant to this interpretation are the 'Beach Beds' that occur at the base of Winnats Pass (Fig. 10.5). These coarse limestones with rounded, reworked fossils show significant depositional dips and have been interpreted variously as beach deposits and as deeper-water deposits resedimentated from the platform. Their fauna is early Brigantian in age, making it possible that Beach Beds and the talus breccias are broadly contemporaneous but from different sources. If the Beach Beds are, indeed, beach deposits, that would require a very large fall in relative sea level at the end of the Asbian and help support a case for the talus breccias being subaerial.

At the south-eastern margin of the massif, in the southernmost fault block, tectonic tilting seems to have been less intense, and eustatic changes were, therefore, more important in determining the Brigantian stratigraphy. Monsal Dale Limestones are quite well developed and intra-Brigantian erosion was confined to karstic dissolution during a fall in relative sea level. However, the local complexity and uncertain stratigraphy at Baileycroft Quarry seem to indicate that the Monsal Dale Limestones are locally removed so that Eyam Limestones rest directly on Bee Low Limestones, possibly within a channel. The lower gradient of the platform margin might have favoured such erosion during low sea level.

A major consequence of ongoing tilting was that the western part of the platform was never reached by the late Brigantian transgression, and Eyam Limestones are confined to the eastern side of the platform.

In the Ashford intra-platform basin, the Eyam Limestones conformably overlie the Headstone Laminite. Ongoing tectonic activity initiated the Stanton Basin, and this persisted into the early Namurian as an embayment on the eastern side of the platform. On the north-east corner of the platform, around Eyam and Middleton Dale, mud mounds within the Eyam Limestones are a final development of fringing mounds before carbonate production died out and the platform was draped by clastic sediments, first by the latest Brigantian Longstone Mudstone and then by the early Namurian Bowland Shale.

4.3 Namurian sedimentology

The end of the Viséan was marked by a reduction in tectonic extension and by a major change in depositional regime. Widespread carbonate deposition came to an end, and the supply of siliciclastic sediment extended dramatically southwards. In addition, it seems that the unconstrained connection to the open ocean that prevailed throughout the Viséan became more restricted, with fluctuating exchange of sea water between a distant ocean and the Pennine basins. The increasing supply of clastic sediment, the continuing glacio-eustatic fluctuation in sea level with its implications for accommodation space and water salinity, and the inherited bathymetry of end-Viséan platforms and basins were the main controls on Namurian sedimentation. The progressive infilling and elimination of the end-Viséan bathymetry led eventually to widespread, low-relief, delta-plain conditions that permitted deposition of the Westphalian Coal Measures.

Sedimentation in the various Pennine sub-basins during the Namurian can be simply characterized as having four main phases; basinal mudstones; turbidite sandstones; turbidite-fronted, basin-filling delta sequences; shallow-water deltaic successions, commonly stacked in cyclothems.

Basin mudstones

Basin mudstones mainly record sedimentation prior to the arrival of coarser sediments in a basin. Their age range varies from sub-basin to sub-basin. In the Bowland Basin, the Bowland Shales are of Pendleian age only. In the Edale Basin, further south, basin mudstones extend from Pendleian to mid-Kinderscoutian. West of Blackburn, basin mudstone deposition persisted to the early Marsdenian, whilst east of the Derbyshire Massif and in the Staffordshire Basin, coarse feldspathic sands did not arrive until mid-Marsdenian times. The Sabden Shales of east Lancashire are sandwiched between the Pendleian Pendle/Warley Wise Grits and the Kinderscoutian Parsonage Sandstones. These mudstones record a period of sand-starvation that allowed subsidence to re-establish deep-water conditions.

In basin areas, deep-water mudstone intervals are typically several hundreds of metres thick, although they are thinner where they drape platform margins, smoothing, but not eliminating, the end-Viséan bathymetric relief. Over the limestone platforms, the mudstones are locally preserved, but are thinner and condensed, as shown in boreholes near Ashover.

Basin mudstones are dark coloured and usually fissile. Most intervals are apparently barren of macro-fossils, but successions are punctuated by thin fossiliferous marine bands, often only a few centimetres thick. The fossils include bivalves, thin-shelled brachiopods and goniatites, indicating fully marine salinity and are commonly flattened through burial compaction. However, calcite-cemented concretions ('bullions') preserve the fossils in three dimensions, and some yield the delicate silica skeletons of radiolaria. The shales have a variable content of organic carbon, which tends to be higher within marine bands. The contrast between the fossiliferous marine bands and the fossil-poor interbeds is thought to reflect salinity changes although geochemical results are inconclusive. Some evidence suggests permanent fully marine conditions, but this may only reflect the deepest, near-bottom water. With density-stratified water columns, salinity changes related to sea-level changes may have been confined to the upper parts of the water column where any pelagic fauna lived. A stratified water column would have favoured anoxic conditions at the basin floor, encouraging preservation of organic matter and enhancing the potential of the sediments to be source rocks for hydrocarbons.

Turbidite sandstones

Turbidites in the Namurian are most commonly associated with turbidite-fronted deltas but for a few examples linkage to a delta is not obvious. The main examples are in the Staffordshire Basin, where deep-water conditions persisted until the Marsdenian. There, the pre-Marsdenian (Morridge Formation) succession is dominated by mudstones with marine bands, but it also includes variably developed intervals of quartzitic turbidites that comprise the Minn, Hurdlow and Lum Edge Sandstones. These vary from being very thin beds of fine sandstone in mud- and silt-dominated intervals to thicker-bedded sandstones in sand-dominated intervals. Mapping suggests that turbidite lobes built out into the basin from a southern source, probably arriving through small delta systems, for which there is limited direct evidence. The first turbidites to arrive in the Staffordshire Basin from the northern feldspathic source comprise the Longnor Sandstone, which records spill-over from the Edale Basin to the north of the Derbyshire Massif (see Fig. 10.9).

Turbidite-fronted deltas

These major delta systems (Fig. 4.11) were largely responsible for eliminating the accommodation space inherited as end-Viséan bathymetry. These basin-filling successions overlie basin mudstones and vary in age across the basin complex, with progressively younger successions from north to south. The earliest basin fill, in the Bowland Basin, is Pendleian in age, the basin fill of the Edale Basin in North Derbyshire is late Kinderscoutian, and that of the Staffordshire Basin is Marsdenian, as is the sequence west of Blackburn and at Ramsbottom (see Figs 4.12; 4.13).

These successions comprise three main elements: an interval of turbidite sandstones; a coarsening-upwards-unit, mainly consisting of siltstones and fine sandstones; and an interval of coarse-grained channel sandstones. Successions are typically several hundred metres thick, reflecting the depth of water that was infilled.

The turbidite element interval is quite variable. In the Bowland Basin, turbidites (Pendle Grit) are dominantly channelized from their very base (Fig. 7.8). Their deposition was probably strongly influenced by complex basin-floor topography and by on-going tectonic activity, as suggested at Waddington Fell Quarry (Fig. 8.4). However, the base of the Pendle Grit at the Trough of Bowland shows a gradual incoming of turbidite sands with no conspicuous erosion. In the Kinderscoutian of the Edale Basin, the earliest turbidites (Mam Tor Sandstones) appear

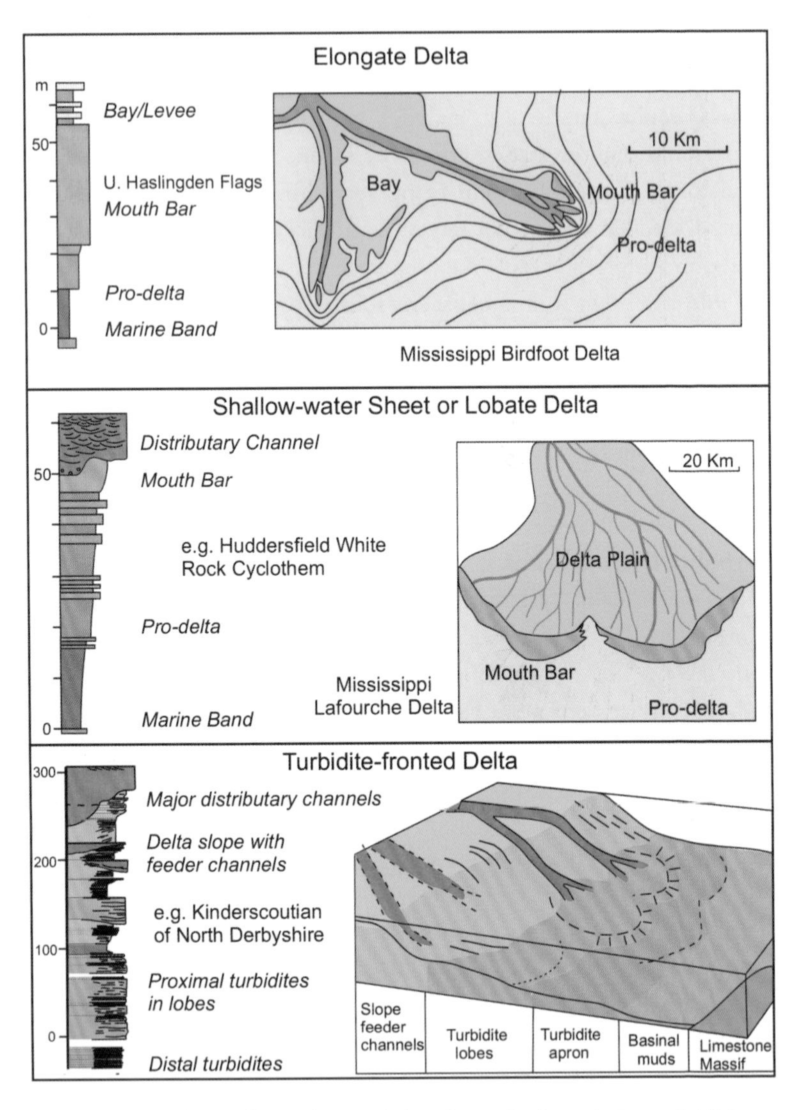

Figure 4.11 Vertical profiles and suggested analogues for the main types of deltaic succession in the Pennine Namurian. The sheet deltas can generate quite variable successions depending on prevailing subsidence and on changing sea level.

quite abruptly above basin mudstones but show no obvious channelling (Fig. 10.6). Neither do they show a gradual incoming of thin sandstone beds as seen at Trough of Bowland but, from the base, they comprise quite thick sandstones that include hybrid event beds. The Marsdenian

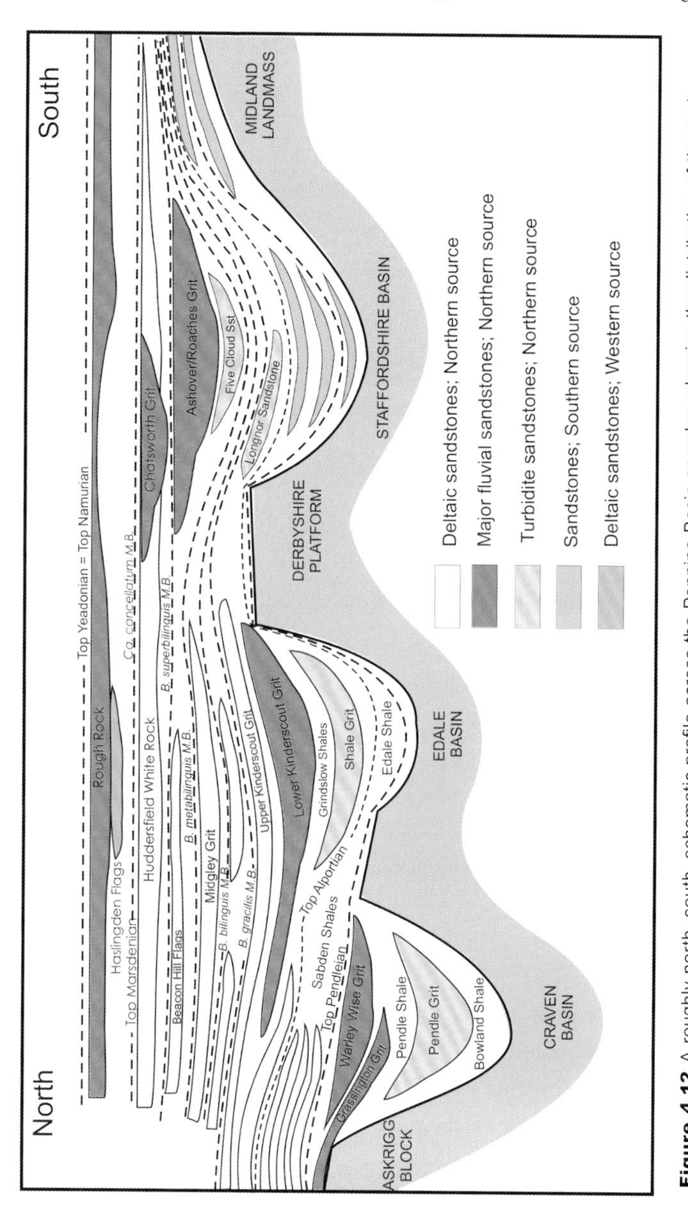

Figure 4.12 A roughly north–south, schematic profile across the Pennine Basin complex, showing the distribution of the major deltaic successions. The turbidite-fronted deltas fill the sub-basins progressively from north to south, illustrating the fill-spill pattern of sedimentation (adapted from Collinson, 1988). A downloadable pdf file is available for this figure.

Alum Crag Grit of Lancashire comprises around 200 m of mainly thick-bedded turbidites with abundant erosion surfaces and associated mudclast conglomerates. These are interbedded with thin units of thinly bedded turbidites, and the succession is split by a mudstone unit containing the R_{2b} (*B. bilinguites*) Marine Band, probably reflecting a rise of sea level which cut off sand supply temporarily. Sadly, this important section was only seen during excavation of a sewer tunnel.

In general terms, the turbidite intervals, which are up to hundreds of metres thick, tend to show increasingly common erosive contacts upwards. Many thick beds are inferred to be amalgamated, although a lack of grading can make this difficult to demonstrate. Patterns of systematic bed-thickness change such as thickening-upward and thinning-upward sequences are uncommon. However, in the Shale Grit (Kinderscoutian) of Derbyshire the overall succession comprises sand-rich packages, tens of metres thick, separated by thinner sand-poor units, mainly mudstone. The contrasting erosion of these harder and softer lithologies is expressed as steps in present-day hillsides that can be traced for kilometres (Fig. 10.10). This pattern of lithological change may have resulted from sea-level fluctuations that switched sand supply on and off, or from lateral switching of sand supply. As a turbidite-fronted delta complex prograded, feeder channels higher on the advancing slope would have switched position over time, feeding sand-rich lobes during active sand supply and leading to mud deposition when supply switched elsewhere.

In their upper parts, turbidite intervals become channelized. Channels are typically of the order of 10 m deep and hundreds of metres wide and are filled with thick-bedded, massive and commonly amalgamated sandstones (Fig. 10.12). In the Pendle Grit, channels can dominate the succession, as seen at Skipton Moor (Figs 7.7. & 7.8). In the Shale Grit, channels are more isolated within a background of sheet-like turbidites. The overall Shale Grit interval (Fig. 10.11) is thought to have been deposited on a pro-delta ramp fed by turbidity currents that by-passed sand from the delta front through channels that crossed the prograding slope. The ramp comprised both lobe and channel elements, with lobes switching in response to shifting of channels up slope. By-passing of sand from the delta front would have been aided by reduced salinity of the basin water, promoting hyperpycnal flows. However, major desalination did not always occur, since trace fossils in the Pendle Grits suggest near-normal marine salinity in the Pendleian.

The coarsening-upward units that overlie turbidite intervals are typically of the order of 100 m thick and mainly comprise siltstones and fine-grained sandstones. In their lower parts, they commonly show

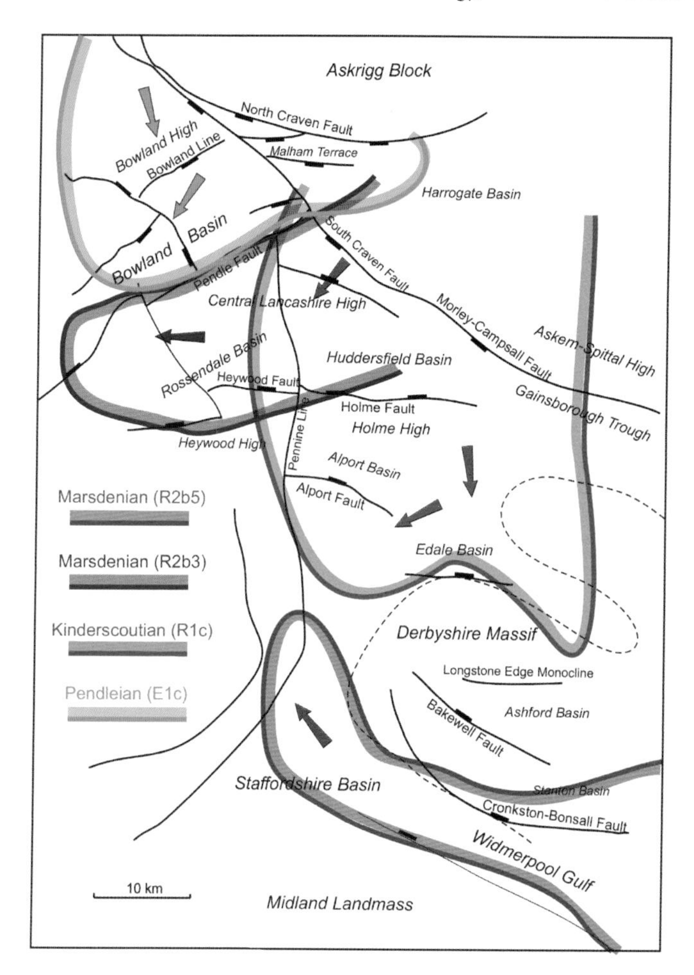

Figure 4.13 The extents of successive basin-filling, turbidite-fronted delta successions across the basin complex through the Namurian. All these systems were supplied from the northern feldspathic source area, despite the palaeocurrents to the NW in the Staffordshire Basin. Their distributions were controlled, in large measure, by basin topography related to major extensional faults and to the distribution of Dinantian limestone platforms.

gradational thin bedding, which may be disturbed by burrowing. Ripples and small-scale cross-bedding are present towards the top in some cases. In addition to the finer-grained sediments, the units include large, widely scattered channel sandbodies made up of generally massive sands. These channels are thought to be the conduits through which sand by-passed the slope to build the base-of-slope turbidite apron. Some channels may have linked back to delta distributary channels, suggesting that hyperpycnal flows were initiated at the river mouths during flood events. Such a model is supported by the scarcity of mouth-bar sandstones, although deep erosion by delta-top channels may have removed such mouth-bar sands. In the Roaches Grit (Marsdenian) in Staffordshire, inferred mouth-bar sands occur down current of large fluvial channels in a zone where delta advance appears to have come to a halt.

The uppermost components of the turbidite-fronted delta successions are major channel sandstones that coalesce laterally to form extensive, commonly multi-storey sheets. Individual channel units are typically up to 30 m thick whilst multi-storey sheets may be up to around 100 m thick with erosional relief on the base up to 60 m. The sandstones are usually coarse and pebbly and in the deeper parts of channels, they tend to be structureless. The higher parts of the channel sandstones are cross-bedded with giant tabular sets commonly overlain by cosets of medium-scale sets. The giant sets, which can be up to 30 m thick, are unique to these channels, occurring in the Warley Wise Grit (Pendleian), in the Lower Kinderscout Grit (Kinderscoutian) and in the Roaches Grit (Marsdenian) of the Staffordshire Basin, and in the Fletcher Bank Grit (Marsdenian).

Channel erosion at this scale makes it impossible to know what facies may have been removed. The cause of such erosion remains a matter of discussion. It may have been an inevitable product of the scale of the river channels feeding the deltas, possibly augmented by large changes in flood discharge. However, the depth of erosion may have been enhanced by falling sea level, which could have promoted rapid delta advance and the by-passing of sands to deeper water. Whilst falling sea levels were probably active, the mechanisms are less clear-cut.

The origin of the giant sets of cross-bedding is still debated. They only occur where the channels are underlain by thick slope deposits. This suggests that they relate to delta progradation into deep water where progradation rates would have been relatively slow, channel gradients would have been relatively high, and channels would have been less prone to splitting into distributaries. Subsidence rates (and hence gradients) may have been enhanced by the rapid compaction of underlying thick basin mudstones. In the Kinderscoutian, giant cross-beds are present in channel sandstones at Earl Crag on the south side of Airedale, whilst they are absent in the correlative Addingham Edge Grit at Ilkley, 6 km to the north (up current). This change coincides roughly with the southern extent of the small deltaic cyclothems that developed at the northern edge of the basin between Arnsbergian and Kinderscoutian times, maintaining relatively shallow conditions in the area. Rapid, compaction-enhanced subsidence may also help to explain the multi-storey nature of the major Kinderscoutian channel complexes (Figs 9.9; 10.11), although sea-level fluctuations may also have played a role.

Whilst these considerations constrain the context for the giant cross-beds, they do not fully explain their development. Before their occurrence within channels was recognized, they were interpreted as sandy (Gilbert-type) deltas that built out following a rise in sea level. Once their channelized setting was established, they were explained as the products of large bar forms, with high slip faces. More recent interpretations suggest they are indeed Gilbert-type deltas, but ones formed in the lower parts of channels during periods of base-level rise. Deep channelling and the giant cross-beds have also been ascribed to a climatic regime that involved major floods, capable of deepening the channel during high flood discharge, with large bars or in-channel deltas developing as flow began to reduce. Such a process may help to explain the massive, structureless sandstones that typically lie between the channel erosion surface and the giant foresets. The massive beds most likely resulted from the very rapid deposition immediately after a major flood peak. Medium-scale cross-bedding that always overlies the giant sets has palaeoflow directions identical to those inferred for the underlying giant foresets and records the migration of dunes over the larger form, feeding sand to the giant slip face.

The widespread extent of the 'Lower Kinderscout Grit', extending from Airedale to North Derbyshire, with giant cross-beds in stacked channels at many exposures is not the result of a single progradation. Several biostratigraphically constrained sandstones, all labelled 'Lower Kinderscout Grit' on BGS maps, occur in progressively younger cyclothems from north to south. This complexity suggests episodic delta advance, with the channel sandstones reflecting first falling and then rising sea level in each cyclothem. There has been considerable discussion of the role of sea-level changes in developing these channel systems. Whilst there are arguments for them being incised palaeovalleys related to low stands, it seems likely that intrinsic properties of the depositional setting, particularly basin water depth, were also important.

Shallow-water deltas

These intervals mainly occur as a succession of cyclothems overlying the turbidite-fronted deltaic basin fills. On the Askrigg Block, which was never under deep water, shallow-water deltaic cyclothems closely overlie Dinantian shallow-water limestones.

Individual cyclothems, typically several tens of metres thick, have a marine-band mudstone at the base. This is commonly separated from the underlying flooding surface at the top of the underlying cyclothem by a thin unit of fine-grained sediment (Fig. 4.2). Above the marine band, mudstones pass gradationally upwards into fine- or medium-grained sandstones with increasing ripple cross-lamination, mainly current ripples but some with wave influence, and often with bioturbation. However, some cyclothems are only a few metres thick and, in some cases, the interval between successive marine bands may be just a thin unit of mudstone, showing no deltaic advance during that time interval. The repetitive stacking of cyclothems was driven by glacio-eustatic sea-level changes, but the variability amongst individual cyclothems resulted from the interaction of sea-level changes, subsidence (which control accommodation space) and the volumes and grade of sediment that was supplied.

These cyclothems result from prograding, fluvial-dominated deltas. Their sandy upper parts are either mouth-bar deposits, lobate or sheet-like, or channel sandbodies that may be of restricted width or widespread multilateral complexes. Thicker mouth-bar sandbodies occur as wide lobes, probably tens of kilometres across. There may be more than one such lobe within a cyclothem (Fig. 4.14), suggesting large-scale switching of sediment supply by avulsion of river channels. The only examples of more elongate mouth-bar sandbodies (Fig. 4.11) are confined to the Lower and Upper Haslingden Flags. Mouth-bar sandstones show variable degrees of wave modification but there is no convincing evidence of tidal influence.

Delta-top channel sandstones typically have erosional contacts with mouth-bar sands, so that mouth-bar sands are variably preserved. The extent of such erosion probably depended on the size and gradient of the active channels, the extent to which they migrated steadily or shifted position suddenly and the prevailing conditions of sea-level change. Variations amongst channel sandstones are described later.

Variations in thickness both within and between cyclothems probably relate to the available accommodation space and to the level of sediment supply. Whilst most cyclothems are largely controlled by externally imposed sea-level changes, some thinner ones may be autocyclic, due to delta switching or the development of sub-deltas, perhaps as bay fills on a delta plain. Distinguishing autocyclic and allocyclic controls is not always easy, although the presence of a marine band above a flooding surface favours an external (allocyclic) control.

The uppermost sediments of many coarsening-upward cyclothems record emergent conditions, with development of palaeosols and thin coal seams. These are typically overlain by a flooding surface, above which are mudstones with a marine band marking the base of the next cyclothem. Mature palaeosols, which may be fireclays or ganisters, depending on whether the host sediment is mudstone or sandstone, probably relate to sustained periods of lowered sea level. Other cyclothems end at a flooding surface characterized by intense bioturbation, reflecting a period of non-deposition when organisms had time to thoroughly rework the sediment. Flooding surfaces record maximum condensation associated with a rapid rise in relative sea level. The overlying marine bands record periods when sea level approached its highest, when water depths were greatest, and the basin waters were self-evidently fully marine (Fig. 4.2).

In many cyclothems the coarsening-upward sequence is truncated by the erosive base of a channel sandstone with mouth-bar sands being at least partially removed. Channel sandbodies, between 5 and 30 metres thick, vary considerably in width, shape and facies of the fill. They are described here in terms of four loosely drawn groups which have different relationships with their associated cyclothems.

Thin channel sandbodies, less than around 8 m thick, and comprising trough cross-bedded, medium-grained sandstone are typically around 1 km wide. They usually preserve some mouth-bar sands beneath the basal erosion surface and are interpreted as distributary channels. In other words, they are envisaged as integral components of the delta progradation, recording the erosion and removal of the upper parts of mouth-bar sands. Palaeocurrents are quite variable, reflecting the low gradients of the channels and their tendency to meander and switch position. They may record delta advance when sea level was relatively stable or only falling slowly.

Some channel sandstones are coarser-grained, with pebbles both scattered and concentrated as lags. They are up to around 15 m thick and

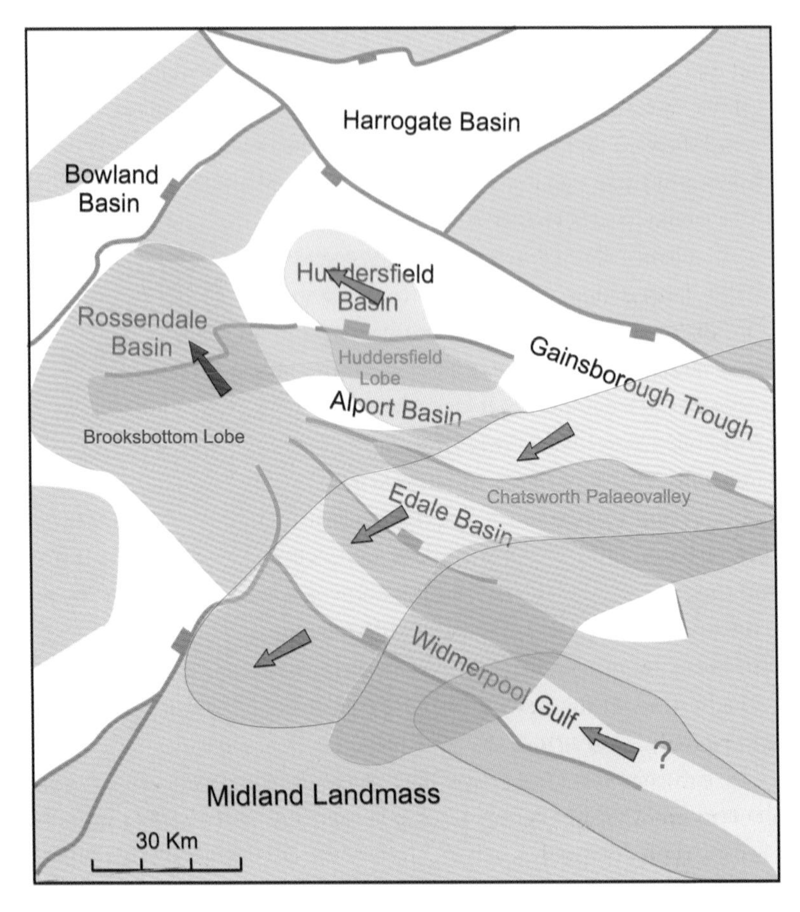

Figure 4.14 The distribution of sub-delta lobes and the incised palaeovalley fo the R_{2c} cyclothem. The deltaic facies comprise the Huddersfield White Rock and the Brooksbottom Grit of Yorkshire and Lancashire whilst the fill of the palaeovalley comprises the Chatsworth Grit, recording a fall and then a rise of sea level (after Waters *et al.*, 2008).

typically preserve thin mouth-bar sands beneath them. They form multi-storey, multi-lateral sheets within which individual channel elements may be up to several kilometres wide. Palaeocurrent directions are strongly unidirectional. This type of channel system is probably the result of delta progradation during conditions of significantly falling relative sea level, leading to a forced regression, whereby more up-current, high-energy fluvial facies advanced rapidly over earlier delta sediments, although without significant incision. The Lower Rough Rock is an example of this type.

A third type of channel sandbody is the very extensive, multi-lateral and typically multi-storey sheet body, typically coarse-grained and with little or no mouth-bar sandstones below. Such sandbodies probably record the very rapid advance of a wide, braided river system. The braid plain would have comprised many shifting and mobile channels, as there would have been little finer-grained cohesive sediment to stabilize channel banks. The Upper Rough Rock is an example of this type.

The largest channel sandbodies are up to 30 m thick and show a high degree of erosion at their base, typically removing all mouth-bar sands and, in some cases, cutting down into sediments of an underlying cyclothem. These channel bodies are commonly multi-storey and up to tens of kilometres wide. The Chatsworth Grit is the best example (Figs 4.14 & 10.16). Channel fills mainly comprise coarse pebbly sandstone, with pebbles both scattered and concentrated in lags. They typically display large-scale tabular cross-bedding, with sets up to 3 m thick, but they lack the giant cross-bedding of the channels associated with turbidite-fronted deltas. Their margins are nowhere exposed, but mapping suggests that some are quite sharp and steep (Fig. 12.11). These sandbodies are thought to be infills of major palaeovalleys, whose incision records large and possibly rapid falls in sea level and their fills record the early stages of the subsequent sea-level rise. One consequence of deep incision is that flanking areas became subaerially exposed for long periods, allowing very mature, typically highly leached palaeosols to develop (Fig. 9.10).

All these delta systems were very extensive, probably comparable in scale to the present-day Lena Delta of northern Siberia.

The Haslingden Flags (Yeadonian) are localized variants of shallow-water delta cyclothems. They are found only around Rossendale (Fig. 4.11). The Lower Haslingden Flags cap a coarsening-upwards unit, about 35 m thick, above the *Ca. cancellatum* (G_{1a}) Marine Band, whilst the Upper Haslingden Flags cap a similar coarsening-upwards unit, up to 55 m thick, above the *Ca. cumbriense* (G_{1b}) Marine Band. Mapping shows that both sandbodies are highly elongated in an east–west direction (i.e. parallel with palaeoflow), thinning away north and south over a few kilometres from their maximum axial thicknesses (Fig. 9.6). They also become thinner down current to the east, dying out over around 15 km. Both comprise dominantly ripple cross-laminated, fine–medium, micaceous sandstones with minor parallel lamination and small-scale cross-bedding. There is little change in lamination style throughout their thickness, but some localities show large-scale, low-angle accretion surfaces, suggesting lateral accretion. The sandstones commonly show vertical escape burrows of surface-resting bivalves (*Lockeia*; formerly *Pelecypodichnus*), suggesting rapid deposition.

Both the Lower and Upper Haslingden Flags are interpreted as elongate mouth-bar deposits (bar-finger sands), similar to those inferred for the present-day Mississippi 'birdfoot' delta. It seems possible that reduced salinity led to underflows down the front of the mouth-bars, creating a uniform boundary layer so that current ripples were active across a range of water depths. The lateral accretion surfaces, once recorded at Heys Britannia Quarries, are thought to reflect shifting of local bars and channels on the delta front. It is interesting that the two Haslingden Flags sandbodies are vertically stacked, with little offsetting (Fig. 9.6). If differential compaction had been the dominant control on subsidence, then a greater lateral offset of the Upper Haslingden Flags might have been expected. The vertical stacking suggests that local tectonic activity over-rode differential compaction, probably movement on a nearby syn-depositionally active fault.

Importantly, the Haslingden Flags contrast markedly in composition and in palaeocurrent direction with other sandstones in the Pennine Namurian. They were deposited by currents flowing to the east and are micaceous, with a distinctive greenish colour. Other sandstones in the Millstone Grit, including finer-grained sandstones, are arkosic and show palaeocurrents mainly directed to the south and southwest. These were derived from a Caledonian source to the north, whereas the Haslingden Flags are the earliest examples of sands derived from an unlocalized western source.

Growth faults

Rapid deposition inferred for the progradational parts of deltaic cyclothems, in some cases, led to instability. Muds and silts, low in the coarsening-upwards unit, became over-pressured, leading to temporary loss of strength and to physical failure, often on discrete slip surfaces. This occurred soon after deposition (Fig. 4.15). Where failure took place close to the sediment surface, slumps and slides occurred on the delta slope, some superficial and some slide surfaces more deeply penetrating

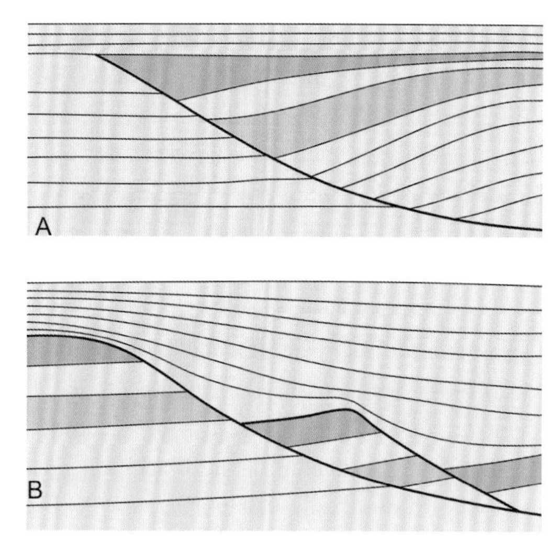

Figure 4.15 Cartoons to show the distinction between (**A**) sedimentary growth faults, where fault displacement persisted during a period of sedimentation, causing thickening and roll-over in the hanging wall of the fault, and (**B**) rotational sliding, whereby a coherent mass of sediment became detached and slid down slope, rotating as a result of the curvature of the slide surface. After movement stopped, the rotated block was draped by later sediment.

(Figs 4.15B, 11.30 & 11.31). Where the failure is steeper and/or deeper, it could take the form of a growth fault. When active below the mouth bar, these would cause anomalous thickening of mouth-bar sediments in the footwall of the fault and the development of a roll-over anticline (Figs 4.15A & 7.9). In areas with solifluction and heavy vegetation, like much of the Pennines, seeing clear field evidence of these features is problematic. Only exceptional quarry outcrops, the presence of anomalous dips or detailed geological mapping allow their detection.

Regional controls on Namurian deposition and palaeogeography

The Namurian successions of the Pennines record the progressive infilling of the basin topography created in Holkerian to Brigantian times by sediment supplied to the basin from at least three source areas at different times during the Namurian (Fig. 4.1). The southern source, the Midlands Landmass, provided small volumes of quartz-rich sand which accumulated, mainly as turbidites, in the Staffordshire Basin and the Widmerpool Gulf during the Pendleian to Marsdenian interval. A second source supplied sand from the west and gave rise to the Haslingden Flags in central Lancashire during the Yeadonian.

By far the most important source area lay far to the north and supplied coarse, often pebbly, feldspathic sands. These sediments were supplied to the Pennine Basin Complex intermittently throughout the Namurian in huge volumes that progressively infilled the inherited bathymetry. This basin-floor relief compartmentalized sedimentation so that sub-basins were filled successively from north to south. Once a sub-basin had been filled and the containing topographic high had been overtopped, sediment then fed into the next sub-basin down current, with a 'fill-spill' process. However, the detailed story was more complex. Between the Pendleian and the Kinderscoutian only minor deltaic advances took place in the northern part of the Craven Sub-basin, a period of time when sediment supply was diverted elsewhere.

Basin-filling involved progradation of turbidite-fronted deltas (Fig 4.12). At least four such episodes are known: the Pendleian of the Bowland Basin, the Kinderscoutian that extends to the northern edge of the Derbyshire Massif, the early Marsdenian of the Blackburn area and the mid-Marsdenian of the Staffordshire Basin (Fig. 4.13). The Pendleian succession in the Bowland Basin was dammed back by the Central Lancashire High. Similarly, the thick Kinderscoutian turbidite succession of the Alport Basin was dammed against the northern margin of the Derbyshire Massif so that no sand escaped to the Staffordshire Basin (Fig. 10.9).

Examples of fluvial sandstones passing down current into turbidites are rare. The fluvial Upper Kinderscout Grit of North Derbyshire (the cyclothem above the main basin-fill) dies out just north of Buxton and a correlative turbidite unit, the Longnor Sandstone, comes in 3 km to the south and thickens into the Staffordshire Basin (see Fig. 10.9).

Once a sub-basin was filled and inherited accommodation space was eliminated, new accommodation space resulted from ongoing subsidence and eustatic sea-level changes. Such conditions favoured a succession of shallow-water deltaic cyclothems. Examples of turbidite components in thicker cyclothems that post-date the main basin fill occurred when, following the main basin fill, during one or more marine band cycles there was no local deltaic progradation. This created fresh accommodation space so that the next deltaic progradation took place into deeper water.

In the north of the Bowland Basin, the Pendleian basin-filling succession (Pendle Grit to Warley Wise Grit) is overlain by a series of small deltaic cyclothems that advanced only a short distance into the basin during Arnsbergian to Alportian times. This allowed a thick mudstone succession (Sabden Shales) to develop as the basin deepened beyond the limits of these deltas. When abundant sand supply was re-established in the Kinderscoutian, the resulting progradation generated turbidites (Parsonage Sandstone and Todmorden Grit) at the base of a turbidite-fronted delta succession capped by the fluvial Lower Kinderscout Grit. This progradation extended mainly to the south but not far to the west as, around Blackburn, the Kinderscoutian comprises deep-water mudstones and the main basin fill is early Marsdenian.

Around Glossop, the main basin fill succession is Kinderscoutian, ending in the fluvial Lower and Upper Kinderscout Grits. However, these units are overlain by a mudstone interval that spans the time between the *B. gracilis* and *B. superbilinguis* Marine Bands (i.e. early Marsdenian). In Yorkshire and Lancashire, these marine bands are separated by shallow-water delta cyclothems, and in the Staffordshire Basin by a major basin-filling succession (Five Cloud Sandstone and Roaches Grit). However, early Marsdenian deltas appear to have stopped short of the Glossop area, leading to a significant deepening. When sand supply resumed, the resulting late Marsdenian R_{2c} cyclothem (Huddersfield White Rock/Chatsworth Grit) is not only unusually thick but also has a thick interval of turbidite sandstones in its lower part (Fig. 9.11).

4.4 Triassic

Triassic red, pebbly sandstones are exposed in road cuttings to the south and southwest of Leek town centre and in areas to the north and south of the town. They show large-scale trough cross-bedding indicating palaeoflow to the north. The facies association indicates sediment supply by a powerful river system, probably ephemeral and in a semi-arid setting. The occurrence of these sediments in an area of low topography compared with surrounding hills of Carboniferous strata suggests that the Triassic rivers flowed northwards in a palaeovalley that was incised deeply into the post-Variscan landscape. It would have been similar to a present-day wadi, subjected to intermittent floods. The sediments themselves are derived from areas far to the south, some possibly from as far afield as the main Variscan deformation belt in southern England or northern Europe.

Chapter 5

Mineralization

The limestones of both the Askrigg Block and the Derbyshire Massif are extensively mineralized and both areas have supported major mining industries, some dating back to Roman times (Fig. 5.1). Most of the mining exploited galena and sphalerite, the sulphides of lead and zinc respectively, although copper was also important locally. Early lead mining may also have exploited the subsidiary content of silver associated with galena. These ore minerals occur in veins along with calcite, fluorite and barite, which were initially regarded as gangue. More recently, barite and fluorite have been exploited both through surface mining and through the reworking of spoil heaps as new industrial demand emerged. Fluorite is important as a flux in iron smelting whilst the high density of barite makes it important in the formulation of drilling muds. Whilst industrial subsurface mining is now finished, Blue John, a purple-coloured variety of fluorite, is still mined for its decorative properties in a few small mines at the northern end of the Derbyshire Massif. Several mines broke through into natural cave systems and some of these now have a second life as show caves. The copper deposits at Ecton, on the western edge of the Derbyshire Massif in Staffordshire, stand outside the main pattern of mineralization but, at one time, their exploitation was very important.

The various minerals that have been exploited occur mainly in veins that follow faults or fractures generated during Variscan deformation and often with strongly preferred orientations. In both Yorkshire and Derbyshire, the wider veins are commonly known as 'rakes' whilst in Derbyshire, narrow veins are called 'scrins' (Figs 5.2; 5.3). Other mineral concentrations occur as linear pipes that broadly follow bedding, and as more irregular flats, where the diagenesis or dissolution of the limestone host seem to have exerted significant control. Thin clay partings (clay wayboards), commonly associated with palaeokarst, appear in some

Figure 5.1 A series of closely spaced bell pits close to Dream Cave, near Wirksworth, Derbyshire. The pits tend to be aligned along veins. In the distance is Site 2 of the Dream Cave locality (see Chapter 11), exposing part of a mud mound within the Bee Low Limestones.

cases to have acted as barriers to the movement of the mineralizing fluids, so that mineralization is concentrated below them. Interbedded units of volcanic lavas also appear to have acted in this way. Some of the Blue John deposits were precipitated in cavities between breccia blocks of submarine talus deposits.

The fluids that precipitated the various mineral phases are thought to have been expelled from thick mudstone intervals (Bowland Shale equivalents) in the basin areas that surround the limestone platforms of both Derbyshire and the Askrigg Block. Deep burial and compaction of these mudstones caused warm acidic hypersaline brines to be expelled, carrying a cocktail of dissolved ions. On the Askrigg Block, and on the eastern side of the Derbyshire Massif, the cations were dominantly lead, zinc, barium and calcium along with appropriate anions. In Derbyshire, there is a horizontal zonation of dominant mineral types that relates to temperature of crystallization. This shows that migration and progressive cooling occurred as fluids moved from east to west, suggesting that the brines were expelled from basin mudstones that lay to the east. The migration of these brines took place after the fluid movements that caused local dolomitization of the limestones. Peak burial of the basin mudstones occurred in late Westphalian times. Expulsion of brines probably began at that time and persisted through the Variscan

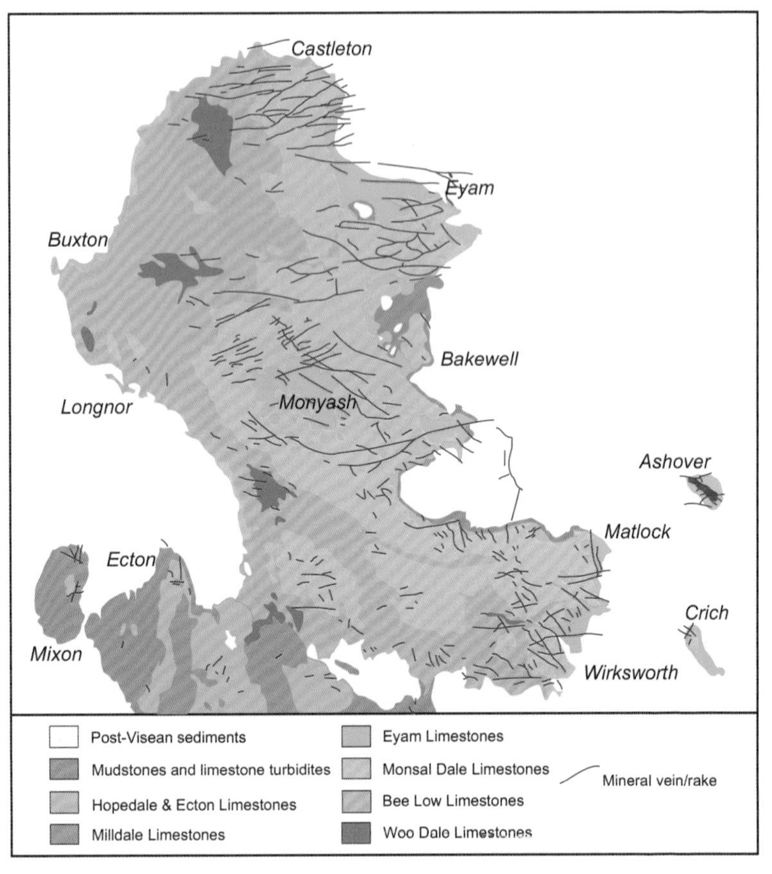

Figure 5.2 The distribution of mineral veins across the Derbyshire Massif. With the exception of the copper mines at Ecton, on the western bourdary, the veins comprised mainly galena and sphalerite (based on Ford, 2002).

deformation when the faulting and jointing exploited by some veins would have developed (Figs 5.3 & 5.4). This origin of the mineralizing fluids differs from earlier suggestions that they were late-stage products of cooling Devonian granites. Whilst such a now-discredited hypothesis is understandable for the Askrigg Block, with its underlying Wensleydale Granite, it was never sustainable for Derbyshire, as no nearby granite is inferred from borehole or geophysical data.

On the western side of the Derbyshire Massif, copper was locally abundant at Ecton where the mines were active until the late nineteenth

Figure 5.3 The mineralized Putwell Hill Vein within the Monsal Dale Limestones exposed on the Monsal Trail near Monsal Head. Columnar masses of calcite occur in several parallel veins with the long axes of crystals normal to the steep vein wall. A mine adit entrance into the vein coincides with where ores, mostly lead, have been extracted. Scale in cm.

century. The main copper mineral was chalcopyrite (copper sulphide) with subsidiary galena and sphalerite, as well as some fluorite and barite. The dominance of copper minerals suggests a different source of mineralizing fluids compared with the lead mineralization further east. A brine source to the west seems most likely, although exactly where remains uncertain. It seems probable that the same source fed the copper mineralization hosted in the Permo-Triassic sandstones of Cheshire (Alderley Edge) and Shropshire (Clive).

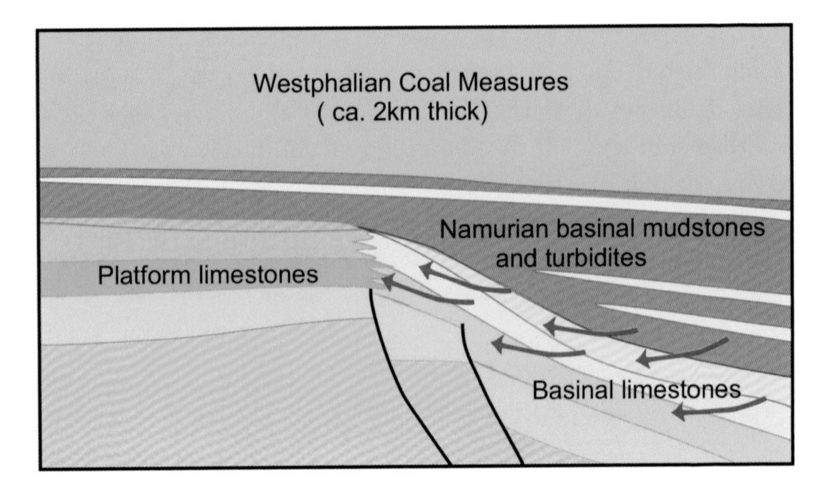

Figure 5.4 A highly schematic model for the mineralization of the Derbyshire Massif by brines derived from the basin sediments of late Dinantian and early Namurian age. This model is thought to best apply to latest Westphalian times when burial was at its maximum.

Chapter 6

Field trips and localities: prelude

6.1 Field trip planning

Most field trips and courses have their own agendas, with different durations, and start and finish locations. Some are focused on specific topics whilst others may wish to concentrate on particular geographic areas or stratigraphic intervals. In this book, therefore, localities are mainly listed and described individually so that field trips can be planned on a 'pick-and-mix' basis. At the end of the locality descriptions, some combinations are suggested as itineraries for topic-based, local or sub-regional trips that could be variously carried out by motor vehicle, cycle or on foot.

In the case of the Derbyshire Platform and its southern and eastern margins (Chapter 11), some localities are organized into groups that make up integrated geo-walks and geo-biking trips to demonstrate the variations in stratigraphy and facies of the limestones on a very local scale.

The localities are organized into chapters in geographic groupings, listed broadly from north to south. The description of each locality has details of access, permissions, parking, walking distances, nature of terrain and safety considerations.

Whilst the chosen localities span a wide variety of features and involve strata of various ages, there is considerable overlap. This could help reduce the amount of travelling needed to provide reasonable coverage of a particular topic. For example, depending on the degree of detail required, it would be possible to have a perfectly satisfactory trip focused on turbidites either in north Derbyshire or in the Skipton/Pendle area, or examples of the Rough Rock could be studied in almost any part of the wider area. The full complexity of the Derbyshire limestone platform and its margins may draw on localities from Chapters 10, 11 and 12. In planning a trip, it can often be useful to carry out a virtual reconnaissance using Google Earth and Google Street View. Indeed, the latter app can be used to see quite a lot of the geology where localities are close to the roadside, allowing preliminary scoping or even an armchair field trip. Localities that can be looked at in this way have the symbol (▶) next to locality name. To allow higher magnification for use in the field either on a device or as hard copy prints, three complex figures (4.12; 10.11 & 11.15) may be downloaded as .pdf files from this book's page on the Liverpool University Press website.

It may often be helpful to make use of smartphone apps that relate to geology and fossils. The British Geological Survey (BGS) app is called *BGS Geology Viewer.* It can be downloaded for free.

6.2 Access and safety

The localities have been visited by the authors over many years and some have been revisited specifically to provide the most up-to-date information about parking, access, permission and safety. Many localities lie within National Parks, some are on National Trust land, some are on open access land and others are on private land. In many cases, it is not evident who owns the land or where permission should be sought. Where that information is available, the locality descriptions indicate the status as presently understood. These should not be construed as definitive statements, and it is for individuals to make their own judgements.

Many of the localities involve significant walks over rough terrain and disability access may be limited. The level of disability access is noted in the introduction to many localities.

Some localities are natural crags and stream sections, but many are quarries, mostly disused but a few still active. Many disused quarries are readily accessible as they fall within areas where there is formal or informal public access and quite a few are used for rock climbing. For quarries where there are physical disincentives to entry (fences etc.) it is sensible to seek permission when it is clear where that should be sought.

For active quarries, it is essential to seek permission, usually in advance of the visit, and to fully understand and comply with any HSE conditions that the quarry management might impose. It should always be assumed that boots, helmet, safety goggles and hi-vis vest are required. Some quarries may restrict sectors that may be visited, and some may require that visitors be accompanied by quarry staff. Do not attempt to visit working quarries without permission as you could be putting yourself in danger. Such attempts also antagonize quarry managements and make it more difficult for others to visit in the future.

Whatever the status of a locality, it is important to follow the Fieldwork and Countryside Codes. Avoid damage to footpaths, walls and fences; close gates; park vehicles so as not to obstruct gateways and other access; do not disturb livestock and keep dogs under control.

Exercise vigilance and be aware of hazards for both yourself and anyone else in the area. Many localities are in open countryside, often approached over rough and slippery ground, and are frequently exposed to wind and rain. It is important to go prepared for the worst conditions, appropriate for the time of year. Remember that conditions can change quickly. Temperatures can fall with altitude and streams can rapidly become impassable after heavy rain. For localities that involve walking long distances from parking places, it may be safer not to go alone, but that decision is for the individual. If you do go alone, ensure that someone knows your plans, and be sure to follow them. Whilst mobile phones will usually permit you to summon help in the event of an accident, remember that remote moorland areas or deep valleys may not have a good, or indeed any, signal. If venturing into high country away from roads, it is sensible not to rely on mobile phone GPS. Try to carry a 1:50 000 or 1:25 000 map that shows vital information (contours, positions of crags etc.).

At crags and disused quarries, helmets are strongly advised. Do not approach the base of rock faces in active quarries or anywhere where there is evidence of recent instability, especially where areas are fenced off pending remediation work. Do not climb on steep outcrops unless you are a competent rock climber and be aware of any people that may be below you. If you dislodge material, always give a warning shout. Equally, if there are people climbing above you, be very aware of their positions and avoid moving directly below them. Remember also that sheep are well able to dislodge debris on steep slopes and tend not to give a warning shout.

Use a geological hammer very sparingly, and only if wearing goggles or toughened glasses. Indeed, at most localities, there is no compelling reason for using a hammer at all. At only a few of the localities included in the Guide would it be appropriate to collect fossils. Accepted sites for fossil collecting are described in dedicated websites and are not included here. If material is needed for petrographic purposes, ideally use fallen blocks. If collecting from an *in-situ* exposure, be discreet in selecting a sample point, avoid sampling material that makes a distinct contribution to the geological context and avoid removing any rock that contributes to the holds of a climbing route. Use of a hand lens can add a lot of useful information and sometimes obviates the need for sampling. Wherever possible, use a hand lens directly on the outcrop or on detached blocks, not on extracted samples.

For landscape views and for inaccessible parts of large exposures, binoculars can make a huge difference to the clarity and detail of observations.

6.3 Field localities

Most of the localities suggested in this Guide are stand-alone exposures or viewpoints that are separated from one another by considerable distances (usually kilometres). In Chapter 11 (Derbyshire Platform), however, some localities are combined into 'geo-walks' that include several closely spaced sites.

Localities are described in subsequent chapters and most are located by a six-figure grid reference and by GPS co-ordinates. An Excel file listing the GPS co-ordinates of all localities can be downloaded from the book's page on the publisher's website. Grid references and GPS co-ordinates commonly coincide but, in some cases, they diverge slightly. Where this happens, the grid reference relates to the actual exposure whilst the GPS co-ordinates relate to the relevant parking place. In the Chapter 11 geo-walks and cycling routes, GPS co-ordinates are used for individual sites within the route. For each locality, we provide basic instructions about access and safety. These can change over time and so everyone should make their own judgement if they plan to visit.

For convenience, the localities have been grouped into six sub-regions that are presented as chapters. These sub-regions have some geological integrity, but some localities could reasonably have been allocated to an adjacent sub-region. Each chapter gives a brief introduction to the sub-region, suggesting some of the key relationships between localities.

These are suggested itineraries at the end of each chapter, which illustrate how a trip might be organized. We know that these suggestions are viable as one-day trips, but they are certainly not prescriptive. They should be regarded as starting points for creative adaptation, incorporating localities described here or including other localities of your own choosing, geological, scenic, heritage or cultural. Most of these itineraries involve transport by motor vehicle between localities, although some, especially in Chapter 11, are more suitable for walking or cycling.

Chapter 7

The Northern Margin: Nidderdale to Airedale

The localities in this chapter mainly illustrate the distribution of sedimentary rocks across the Craven fault zone (Fig. 7.1). Three localities are on the Askrigg Block, where relatively low rates of subsidence in the Carboniferous led to a thin succession of shallow-water sediments punctuated by unconformities. The Viséan limestones, which rest unconformably on Lower Palaeozoic sediments, are of rather uniform shallow-water facies whilst the Namurian succession comprises thin and incomplete clastic-dominated cyclothems. The Craven fault zone separates the Askrigg Block from the Bowland Basin and is associated with a major change in thickness, facies and stratigraphic continuity. In the Viséan, the platform limestones pass southwards through a major 'reef' belt into a deep-water carbonate setting, best seen further south in the Clitheroe area. In the Namurian succession, there is a rapid expansion from a thin, discontinuous sequence on the block into thicker, basin-filling successions involving deep-water sandstones in the basin. This change is well illustrated by the Pendleian succession between the Askrigg Block and the area around Skipton (Fig. 7.5). Later Namurian deltas were rather localized until a second major basin-filling succession was initiated in the Kinderscoutian (Fig. 7.2). The area also provides good examples of the Rough Rock (Yeadonian, G_1), which extends basinwide to Staffordshire as an extensive braided river deposit.

Brimham Rocks (SE 210650: 54.0769, -1.6828) (►)

This locality can be approached by a variety of routes, and it is then possible to follow the tourist signs, once in the vicinity. The locality is a popular, heavily visited tourist attraction that is managed by the National Trust. Parking on the main car park is subject to a charge, (free for NT members),

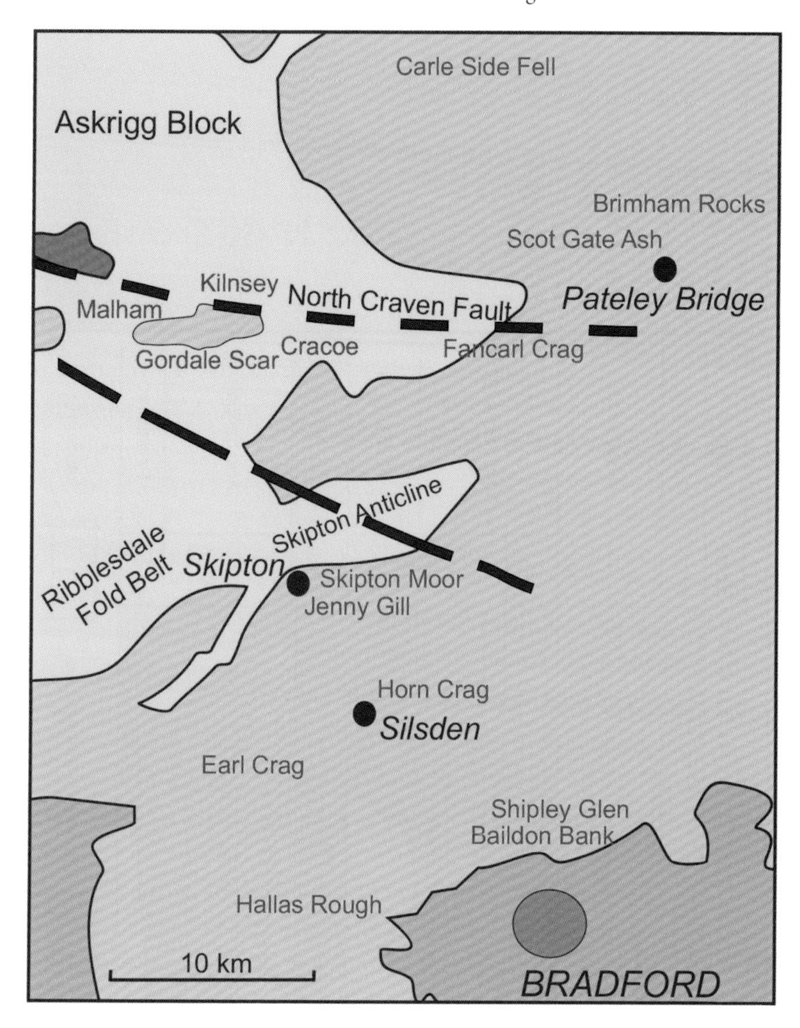

Figure 7.1 Geological map of the northern area, showing the positions of the localities described.

but there is no charge for access to the site. Minibuses should book a place in advance. From the car park, follow the main footpath for around 400 m through the woods which leads into the more open area where the spectacular, tor-like exposures of the sandstone are well seen. Disabled access is mainly limited to the main footpaths, but the National Trust have a mobility scooter for rugged ground which can be borrowed, free of cost, during main opening times. This must be pre-booked by telephone (01423 780688).

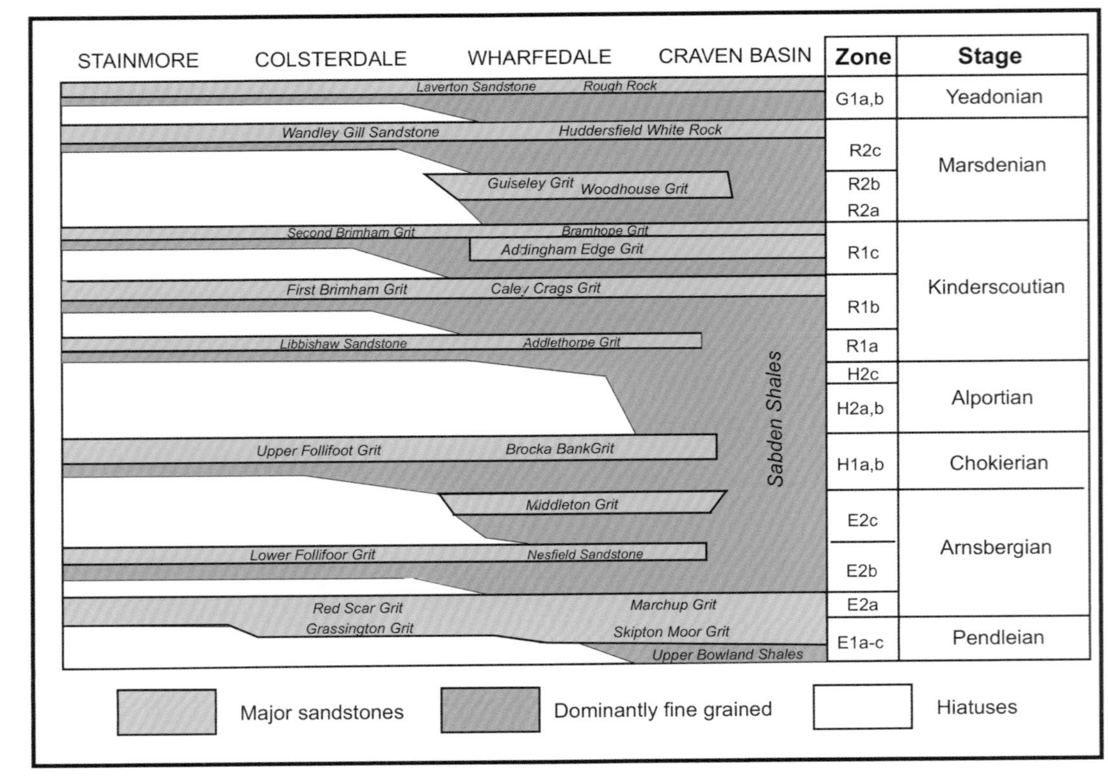

Figure 7.2 The names and southerly extents of Namurian sandstones across the Craven Fault belt. Many of the sandstones, particularly in the Arnsbergian–Alportian interval, do not extend far to the south. Note the abundance of hiatuses in the succession on the Askrigg Block compared with the more complete succession in the basin (after Ramsbottom,1977).

These tors are up to around 10 m high and can be examined by walking around them and by some gentle scrambling. The popularity of the locality means that there are sometimes serious climbers and often children climbing around on the rocks. ***Geological hammers are banned.***

These spectacular crags and tors are in the Lower/First Brimham Grit of Kinderscoutian (R_{1b}) age and are located near the southern edge the Askrigg Block where the overall Carboniferous succession is thin compared with that in the Bowland Basin to the south (Fig. 7.2). The three-dimensional nature of the exposure at Brimham Rocks allows the geometry of the bedding to be examined in detail (Fig. 7.3). The exposures comprise coarse, pebbly sandstones, with lag layers and lenses of fine-pebble conglomerate. They are extensively cross-bedded in medium-scale sets, with both tabular and trough geometries. Cross-bedding directions are variable both within and between individual crags but the predominant palaeocurrent is to the west.

The succession is subdivided by laterally extensive erosion surfaces, separating major channel units which can, in turn, be further subdivided into a hierarchy of cosets and sets. The inclinations of set boundaries in relation to the cross-bedding dip directions indicate that major bars were growing in the channel complex by both lateral and downstream accretion. The sandstones also show several examples of soft-sediment deformation suggesting rapid deposition and temporary loss of strength through liquefaction.

The sandstones represent deposits of an extensive, low-sinuosity river system of unconfined channels that derived both water and sediment

Figure 7.3 A general view of Brimham Rocks, an exceptional exposure of the Lower Brimham Grit (Kinderscoutian) in tor-like columns. The sandstones show complex medium-scale cross-bedding produced by bedforms on a wide, braided sandy river bed (photograph courtesy of Roman Soltan).

discharge from a mountainous catchment in Scandinavia or Greenland. These same river systems fed the major deltaic advances that took place throughout the Namurian in the basin area south of the Askrigg Block.

The crag and tor morphology that characterizes Brimham Rocks is unusual in Britain, particularly in sandstones. It seems that weathering exploited joint systems and that weathered debris was then removed by a combination of running water and gravity, possibly with some help from strong winds, to slowly widen the gaps between less weathered cores. However, the conditions under which the weathering occurred are uncertain. Some favour a dominantly chemical weathering in warm humid conditions, whilst others argue for a more physical process, driven by freeze-thaw activity. These processes are not mutually exclusive and, over time, some combination may have been involved.

Scot Gate Ash Quarries, Pateley Bridge (SE 1607; 54.097542, –1.7586)

From the High Street in Pateley Bridge, take New Church Street and follow it to the edge of the village where it becomes Wath Road. After about 1 km from the High Street, the road splits. Take the righthand branch and follow it uphill to an acute road junction. Turn right along Wath Lane for around 400 m and park on the roadside. Go through gate on south side of road and follow the footpath that skirts overgrown quarry spoil heaps for about 400 m. Turn left off the path and go over spoil tips to the main quarry face. The face is steep, but it can be accessed along the base. The higher part of the section can be accessed (with care) by scrambling up the steep slope to the right of the main face.

These long-disused quarries are located 1 km north of Pateley Bridge. When active in the nineteenth century they were a major local industry, with the sandstone being exported throughout Britain, especially for kerbstones and railway platforms on account of the rock's very high crushing strength. The quarries were connected to the railway in Pateley Bridge by an incline. The locality lies a few kilometres north of the North Craven Fault and the succession was deposited, therefore, near the southern edge of the Askrigg Block. This subsided more slowly than the Bowland Basin to the south, resulting in an overall thinner Namurian succession.

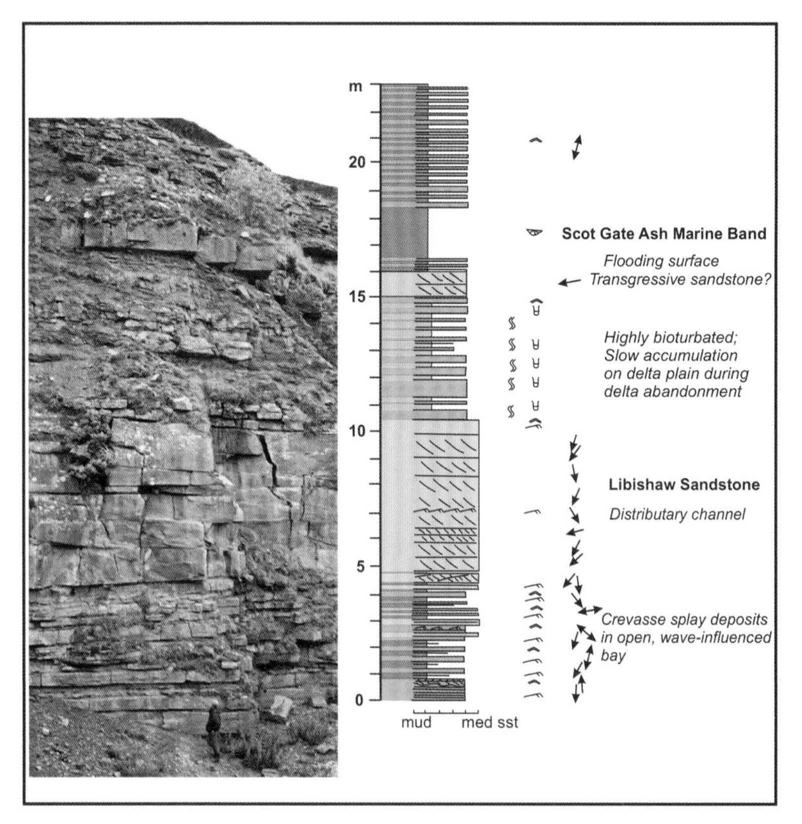

Figure 7.4 Photograph and sedimentological log of the Libishaw Sandstone (Kinderscoutian) and its neighbouring sediments at Scot Gate Ash Quarry, Pateley Bridge (log after Martinsen, 1990a).

The quarries exploited the Libishaw Sandstone (Kinderscoutian $R_{1a1?}$) and now expose that unit and the adjacent sediments (Fig. 7.4). However, a lack of diagnostic marine fossils makes the confident assignment to a minor cycle impossible. About 23 m of strata are exposed in the quarries. The lowest 5 m (exposed at the south of the quarry) is a heterolithic sequence of tabular, sharp-based sandstones, often cross-laminated with wave-rippled tops, interbedded with silty, burrowed mudstones. Drifted plant material is common on bedding planes. A tabular sandstone unit, some 5–6 m thick, with a flat base and internal cross-bedding overlies these beds. Cross-bed sets are up to 1 m thick, some have cross-laminated tops. The palaeoflow appears to have been to the southwest. At first sight, this sandstone might be interpreted as a channel sandbody, but the sharp base is very flat with no evidence of erosion, and there are no large clasts that might suggest a channel lag. Therefore, it seems equally reasonable to interpret this sandbody as a proximal mouth bar that advanced suddenly into the area either by delta switching or through a forced regression driven by falling sea level.

The sandstone unit is overlain by a thoroughly bioturbated interval with few preserved primary structures, suggesting lower rates of sedimentation and possible marine influence. This is, in turn, overlain by several metres of sharp-based, cross-bedded and cross-laminated sandstone beds with wave-rippled tops. Some cross-beds are low-angled and have current ripples migrating on their foresets. These sands may be the deposits of floods or storms that brought coarser sediment into a delta-top embayment or floodplain.

The so-called 'Scot Gate Ash Marine Band', which overlies these beds, is a 2 m thick fossiliferous mudstone containing mainly brachiopods, but with no age-diagnostic goniatites, reflecting the shallow-water setting. Numerous sharp-based sandstone beds up to 40 cm thick, some with wave-rippled tops, overlie the marine band and are interbedded with mudstone. Again, storm or flood events within a delta-plain embayment seem a reasonable interpretation. The entire sequence shows palaeoflow generally towards the south whilst the wave approach direction appears to have been broadly from the south.

Fancarl Crag (SE 065627; 54.05792, -1.8991)

This small crag is best approached from the lane from Appletreewick (New Road), about 600 m from the bend in the road and about 500 m from the junction with the B6265. Park on the roadside near the gate. Follow the path along the top of the feature to the small crag. This is a natural crag and is quite stable. It is, however, quite steep and exposed so that care is needed, especially in wet and windy conditions. The main purpose of a visit is to look at the view, and so it is not so useful if visibility is poor.

Fancarl Crag is located just south of the North Craven Fault, the southern boundary of the Askrigg Block. The fault runs close to the B6265, about 250 m north of the crag, and there are Viséan limestones in the field just to the north of the road. Fancarl Crag itself is a small exposure of Grassington Grit, a fluvial channel sandstone of Pendleian (E_{1c}) age that is incised into Viséan limestones of the Askrigg Block to the north. At Fancarl Crag, the sandstone rests on a unit of (Pendleian, E_1) mudstone, a few tens of metres thick, which gives rise to the rather boggy ground below the crag. The mudstone, in turn, rests on Viséan limestones exposed in the Skyreholme Anticline lower down the hill, across the road, at Troller's Gill. This mudstone unit equates broadly with the Pendle Shales that are encountered further out into the Bowland Basin to the south. The succession between the Viséan limestones and the Grassington Grit or its equivalent show important changes in facies and thickness when traced out into the basin from this marginal position (Fig. 7.5).

Looking south from the crag, the Grassington Grit caps the fell on the south side of Wharfedale. The hillside below the Grassington Grit is made up of an expanded mudstone interval and to the west this is complicated by rounded hills that penetrate upwards into this mudstone

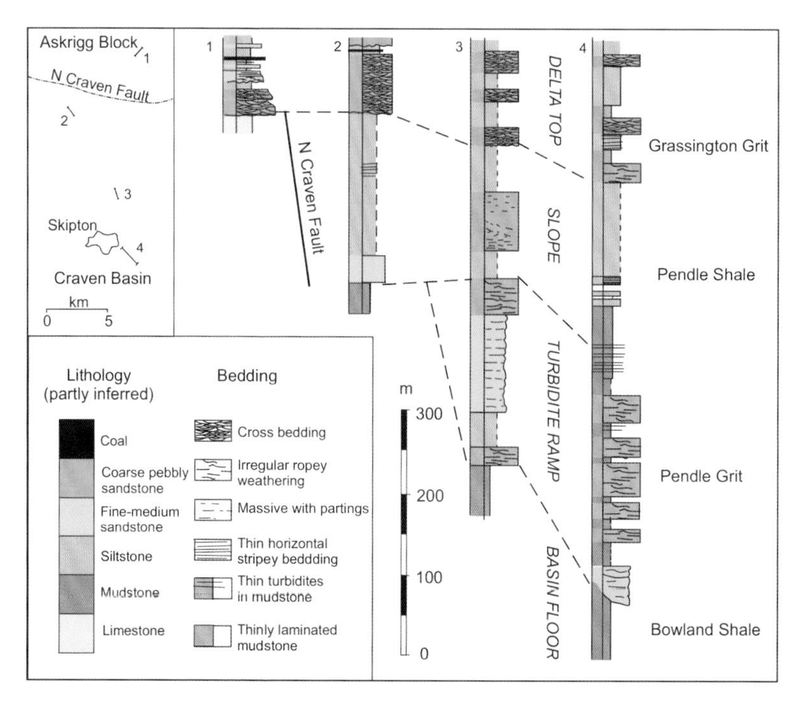

Figure 7.5 Correlation of the Pendleian succession between the Askrigg Block and the Bowland Basin near Skipton showing the rapid expansion of turbidite and slope deposits into the basin. Fancarl Crag lies between Profiles 1 and 2 (after Baines, 1977).

unit. These are examples of the 'reef knoll' mud mounds that occur in a belt stretching eastwards from Malham, in a zone close to the Middle Craven Fault (Fig. 4.6). They are better seen from the Linton to Cracoe road locality and are discussed further there.

The Grassington Grit at Fancarl Crag is coarse, pebbly sandstone, typical of much of the Millstone Grit. It is cross-bedded in medium-scale, trough sets, indicating palaeoflow to the south, into the Bowland Basin. It is the deposit of an extensive, high-energy and laterally mobile river system that produced a multi-storey, multi-lateral sheet sandbody. It is the earliest major pulse of coarse sand from the northern, feldspathic source area that dominated sediment supply through most of the Namurian.

Kilnsey Crag (SD 974683; 54.016; 54.1062, –2.0402) (▶)

The crag lies just north of the small village of Kilnsey on the B6160 and is best approached on foot after parking in the village. The crag is a popular climbing face, and it is best viewed from a distance before approaching closer. The distant view is viable for those with disabilities. Approach the base of the crag by the footpath to see the lowest part of the succession. A close view of the upper part of the succession can be had by climbing up to the scars around the southern side of the main crag.

Kilnsey Crag provides an example of the Viséan limestone succession (Great Scar Limestone) to the north of the North Craven Fault. It has been suggested that the basal unconformity of the Kilnsey Formation with underlying Lower Palaeozoic sediments may be present below Quaternary sediments in the floor of the valley. Glacial over-deepening at the confluence of Wharfedale and Littondale may have led to deep erosion. This suggestion is based on the presence of erratics of cleaved Lower Palaeozoic sediments which are not seen *in situ,* but which are present as blocks in drystone walls. Morphologically, the crag itself largely reflects erosion at the margin of a valley glacier.

The crag exposes the full local thickness (29 m) of the Kilnsey Limestone (Holkerian), the upper member of the Kilnsey Formation (Arundian–Holkerian), resting on a small thickness of the underlying Kilnsey Limestone-with-Mudstone Member (Arundian) (Fig. 7.6). This

Figure 7.6 The base of Kilnsey Crag, Wharfedale, showing the Kilnsey Limestone-with-mudstone Member (Arundian) overlain by the more massive Kilnsey Limestone Member (Holkerian), both part of the Kilnsey Formation (photograph courtesy of Nigel Mountney).

comprises mainly bioclastic limestone with thin interbeds of mudstone. The Kilnsey Limestone consists of dark limestones that are well bedded and are mainly bioclastic with variable proportions of micritic matrix and with some large intact fossils, corals, brachiopods and molluscs. The unit forms the main crag and extends about 14 m above the overhang. Above that, the face comprises the lower part of the Malham Formation (Holkerian–Asbian) which is paler in colour and made up of mainly bioclastic limestones with variable micritic matrix.

The limestones were deposited in relatively quiet, low-energy conditions with little winnowing of lime mud, suggesting a setting that was protected from waves and tidal currents, probably by the mud mound 'reef' belt that developed just south of the Middle Craven Fault, at the boundary between the platform and the Bowland Basin.

Linton–Cracoe Road (*c*.SD 993625; 54.0596, –2.0109)

Park on Lauradale Lane between Linton and Cracoe, about 800 m from Linton village and 700 m from the junction with B6265. If travelling from Skipton, take the right turn about 150 m north of the entrance to the limestone quarry. Park on roadside with caution and keep a good lookout for traffic. An entry to a field on the southeast side of the road gives a safe place for viewing. Suitable for disabled.

The view to the east shows several Asbian mud mound 'reef-knolls' (Carden and Skelerton hills), exposed as topographic features up to around 100 m high in the landscape. These microbial build-ups have been partially exhumed from the overlying Bowland and Pendle Shales (Pendleian, E_1), which are much thicker here than they are below the Grassington Grit at Fancarl Crag. The mud mounds are developed about 1.5 km south of the Middle Craven Fault from where the top Viséan surface steps down into the Bowland Basin (Fig. 4.6). The mounds, therefore, occur close to the margin of the block, in a zone between shallow-water limestones of the Askrigg Block and the deeper-water limestone-mudstones of the basin.

The Pendle Shale is the deposit of a prograding muddy slope, here resting directly on the Viséan limestones. The shale unit thins out to the north so that 3 km away, on the Askrigg Block, the unit is missing and the overlying Grassington Grit rests directly on limestones. The Grassington Grit, which caps the fell, is thought to be the major fluvial channel deposit, part of the prograding Pendleian delta system. To the south, the Pendle Shale thickens into the basin (Fig. 7.5). Two kilometres south of this locality, turbidites of the Pendle (Skipton Moor) Grit come in between the Bowland Shale and the Pendle Shale and thicken southwards into the basin.

The large quarry (Swinden) on the other side of the main road has mostly removed a large mud mound, with limestone being taken out by a dedicated rail line. The limestone is now crushed for aggregate for general building and construction, but at one point some was used for acid scrubbers in coal-fired power stations.

Skipton–Bolton Abbey Road (SE 003524; 53.9728, -1.9706) (►)

The stopping point is a large lay-by on the A59, about 700 m from the roundabout where the A59 and the A65 diverge. There is usually ample parking, set well back from the road. The lay-by has a truck-stop café, and it is often possible to enjoy refreshments along with the view. Suitable for disabled.

The lay-by itself is on the southern flank of the Skipton Anticline, a tight northeast–southwest fold, which exposes Courceyan limestones in its core. These Haw Bank Limestones are muddy, quite thinly bedded and with muddy partings and are probably of rather deep-water origin. However, they are not accessible to view at present, as the extensive quarry is now a landfill site. The Skipton Anticline is one of several folds that make up the Ribblesdale Fold Belt.

The locality provides a viewpoint to the south, where the extensive hillside below Skipton Moor is underlain by a succession that spans the latest Viséan to the Pendleian. The valley floor and the lower part of the hillside are occupied by the Bowland Shale, basin mudstones of both late Viséan and earliest Pendleian age. The Base Namurian boundary does not create a conspicuous feature, but the diagnostic marine band (*C. leion*, E_{1a}) is found in some of the stream sections that cut the hillside. Most of the hillside is the outcrop of the Skipton Moor/Pendle Grit (E_{1c}) and the stepped topographic features in the landscape reflect the underlying geology (Fig. 7.7). The discontinuous ridges on the hillside are formed by more resistant sandstones and the intervening recessive

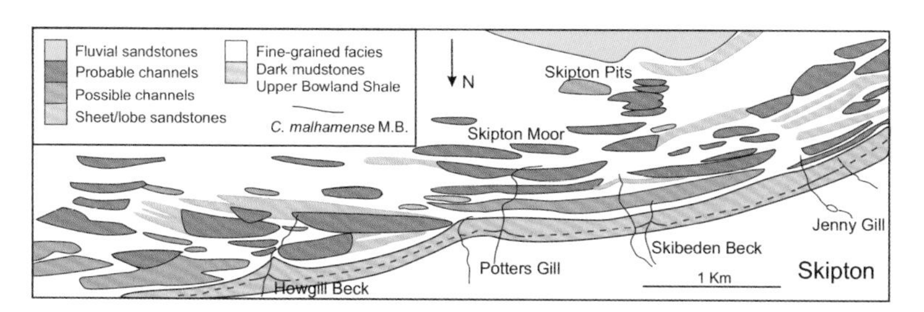

Figure 7.7 Interpreted facies of the Pendleian succession on Skipton Moor, based largely on feature mapping. The map is orientated North downwards to facilitate understanding the view from the Embsay road. Note that this is a map, not a drawing of the view (after Baines, 1977).

areas are dominantly finer-grained sediments. These discontinuous lenses are thought to reflect channel sandbodies within a turbidite ramp setting. The overlying Pendle Shales, towards the top of the hill, are slope deposits that separate the turbidites below from the fluvial channel sandstones of the Grassington/Warley Wise Grit that caps the hillside. The turbidite sandstone interval thins out to the north towards the basin margin so that the Pendle Shales come to rest directly on the Upper Bowland Shale (Fig. 7.5).

Jenny Gill Quarry, Skipton; (SE 005510; 53.9567, –1.9992)

Coming from Skipton centre, on A6069 (Newmarket Street), take a right turn where the road turns slightly to the left, about 350 m from the Market Place. Take the road that turns only slightly (Shortbank Road), not the one at right angles to Newmarket Street. Drive to the very top of Shortbank Road where surfaced road passes into a rough footpath. Park in front of garages opposite houses unless the garages appear to be in use. Walk up the path straight ahead for around 200 m and, where the track opens into woodland, turn half right, and walk up the slope for around 100 m to the floor of the quarry.

Walking up the track, small exposures of mudstone can be seen in the floor of the path and in the banks on either side. These sediments are in the upper part of the Upper Bowland Shale (Pendleian, $E_{1a\text{-}b}$), which comprises deep basin mudstones deposited between the *C. leion* and *C. malhamense* Marine Bands before the arrival of the first turbidite sandstones. These mudstones thicken southwards into the basin, where they were the target for fracking operations in Lancashire.

The quarry exposes Pendle (Skipton Moor) Grit (Pendleian, E_{1c}), which are turbidite sandstones that record the first arrival of coarse sands from the northern source area (Fig. 7.8). The base of the sandstone unit is not seen, but it seems likely that it is not far below the base of the exposed section. The sandstones are mostly thickly bedded, and there is large-scale erosional relief on the bases of some beds, as well as smaller erosional sole marks such as flute casts. Two channel units are present in the quarry. The lower one is inferred from the thinning-upward sequence seen in the northern face of the quarry, whilst the second shows an erosional base with relief of several metres. Large-scale sole marks occur on this surface.

These features suggest a 'proximal' setting even though these sandstones are the first arrivals into the basin. This is best explained by the proximity of the basin margin so that most of the turbidity current flows were confined to channels and were not spreading out widely over the basin floor.

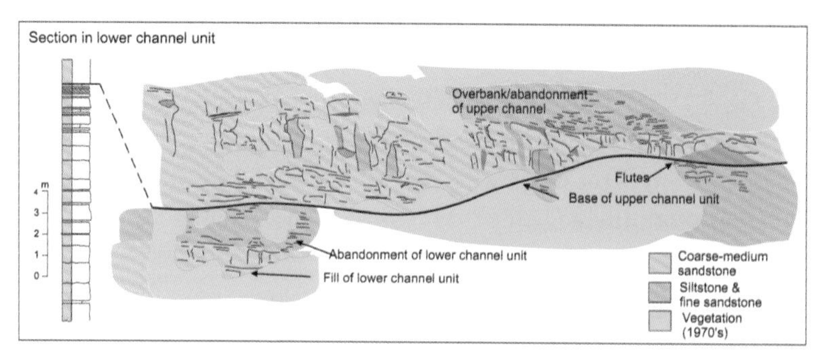

Figure 7.8 Drawing from photographs of the main face at Jenny Gill Quarry, Skipton. The sandstones are close to the base of the Pendle Grit which rests sharply, and probably erosively, on the Upper Bowland Shale.The face comprises two channel units filled with massive turbidite sandstones that are separated by a mud-rich interval. The vertical log on the left is measured in the lower channel unit at the left hand side of the drawing (after Baines, 1977).

Horn Crag Quarry, Silsden (SE 053480; 53.9262, –1.9222)

This locality lies to the east of the A6034, one mile north of Silsden. Approaching from Silsden, turn right along Fishbeck Lane, just before the plantation on the right. Follow the lane beyond the houses until it opens into pasture. Park on a track on the left-hand side of the lane, about 100 m beyond the last house. Walk up the track to the gate and cross the wall at the stile. Walk through old workings to the quarry. The quarry is old and appears stable and there is no need to climb around on the face. If you have the option, try to visit in the afternoon when the sun is in the best position.

This small, abandoned quarry exposes the Middleton Grit (Arnsbergian, $E_{2c2(i)}$ minor cycle), which caps one of several thin deltaic cyclothems present in this area between the Pendleian basin fill and the start of the major Kinderscoutian succession. The outcrop appears to be in the upper part of a coarsening-upwards cyclothem that does not extend much further to the south.

The beds at the top of the quarry show the local tectonic dip so that all other dips, relative to this are of syn-sedimentary origin (Fig. 7.9). The section shows some 26 m of medium-grained sandstones. The lowest 23 m of the exposure shows near-tabular beds 0.15–2 m thick, which tend to become thicker upwards. Beds are commonly cross-bedded and cross-laminated and have wave-rippled tops. Bioturbation is quite common in the lower part of the sequence. These mouth-bar sands appear to have been deposited in the hanging wall of a small growth fault (see Fig. 4.15A). The fault is not actually seen, but its position is inferred to be close to the northern margin (left) of the quarry. The mouth-bar sandstones dip to the north with clear roll-over, and also thicken in that direction. Both these features are typical of the hanging wall area of a

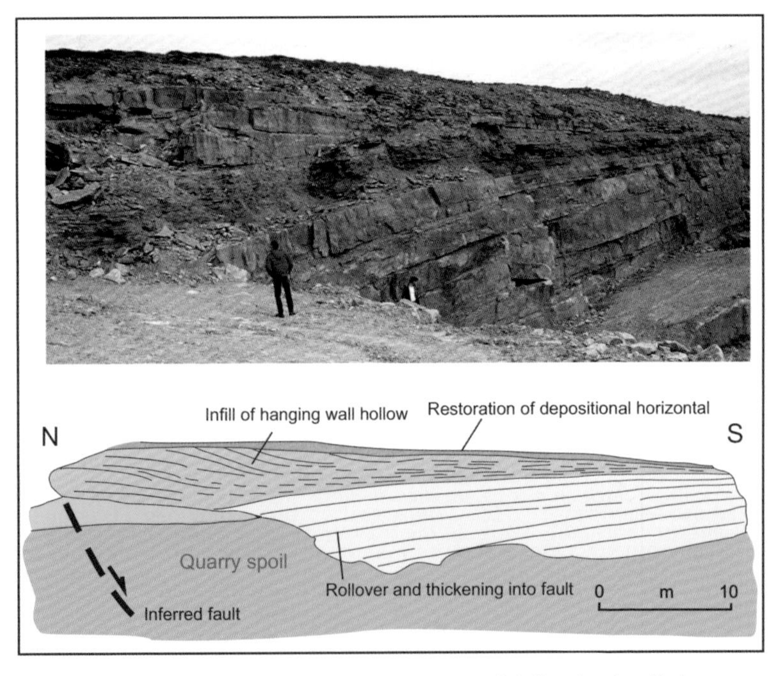

Figure 7.9 Horn Crag Quarry, near Silsden. Middleton Grit (Arnsbergian, E_{2c2}). Mouth-bar sandstones thicken into the hanging wall of an inferred growth fault and show a roll-over structure due to the listric nature of the fault. Large-scale cross-bedding records the later infill of a space created by the continued movement of the fault after initial sand supply ceased (drawing based on Martinsen, 1990a).

syn-sedimentary growth fault where sand supply kept pace with fault displacement.

Beds in the uppermost 3 m form a lenticular unit which seems to have filled the space that resulted from continued rotation of the underlying beds after supply of sand to the mouth bar ceased. The unit is thickest at the northern (upstream end) and thins out to the south within 100 m. Up to 2.5 m thick sandy sigmoidal foresets become finer and thin out into wave-rippled beds down dip. The stacking of these sigmoidal units suggests a subtle interaction of accommodation and sediment supply.

The sediments are thought to have been deposited in a sandy mouth-bar setting, mainly from repeated and sustained (flood?) traction current events. Between events, the bar surface was reworked by wave action. The dips are related to movement on a syn-depositional normal, listric fault located immediately north of the exposure. This fault probably flattened out downwards into the mudstone around an underlying marine band. The quarry exposes one of the few growth-fault geometries known in the Pennine province, and this locality is the only one that shows the detailed form and internal organization of the sediments.

Baildon Bank (SE 150390; 53.8497, -1.7713)

These old quarries cap the escarpment above Green Road, Baildon and are best approached from the big roundabout in Baildon, turning west into Westgate and then left into West Lane at the cross-roads. After about 300 m, turn left into Salisbury Avenue and park towards the far end. From the end of this road, a footpath descends the hill. Turn right at the first bend and follow a footpath along the slope at the base of the exposure. The faces are stable and access along the path is good. There should be no geological reason to climb on the face.

The main sandstone that forms the upper part of the face is the Upper Rough Rock (Yeadonian, G_{1b}) and is exposed over a distance of 250 m. The face is orientated SW–NE, slightly oblique to the dominant palaeocurrent towards the west (Fig. 7.10). The section shows coarse, pebbly sandstone, cross-bedded in a variety of styles. The lower parts are characterized by tabular sets whilst the upper part includes some trough cross-bedding. The tabular sets in the lower part show various

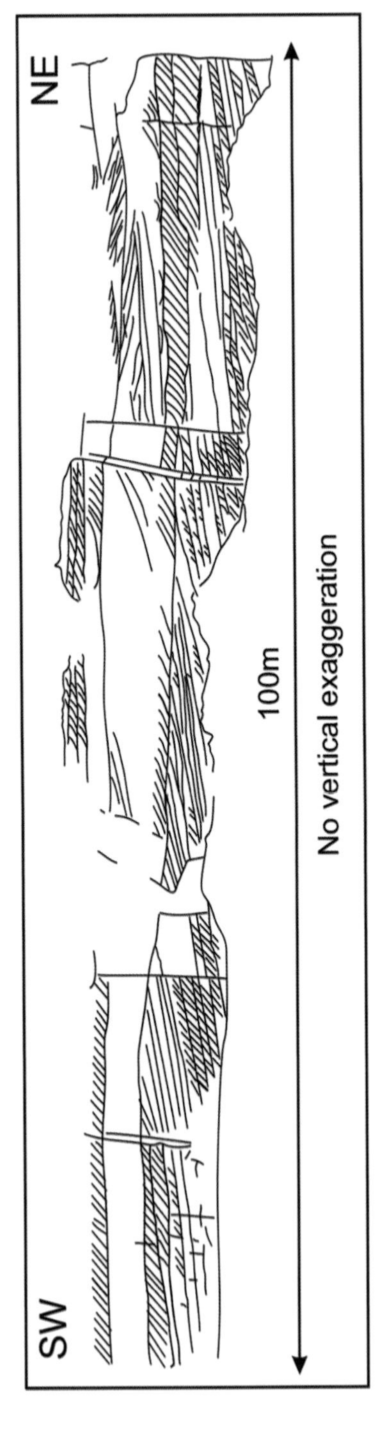

Figure 7.10 Sketch of the southwestern part the Upper Rough Rock (G_{1b}) exposure at Baildon Bank, showing a variety of styles of cross-bedding. These include descending sets, reactivation surfaces and downstream overtaking of bedforms. All these features are indicative of a very active sandy braided river channel. The section is oriented close to parallel with the dominant palaeocurrent direction (after Bristow, 1987).

geometries with descending sets very apparent, indicating that dune bedforms migrated over the downstream-inclined surfaces of larger bar forms. The central part of the unit is characterized by the thickest tabular sets, indicating the migration of large, straight-crested bedforms. Elsewhere there are examples of reactivation surfaces and of one set overtaking an underlying set to merge into a single thicker set. All these features are characteristic of the deposits of large sandy braided rivers where large sand bars and smaller, superimposed bedforms were constantly migrating and shifting position, accumulating a sandbody as rising base level created accommodation space.

A nearby smaller outcrop at Shipley Glen (SE 129392; 53.8471, −1.8020), accessed by going along West Lane to junction at the end of Lucy Hall Drive, and crossing about 150 m open ground to the wooded valley, shows the Upper Rough Rock in an exposure oriented at a high angle to palaeoflow.

Hallas Rough Quarry, Cullingworth (SE 055355; 53.8188, -1.9170)

This quarry is accessed from the A629, Halifax Road between Flappit Springs and Manywells, near Cullingworth. The quarry is currently active as a tip, and permission to visit should be sought from the site manager or from the head office. If access to the deeper part of the quarry is not practical, then on the north side of the quarry is high ramp of tipped material. At its top is a track which gives good distant views of the main faces of the quarry and allows easy access from the western end of the rampart.

The quarry provides one of the most complete and extensive sections through the sandstone-rich upper part of the Yeadonian (G_{1b}), which comprises the Rough Rock Flags, and the Lower and Upper Rough Rock intervals. Around 30 m of sedimentary succession are exposed in the quarry, extending over 300 m laterally (Fig. 7.11). The layout of the quarry has faces at right angles to one another, allowing an appreciation of the three-dimensional organization of the sediments.

The Rough Rock Flags occur in the lowest parts of the quarry and comprise horizontally bedded, fine–medium, sometimes micaceous sandstones. Both parallel and ripple lamination are common and

Figure 7.11 Hallas Rough Quarry near Cullingworth. The exposure is all within the Yeadonian (G_{1b}) and comprises the Rough Rock Flags, and the Lower and Upper Rough Rock sandstones. There is a conspicuous interval of finer-grained sediment directly below the erosive base of the Upper Rough Rock.

there are also lenticular sets of trough cross-bedding with palaeoflow directions towards the west. Larger trough cross-bedding is present towards the top of this unit and some of these sandstones may occupy channels. There is extensive bioturbation by *Olivellites* (*Psammichnites*) and by *Lockeia (Pelecypodichnus)*, both as pits and bumps on bedding surfaces and showing upwards escape traces cutting through tens of centimetres of sandstone. Such examples suggest rapid deposition as the bivalve responsible, which lived at the sediment surface, moved progressively upwards to keep pace with sedimentation. This assemblage of features suggests that these units resulted from rapid progradation of a major, proximal mouth bar.

The Rough Rock Flags sandstones are overlain by around 1 m of micaceous silty mudstone, thinly and horizontally laminated (Rough Rock Shale), which can be traced extensively around the quarry. The unit records the abandonment of the mouth bar, due to delta switching or to a minor rise in sea level. The mudstones are overlain sharply by around 2 m of fine–medium sandstone with ripple lamination and small-scale cross-bedding. Although sharp, the basal contact does not appear erosive, and this suggests the advance of a new mouth bar, possibly accelerated by a minor fall in sea level

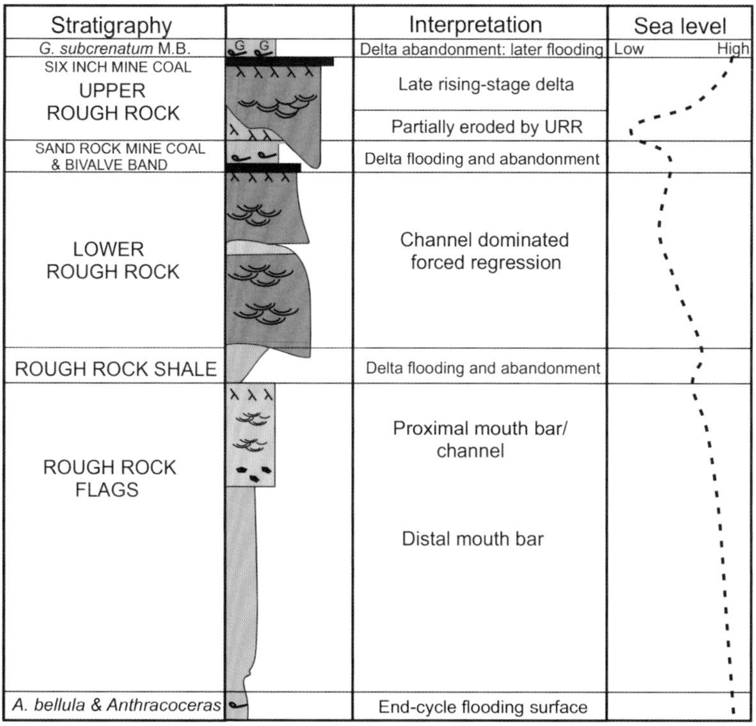

Stratigraphy		Interpretation	Sea level
G. subcrenatum M.B.		Delta abandonment: later flooding	Low High
SIX INCH MINE COAL			
UPPER ROUGH ROCK		Late rising-stage delta	
		Partially eroded by URR	
SAND ROCK MINE COAL & BIVALVE BAND		Delta flooding and abandonment	
LOWER ROUGH ROCK		Channel dominated forced regression	
ROUGH ROCK SHALE		Delta flooding and abandonment	
ROUGH ROCK FLAGS		Proximal mouth bar/ channel	
		Distal mouth bar	
A. bellula & *Anthracoceras*		End-cycle flooding surface	

Figure 7.12 The stratigraphy of the fully developed Late Yeadonian (G_{1b}) succession related to changes in sea level during deposition of the interval. At many localities across the southern Pennines, one or more of the component units may be missing due to variable subsidence and/or erosion (courtesy of Colin Jones).

(Fig. 7.12). Above is an erosively based, rather massive sandstone, around 8 m thick. It is medium- to fine-grained and has medium-scale trough cross-bedding. This unit is the Lower Rough Rock and is a channel deposit that may have been genetically related to the underlying mouth bar.

In the uppermost part of the section, a further sharp erosion surface marks the base of a very coarse sandstone, the Upper Rough Rock. The finer-grained sediments that are locally present between the Lower and Upper Rough Rock (including the Sand Rock Mine coal) in Rossendale are absent here, presumably removed by contemporaneous erosion, as the widespread braided river complex spread out, possibly helped by a falling sea level. The Upper Rough Rock shows large cross-bedding, mainly as broad troughs with foreset dip components to the west.

Suggested Itinerary: Nidderdale to Airedale

This itinerary could be used to illustrate the transition from the Askrigg Block in the north to the Craven Basin to the south. It is best carried out from north to south, as the view from Fancarl Crag then sets the scene for later localities. The following localities would make up the itinerary, but some could be omitted depending on time available and individual interest. It would require a full day, using a motor vehicle.

Brimham Rocks – Scot Gate Ash Quarry – Fancarl Crag – Kilnsey Crag – Linton-Cracoe road – Skipton–Bolton Abbey road – Jenny Gill Quarry.

Chapter 8

The Deeper Basin: Clitheroe–Bowland

This area (Fig. 8.1) illustrates Viséan deeper-water, mid-basin settings that contrast with the marginal setting seen north of Skipton. The Viséan succession comprises limestone turbidites and 'Waulsortian reef' microbial mud mounds. These large mud mounds differ in their setting compared with the Cracoe 'reef knolls' to the north and with the fringing 'reefs' and on-platform mud mounds of the Derbyshire Platform. They appear to have developed as the basin began to deepen in late Courceyan (Tournaisian) times. Slumping in some limestones suggests that local extensional faulting was creating sediment instability through increasing basin-floor gradients or by seismic shocking. The overlying Pendleian succession is entirely clastic in composition and comprises deep-water mudstones (Bowland Shale) and a basin-filling

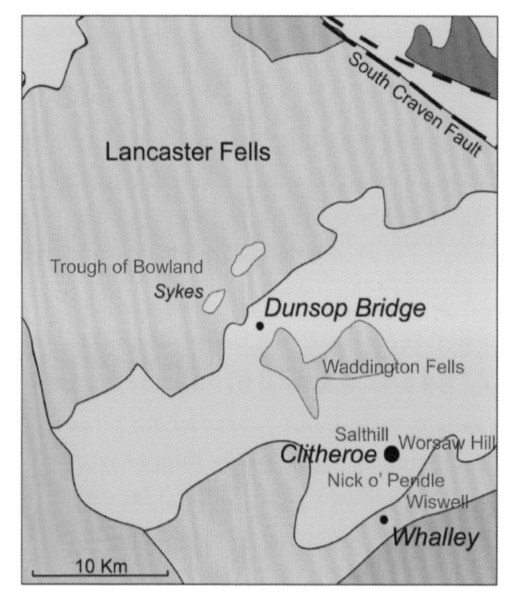

Figure 8.1 Outline geological map of the Clitheroe area, showing the positions of the localities described.

succession of turbidites and slope sediments that are the more distal equivalents of the succession near Skipton. There is evidence that active extensional faulting persisted into the Namurian. The Pendleian basin-fill was confined to the north of the Pendle Lineament which, at that time, was a north-facing fault-controlled ramp that acted as a barrier or deflector to southwards-flowing turbidity currents (Figs 4.12 & 4.13).

Clitheroe by-pass (A59), Worston (SD 768430) (▸)

The erection of fencing and the consequent lack of safe parking means that this important locality now cannot easily be visited. The heavier traffic and the growth of trees in recent years have not helped the situation. It is possible to get some impression of the sediments by driving through the road cutting but, obviously, it is important not to combine driving with looking at the rocks. ***Only try to look at this section if your driver is sworn to geological abstinence.*** *An alternative is to use Google Street View, which provides partial views. It is also possible to get a remote view of this section from the road bridge over the cutting between Chatburn and Downham.*

This long road-cutting in the Hodder Mudstone Formation (formerly Worston Shale: Chadian) lies 3 km to the northwest of the Pendle Monocline. This dipping structure resulted from Variscan inversion of the normal Pendle Fault that had formed the south-eastern margin of the Bowland Basin in the early Carboniferous. The exposure shows a succession of interbedded limestones and mudstones of relatively deep-water origin. The smell on hammering suggests that both the limestones and the mudstones contain significant organic carbon. The limestone beds are probably the deposits of turbidity currents shed from surrounding bathymetric highs. At the southern end of the cutting are rounded hills formed by exhumed mud mounds which directly underlie the Hodder Mudstone (see Worsaw Hill, below).

Worsaw Hill: (SD 779434; 53.8788, –2.3375) (▸)

Worsaw Hill is a clear feature in the landscape. The GPS reference applies to a good stopping point on the lane between Worston and Downham. A gateway allows a good view across fields to the hill. Disability friendly. Further views of the eastern side of the hill can be had by stopping further along the lane towards Downham.

Figure 8.2 Worsaw Hill from the west. A large 'Waulsortian reef' mud mound exhumed from the overlying Hodder Mudstones. It stands about 90 m high.

This spectacular topographic feature is a large, exhumed microbial mud mound, typical of several examples in the area, some of which have been extensively quarried around Clitheroe (Fig. 8.2). The overlying Hodder Mudstone has been stripped away by glacial and later erosion so that the present-day topography partly reflects that of the Courceyan sea floor. These microbial build-ups have rather structureless cores of lime mud, whilst their flank beds are particularly rich in crinoid debris. They are thought to have developed in the early stages of major deepening that led eventually to deposition of the Hodder Mudstone Formation. Their association with deeper-water sediments above has led to them being characterized as 'Waulsortian' by comparison with examples in Belgium, but the water depth under which the mounds grew remains somewhat uncertain.

Salthill Quarries, Clitheroe; (*c.*SD 754425; 53.8794, –2.3731)

From the A519, take the road to the northwest (Pimlico Link Road; A671), close to Worston, which leads towards the currently active quarries. Take the first left, (Lincoln Road) and follow for around 300 m. Park in the car park on the righthand side. There are around 30 parking bays. This important site, which is a designated SSSI, has been developed as a nature trail around the margins of what was once a quarry, and which is now an industrial estate. It extends on both sides of the road and has accessible and well-labelled footpaths. A walk around the whole circuit is around 1.5 km long. None of the exposures are high and the footpaths are good. Ideal for disabled access. Helmets are not necessary. Do not cross fences. Many features are best viewed from a distance. As this is a SSSI, it is important not to hammer the in-situ exposures. Collecting of specimens is discouraged as the site has suffered greatly from uncontrolled collecting in the past, and any necessary collecting should be confined to already broken, small-scale material. There is a series of good, published guides on the 'Ribble Valley Geotrails', published by the Lancashire Group of the Geologists' Association. 'Walk 5; Clitheroe' is relevant to this locality. Beware that there is a slight mismatch between the stratigraphic ages in the guide and the ones presented here.

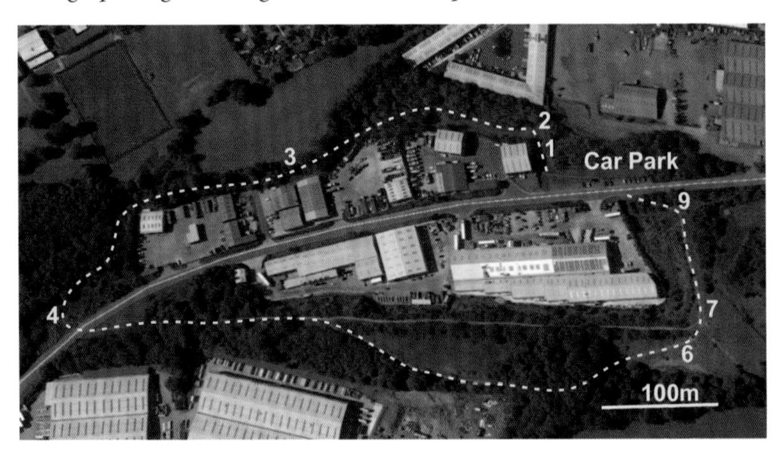

Figure 8.3 Satellite image of the area around the Salthill Trail that follows the margins of a now quarried large mud mound. The positions of the numbered localities are shown. Google Earth image.

The quarry removed the central part of a large microbial mud mound, one of several so-called 'Waulsortian reefs', which are exhumed to form prominent features in the local landscape. They occur within the Clitheroe Limestone Formation of late Courceyan age. Exposures in the old quarry walls that remain show a variety of mound-margin or flank facies. Depositional dips in flank beds are inclined away from the core of the mound. The flank beds themselves are characterized by abundant and diverse crinoidal debris, mainly preserved as lengths of disarticulated stem or as individual ossicles. Crinoids probably flourished in the clear and mobile water around the flanks of the mound after it had stopped growing but did not play a significant part in its growth or stabilization. This locality has also yielded intact crinoid calyxes, suggesting that energy levels were low, as these delicate structures would otherwise have been broken up. Around mud mounds on the Derbyshire Platform, preservation of calyxes is virtually unknown, suggesting that higher (wave) energy broke them up in shallower-water settings. Whilst water depths at Clitheroe were probably greater than those on the Derbyshire Platform, the higher parts of the Clitheroe mounds possibly built up into the photic zone.

This brief resumé is largely based on existing geological guides to the quarry (Fig. 8.3). The description takes an anti-clockwise route around the trail, which follows the perimeter of the industrial estate. Immediately around the corner next to the car park is a section (Points 1 & 2 of the trail) showing pale, rather massive limestone in the core of the mud mound draped by darker, richly crinoidal limestones that are part of an inter-mound facies. There is a boulder bed at the rather irregular junction of the two facies, suggesting a period of erosion, possibly related to the seabed becoming above wave base. This topography is smoothed by the overlying crinoidal limestone. Follow the path for about 250 m (Point 3) where there is a well-exposed face in crinoid-rich limestones that are flanking beds to the mud mound. The crinoid fragments are quite well sorted, suggesting some limited reworking after death. The next point on the trail at the roadside (Point 4) shows flank beds, similar to the last locality, with a depositional dip of around 25°. In addition to abundant crinoid stems, fossils present here include corals and brachiopods.

Crossing the road and following the path to the left for about 500 m leads to Point 6 of the trail, where dipping beds yield abundant crinoidal debris. This mostly consists of crinoid stems, but calyxes have also been found, usually fragmented, along with corals.

At the northern end of the trail there are other numbered points. Point 7 is on higher ground that allows a wider view, setting the mud mound in its regional context, particularly its relationship with overlying strata. To the east is Pendle Hill, capped by Pendle Grit (Pendleian, Namurian E_1), underlain by the Upper (Pendleian) and then Lower (Brigantian)

Bowland Shale, including the turbidite Pendleside Limestone. Point 8 highlights the more recent effects of Pleistocene glaciation. The rounded form of the limestone surface suggests a roche moutonnée, and the striations on some surfaces also point to glacial scour.

Waddington Fell Quarry, north of Clitheroe (SD 719479; 53.9264, -2.4303)

From Clitheroe, take the road north through Waddington village and proceed to the top of the moor. The quarry entrance is on the west side of the road. Park in the car park opposite the weighbridge and report to the office. At the time of writing, the quarry is run by the Armstrong Group and prior permission should be obtained from the head office. The quarry operates on weekdays only. The usual conditions of high visibility clothing, helmets and boots apply, and, especially in windy conditions, eye protection is advised. It may be possible to negotiate disabled access to some areas. If prior permission is granted, check in at the weighbridge office before proceeding further and follow any instructions given at the time. Be sure to look at some of the cut slabs near the car park as these can show clear internal features that are not readily visible on broken quarry blocks and faces.

This is a large working quarry in the Pendle Grit (Pendleian, Namurian E_1). The sandstones are mainly crushed for aggregate. Large blocks are taken elsewhere for sawing into dimension stone. The main active quarry face exposes around 40 m of sandstone over a distance of around 400 m. Older workings extend the useful exposure to the south for another 500 m. The deepest levels of the quarry used to show the sandstones resting sharply on mudstones of the Upper Bowland Shale, dipping to the northeast. The medium-coarse feldspathic sandstones are mainly massive and thickly bedded. There is commonly erosional relief between beds, in places giving thick amalgamated intervals. Intra-formational mud clasts are very abundant, in some cases occurring in high-concentration pods of mudflake conglomerate. The distribution of mud clasts in some sandstone units suggests that they are hybrid event beds (HEB), deposits of flows which were transitional between turbidity currents and debris flows. Within the stacked massive sandstones, lenticular units of fine-grained sediment occur as small erosional remnants. These intervals were probably originally thicker and more extensive, and could have been overbank sediments or the infills of abandoned channels.

Palaeocurrents based on sole marks on the massive sandstones indicate a dominant flow to the southwest, whilst palaeocurrents from thinner-bedded sandstones show flow to the north, down the dip of inclined lateral accretion surfaces. These surfaces are an important feature of the sandstones at this locality. They dip at relatively low angles, in units around 4 m thick, and are thought to record deposition on the side of a laterally migrating channel, probably the inner bank (point bar) of a meander. The inclination of these surfaces to the north and their vertical stacking have led to the suggestion that the migrating channels responded to differential subsidence caused by syn-depositionally active normal faulting (Fig. 8.4). Some faults seen in the quarry may have been

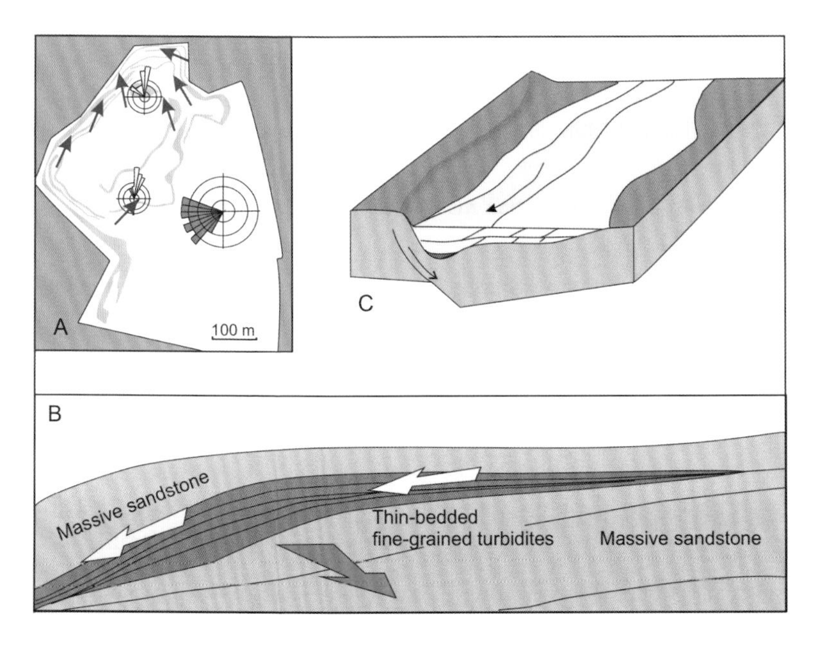

Figure 8.4 Lateral accretion in channelized turbidites in the Pendle Grit at Waddington Fell Quarry. **A**) Plan of quarry showing the distribution of the major sandbodies and associated palaeocurrents: Blue – paleocurrents from thick massive sandstones; Yellow – palaeocurrents from thin-bedded, fine-grained turbidites; Magenta – dip direction of lateral accretion surfaces. **B**) Schematic model for the occurrence and directions of palaeocurrents in laterally accreting unit. **C**) Suggested control on the migration direction of laterally accreted channels by movement on a syn-depositionally active normal fault (courtesy of Ian Kane; see also Kane *et al.*, 2010).

extensional structures during deposition, but they probably underwent some inversion during Variscan compression.

These 'proximal' turbidites are thought to be part of a major deep-water channel complex which extended out into the basin from the margin of the Askrigg Block to the north. In early Pendleian times, the continuing but waning extensional tectonic regime probably led to an irregular basin-floor topography, and this would have channelled turbidity currents into laterally constrained pathways. Unlike the succession at the Trough of Bowland, there are no 'distal' turbidite precursors to the thick-bedded sandstones.

The red coloration that affects some of the sandstones here is thought to be due to deeply penetrating weathering beneath the Variscan Unconformity, which would have been within perhaps 100 m above the present-day land surface, during arid Permian and Triassic times.

Even if you are not planning to visit the quarry, it is worth stopping at the roadside for the view, especially towards Pendle Hill. The small valley that goes up the side of Pendle Hill is Little Mearley Clough, the stratotype for the Pendleian stage.

Trough of Bowland (SD 626528; 53.9726, –2.5775) (▶)

If approaching from Sykes, it is best to park vehicles at the top of the hill, on the parking place beyond the cattle grid. Then walk back down the road (c.300 m) before climbing down into the stream bed and working back up the section. It is quite rugged, and care is needed, especially when the vegetation is high. The stream bed and the gullies on the western bank need care as some are quite steep. Not advised after heavy rain. No disabled access to the stream bed, but a good impression of the section can be had from a car on the road. The road is very steep.

The section is exposed in both banks of the stream and in its small, steep tributary gullies. The sediments are latest Viséan to earliest Namurian (Pendleian, E_1) in age, and record the earliest arrivals of clastic sediments from the northern source area that eventually supplied most of the Millstone Grit. The overall section shows an upward transition from grey silts (Lower Bowland Shale) into thinly laminated dark mudstones (Upper Bowland Shale). Above these shales, the incoming of relatively thin-bedded turbidite sandstones marks the base of the Pendle Grit. The section is, therefore, broadly equivalent to the section seen at Jenny Gill, Skipton and at Waddington Fell. However, here the incoming of sand is gradual, with thin, interbedded sandstones and mudstones, contrasting with the erosional channelling and thick amalgamated sandstones present at Skipton and at Waddington Fell. These differences probably reflect different topographic settings on the basin floor which, in the Pendleian, is thought to have been quite irregular because of on-going tectonic activity. Within the shales there is a unit of sandy breccia, possibly recording seismic activity associated with this tectonic activity. The sandstone turbidites are separated by interbedded shales and show an overall upwards thickening, suggesting progradation of the turbidite system.

Nick o' Pendle (SD772386; 53.8426, -2.3484) (▶)

From the A59 near Clitheroe, take the road towards Sabden. Follow this road for around 2 miles and stop at the top of the hill. There is plenty of space for parking at the roadside. This locality is mainly for the view and does not contribute much if visibility is poor. Disability access fine for view.

This is the type locality for the Pendle Grit and is a good viewing point for understanding the basin context and the present-day geological structure. The locality lies close to the Pendle Lineament, which exercised strong control both on extensional basin development and subsequent Variscan deformation. The steep tectonic dip towards the southeast is an expression of the Pendle Monocline, an inversion structure above the earlier major extensional Pendle Fault (down to the northwest). The southwest–northeast trend is typical of inherited Caledonian structures. To the west, across the Ribble Valley, is Waddington Fell, which overlies the Bowland intrabasin high, and beyond that the Lancaster Fells. To the southeast, the succession dips into the Burnley Coalfield, with Kinderscoutian and Marsdenian deltaic cyclothems picked out by west-facing escarpments. In the further distance to the east is Boulsworth Hill, the site of an early oil exploration well that penetrated red beds at the base of the Carboniferous section.

The Pendle Grit exposure at this locality is somewhat discontinuous. The sediments are rather structureless sandstones, rapidly deposited from large-scale turbidity currents. Some surfaces show prominent scours that may be megaflutes, large-scale flute-like erosional features that occur on the top surfaces of some thick turbidite beds.

Wiswell Quarries, Wiswell (SD 755373; 53.8266, –2.3754)

From the roundabout on the A59, take the A671 for about 1 mile before turning left on the road to Sabden. This road swings left around the end of the golf course. At the junction, go straight ahead where the Sabden road turns right. Follow this road (Clerk Hill Road) beyond the entrance to the golf club for around 1 km and park on the roadside. A footpath from the gate on the left leads to the upper quarry. The quarries are long disused and have an upper and lower part. It is possible to move around freely although the usual care should be taken, particularly if the rocks are wet.

The quarries expose some 80 m of strata in the upper part of the Pendle Grit. The quarry in the higher part of the hill shows extensive bedding surfaces whilst the lower quarry shows dipping beds in a vertical section around 40 m thick (Fig. 8.5A).

These turbidite sandstones, which are arkosic and medium- to coarse-grained, are mainly thickly bedded and structureless, separated by thin beds of mudstone and siltstone. These finer deposits are commonly eroded away so that sandstones occur in thick amalgamated units that resulted from multiple depositional events. Overall, the succession occurs as two thinning-upward intervals separated by a mudstone unit. In each interval, the finer-grained components become more common upwards as sandstones become thinner and show less amalgamation. In the lower parts of the intervals, where amalgamation is common, erosional sole marks are common features and large flute-like erosion features occur on bedding surfaces in the upper quarry. Shale clasts help to define some amalgamation surfaces whilst other shale clasts are scattered within the sandstone beds. Shale-clast debrites, which occur at the tops of some sandstone beds, are regarded as turbidite-debrite couplets, with the two components genetically linked. These hybrid event beds (HEBs) appear to have involved flows that evolved from fully turbulent turbidity currents to slurry-like flows with higher viscosity. It is likely that the erosional collapse of channel banks, upstream of

Figure 8.5 The lower quarry in the Pendle Grit at Wiswell. **A**) Thick-bedded massive turbidite sandstones overlain by thinly interbedded sandstones and siltstones. **B**) Intervals A, B and C of the Bouma sequence of lamination styles in a turbidite sandstone (see Glossary).

the depositional site, generated clasts of cohesive mud which were incorporated into the flows.

Towards the top of the upper interval, where sandstone beds are thinner with more fine-grained interbeds, some sandstone beds show complete Bouma sequences of internal lamination whilst the interbedded siltstones show starved sand ripples (Fig. 8.5B). These are best seen by climbing over the wall into the small cutting.

Suggested itinerary; Clitheroe area

This suggested itinerary illustrates the history of the central part of the Bowland Basin from Courceyan to Pendleian times. The route is a circuit that starts and finishes in Clitheroe town. It should be comfortably carried out by motor vehicle in a day, with a total driving distance of around 90 km.

Clitheroe – Worsaw Hill – Clitheroe by-pass – Salthill Quarries – Trough of Bowland – Waddington Fells Quarry – Nick of Pendle – Wiswell Quarries – Clitheroe.

Chapter 9

Millstone Grit Deltas of the Central Pennines

This is classic Millstone Grit country, made up largely of high moorland and exposing Namurian strata ranging in age from Kinderscoutian to Yeadonian (Fig. 9.1) on both the moor tops and on the sides of the dales. The succession comprises almost entirely deltaic sediments, arranged in a series of coarsening-upwards cyclothems deposited after the main phase of basin filling, which occurred mainly during the Kinderscoutian. The Todmorden Grit, outcropping in the core of the Pennine Anticline, is the only exposed example of the deeper-water facies of that fill succession. The turbidites of the Alum Crag Grit (Marsdenian) which outcrop west of Blackburn are part of a later basin-filling succession that extended beyond the western limit of the Kinderscoutian progradation. The cyclothems that make up most of the Millstone Grit are mainly the products of sheet-like or lobate deltas that developed during periods of fluctuating sea level. Deepening events related to sea-level rise led to flooding surfaces, fossiliferous marine bands and to the development of accommodation space for the next deltaic progradation. The marine bands, with their goniatite fossils, allow detailed correlation of cyclothems across the basin. Differences in subsidence rates, sea-level changes and sediment supply led to differences in thickness and character within and between cyclothems. The youngest sandstone interval of the Namurian, the Rough Rock, extends across the whole area as sheets of high-energy fluvial channel deposits. Subtle differences within certain stratigraphical intervals demonstrate on-going local differential subsidence, close to active faults, which lasted into the Yeadonian.

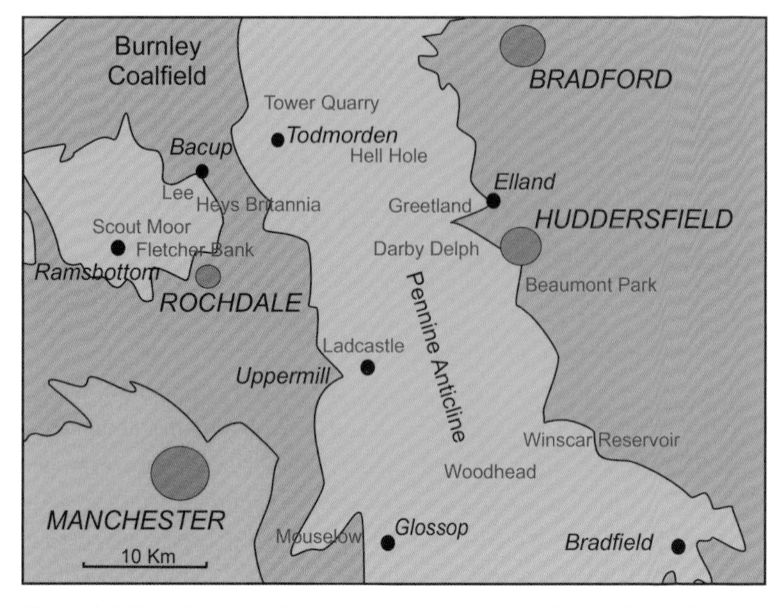

Figure 9.1 Map of the Central Pennine area showing the positions of the localities described.

Greetland Quarry (SE 095216; 53.6894, −1.8600)

From the traffic lights in Greetland, where the B6112 crosses the B6113 (Andy Thornton warehouse), take the B6113 (Rochdale Road) northwards for about 400 m. Turn right into Hoults Lane and then first right after a few metres, to park opposite the Star Inn. Walk up the hill for 50 m and turn right along Dean End. A few metres beyond the old weavers' cottages, a ramp on the left with a gate goes up from the main footpath. Follow this and walk around the gate into the quarry. Disabled access possible but rather challenging.

The quarry is in the Upper Rough Rock (Yeadonian, G_{1b}) of which around 10 m are exposed here. The sandstone is the equivalent of the upper of the two channel units in the nearby A629 cutting. The quarry faces are vertical and quite clean and seem stable. The quarry faces have different orientations allowing three-dimensional appreciation of the sedimentary organization. The sandstones are unusual in showing well-defined, large-scale inclined surfaces that span 6.5 m vertically and extend over 30 m laterally (Fig. 9.2A). Within the inclined layers are sets

Figure 9.2 Rough Rock (Yeadonian G_{1b}) at Greetland Quarry. **A**) large-scale, low-angle cross-bedding and **B**) small-scale trough cross-bedding within the large set, showing palaeoflow along the strike of the large foresets. This suggests lateral accretion on a large bar within a large sandy braided river.

of medium-scale trough cross-bedding showing palaeocurrents directed along the strike of the inclined layers (Fig. 9.2B). Such a relationship suggests lateral accretion of a large bar form that migrated normal to the main flow direction which is indicated by the medium-scale cross beds. This relationship is commonly associated with deposition on the point bar of a meandering channel, and this might initially appear at odds with interpretations of the Rough Rock as the deposit of a braided river. However, lateral accretion and is recorded on the flanks of large compound bars in some sandy braided rivers, for example the modern Brahmaputra of Bangladesh, which is probably of a comparable scale to the river envisaged for the Rough Rock.

A629 Road Cutting, near Elland (*c.*SE 103214; 53.6929, −1.8498) (►)

When freshly excavated in the 1970s, this cutting provided one of the most extensive and informative exposures of the Rough Rock and its immediately underlying sediments. However, over the intervening fifty years, it has become heavily vegetated and the traffic along the road has increased greatly, making crossing the road impossible. In winter, when the leaves are off the trees, it is still possible to see the large-scale organization from the footpath on the opposite side of the road. To access this footpath, park on the old Halifax Road, close to the last buildings on the left. The road is a slip road to the northbound lane of the A629. Walk along to the junction and view the face from the footpath. Disability access. ***Do not attempt to cross the dual carriageway.***

For exceptional reasons, full access to the cutting is, however, possible from Exley Road, which runs behind the cutting and can be accessed from the A629 at either end. Park near the bend in the road at the GPS location above. Follow a path across the field and descend to road level. A footpath runs along the base of the cutting. Take enormous care as traffic on the road is heavy and fast-moving. Small parties should walk strictly in single file.

This road cutting exposes a section 500 m long and around 30 m thick in the Rough Rock Group (Yeadonian, Namurian G_{1b} minor cycle) (Fig. 9.3). The lower part of the face shows finer-grained sandstones of the Rough Rock Flags, dominated by gradational facies changes. They occur above an unexposed coarsening-upward sequence and are interpreted as the mouth-bar deposits of a prograding delta. They are chronostratigraphically younger than the Upper Haslingden Flags of Lancashire and are sourced from the north. The mouth-bar deposits are erosively overlain by channel sandstones that reflect both minor and

Figure 9.3 The Yeadonian (G_{1b}) succession in the Elland road cutting (A629) soon after its excavation in the 1970s. The lower part shows Rough Rock Flags overlain by a channel unit with inclined erosion surfaces, probably the Lower Rough Rock. The uppermost interval is the Upper Rough Rock with a basal erosion surface that cuts down to the left.

major fluvial channels. The lowest channel units may be distributaries genetically related to the mouth bar. Above, a thicker unit of laterally extensive channel sandstones is divided by a major erosion surface with fine-grained sediments locally preserved below it. The lower channel unit is the Lower Rough Rock, with medium-scale cross-bedding. This unit is regionally extensive over virtually the whole of the Pennine Basin province and is associated with a fall in base level.

The Upper Rough Rock, which occupies the upper one-third of the face, is a channel unit of extensive sheet-like geometry, but which is laterally restricted compared with the Lower Rough Rock. Regionally, both Rough Rock units show somewhat variable palaeocurrent directions but, in this exposure, they are dominantly towards the south. Both are interpreted as the deposit of extensive sandy braided rivers with highly mobile channels. They are responses to a lowered sea level, but apparently this failed to cause significant channel incision.

This section, when freshly cut and accessible (Fig. 9.3), was subjected to intensive study. Myers and Bristow (1989) logged the section at 20 m spacing using a portable gamma-ray spectrometer whilst Cameron *et al.* (1992) measured three-dimensional orientations of texturally defined surfaces in a simulated dipmeter study.

Darby Delph, near Rishworth (SE 015161; 53.6421, –1.9712)

This quarry lies just below the A672 (Oldham Road), alongside Booth Wood Reservoir. The parking place is a gateway on the south side of the road, about 700 m west of the Turnpike Inn. There is space for 3–4 cars. The quarry is disused and occasionally used for rock climbing and so there may be other cars parked. If there is no space, there are some small parking spaces on the other side of the road a short distance away. Climb the gate and walk down the track for about 250 m to access the quarry floor.

Access to the quarry is not possible between March and the end of July because of nesting birds*. Conspicuous signs at the gate make the situation clear. These signs are sometimes left in place beyond the prohibited season and can then be safely ignored.*

The quarry, which is around 100 m long, exposes about 10 m of coarse feldspathic sandstone within the Lower Kinderscout Grit (Kinderscoutian, R_{1c}) (Fig. 9.4). It was opened to supply rock for construction of the dam of Booth Wood Reservoir. The quarry section is all within a single channel sandbody which is one element in a widespread multilateral channel complex. It shows an unusual example of the giant cross-bedding that is a typical feature of this interval. The quarry is best read from right to left, as the various components of the complex bedding evolve in that direction. At the right hand (eastern) end of the quarry, rather thick and broadly parallel-bedded sandstone shows large-scale undulations in units separated by low-angle erosion surfaces (Fig. 9.4A). This is the only known exposure of this bedding style within the Kinderscout Grit and its origin is somewhat uncertain. It has been explained by vertical and lateral aggradation of elongate sand-ridges close to the margin of a large river channel, citing images of the channel of the Brahmaputra in Bangladesh as a modern analogue. However, there may well be other explanations. To the left, this bedding style passes, via erosion surfaces, into more typical giant cross-bedding, although here it is highly broken up by inclined reactivation surfaces (Fig. 9.4B). These suggest a rather fluctuating river discharge over the large bar form whose downstream migration within the channel complex during periods of high river discharge led to the cross-bedding, with local shifting and switching of deposition and erosion reflecting possibly seasonal discharge fluctuations.

Figure 9.4 Lower Kinderscout Grit (Kinderscoutian R_{1c}) at Darby Delph near Rishworth. (**A**) Undulatory bedding in distinct packages separated by low-angle erosion surfaces. This underlies and passes laterally into (**B**) giant cross-bedding that is punctuated by several inclined erosional reactivation surfaces. The face is about 10 m high and the ramp at A is 4 m high.

Tower Quarries, Cornholme, near Todmorden (SD 906257; 53.7321, -2.1469)

From the A649 at the western end of Cornholme, turn into Mount Pleasant Street, about 100 m west of the western railway bridge. Follow the street to the end where there is ample parking space. From there, walk along the front of the houses to a second parking space with a large green shed at the end. From the track at the side of the shed, a footpath goes up the hill. This can be followed to the higher part of the quarries, or to a stile a short way up the path, which allows access to the lowermost quarry. Several footpaths criss-cross the hillside and provide access to the higher quarry areas. The quarries are long abandoned and seem quite stable.

This series of quarries, running up the hillside, provides a fragmented section through the middle Marsdenian (R_{2b}) succession in an area where there appears to have been considerable local tectonic control on sedimentation. The *V. sigma* (R_{2c}) Marine Band, present near the top of the section, anchors the succession in the zonal framework both here and at Beater Clough, 1 km to the west. A major down-to-the north normal fault trends across the valley and displaces the section between Tower Clough and Beater Clough. BGS mapping indicates more Marsdenian sandstones to the south of the fault than to the north, suggesting that the fault behaved as a down-to-the-south growth fault during deposition. The fault was then inverted during the later Variscan deformation.

Correlation of the sandstones between the two sections carries some uncertainty, as no marine bands have been recorded between the sandstones. However, a seatearth and thin coal seam within both sections provide a basis for correlation (Fig. 9.5). In the hillside at Tower Quarries, the lowest exposure is of coarse-grained, pebbly sandstone equivalent to the Fletcher Bank Grit (R_{2b3}). These sandstones show large-scale cross-bedding in sets up to 3.5 m thick suggesting a major fluvial channel deposit, similar to the equivalent sandstone at Fletcher Bank, near Ramsbottom. There, the channel sandstone occurs above a major basin-filling, coarsening-upward succession. At Cornholme, BGS mapping indicates a relatively thin interval of fine-grained sediment between the thick fluvial channel succession of the Kinderscout Grit and the base of the Fletcher Bank Grit. This suggests that the Kinderscoutian deltaic progradation that deposited the basin-filling succession stopped between Cornholme and Ramsbottom, some 13 km to the southwest, where deep water persisted into the Marsdenian.

Above the Fletcher Bank Grit, there is a gap in exposure of around 10 m, at the base of which the R_{2b4} and R_{2b5} Marine Bands are likely to be present. The covered interval is probably predominantly mudstone and siltstone with a coarsening-upward trend. The overlying exposure is of about 6 m of very fine-grained sandstones dominated by ripple cross-lamination but with plane lamination and bioturbation, particularly the meandering shallow burrow, *Psammichnites*. These sandstones are interpreted as mouth-bar deposits capping the progradational sequence.

They are erosively overlain by a 4 m thick, sharp-based channel sandbody of coarser sandstone, which has been correlated with the Lower Hazel Greave Grit. It is interpreted as a deltaic distributary channel. Directly above are a seat earth palaeosol and a thin coal seam, formerly exposed in the quarry, which permit local correlation. See Jones (2014) for a full discussion of this locality and its context.

The overlying sandstone is around 10 m thick. It is well sorted and siliceous, with wave-rippled surfaces, and a mixture of trace fossils suggesting a wave-reworked *marine* sand. This marine influence is very unusual in the Namurian of the southern Pennines basin, although comparable examples are found on the Askrigg and Alston Blocks. Traced laterally towards Tower Clough, using the coal seam for correlation, this sandstone thins to around 2 m and is replaced above by finer-grained sediment, above which is the *V. sigma* (R_{1c}) Marine Band. Lower in Tower Clough, the lowest exposed sandstone is very fine-grained and

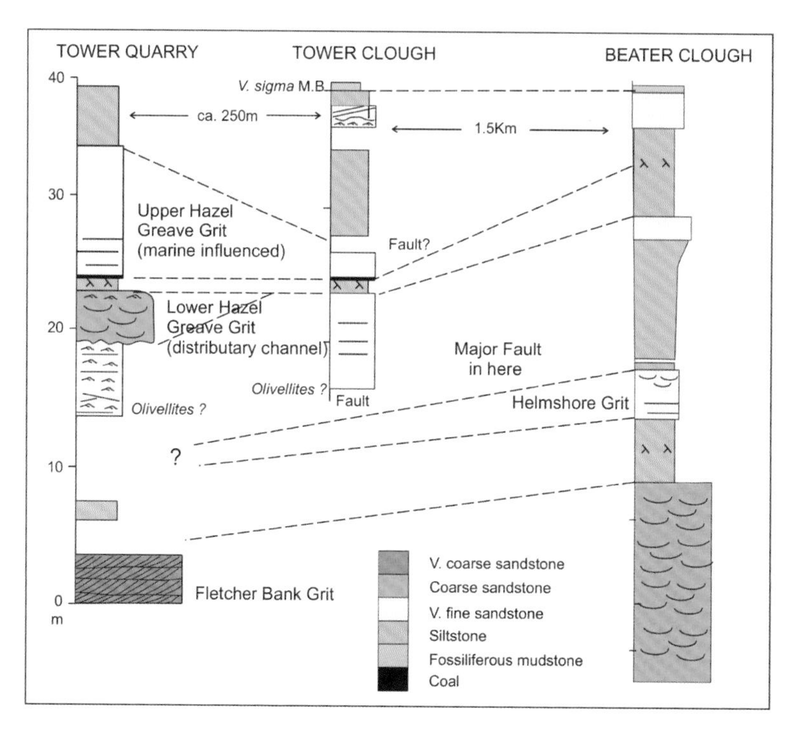

Figure 9.5 Sedimentary logs in the Marsdenian (R_{2b}) section at Tower Quarry, Cornholme, and a suggested correlation to Beater Clough, across a major fault to the north (courtesy of Colin Jones).

micaceous with undulating lamination and trace fossils. It is thought to be a mouth-bar deposit that is eroded in the quarry section by the distributary channel.

Correlations to Beater Clough, 1 km to the west, are less than secure as there is an intervening major fault. The Beater Clough section shows a good development of Fletcher Bank Grit and a thin development of the Helmshore Grit which is difficult to correlate back to the Tower Quarry/ Clough section. The overlying section has much less sand. Higher in the hillside there is a good section in the Yeadonian Rough Rock Flags and the Lower Rough Rock.

The area lies on the southern flank of the Holme High, and the stratigraphic complexity may relate to on-going tectonic movements around this structure.

Heys Britannia Quarry, Whitworth (SD 872203; 53.6780, -2.1874)

From the A671 in Whitworth, turn into Tong Lane, close to the zebra crossing. Follow Tong Lane to its end and along its continuation as a quarry track that climbs the hillside above Cowm Reservoir. Follow all the way to the quarry buildings and report to manager's office. If visiting as an individual or as a very small party, it may be possible to negotiate access immediately. If planning a larger trip, it is advisable to organize permission in advance. This is a large quarry complex, and it is usually best to leave vehicles near the office and to walk in from there. Disability access by negotiation, but rather rough ground. Beware of quarry traffic at all times. Keep well away from any waste pits where the ground may be water-saturated, unstable sand. The quarry faces are best examined at a distance from a good vantage point. Individual lithologies, lamination types and trace fossils are best examined in loose blocks.

This large and complex area of quarries exploits the Upper Haslingden Flags (Yeadonian, G_1) and extends across a large area of the plateau. The complex dates back to the times when the main products were high-quality flagstones. In the presently active quarries, it is impossible to predict exactly what might be visible at any time. The Haslingden Flags are stratigraphically distinct from, and demonstrably just older than the Rough Rock Flags. Importantly, both the Lower and Upper

Haslingden Flags are the products of a totally distinct sediment supply system derived from the west that has nothing to do with the source of the rest of the Millstone Grit (Fig. 4.1). The Upper Haslingden Flags occur at the top of a coarsening-upward unit some 25 m thick that overlies the *Ca. cumbriense* Marine Band (Fig. 4.11). The underlying Lower Haslingden Flags form the upper part of another coarsening-upward interval above the *Ca. cancellatum* Marine Band. Both these sandstones have a distinctive lithology, fine–medium sandstones with conspicuous mica and a greenish colour. This contrasts with the feldspathic sands that characterize the rest of the Millstone Grit. The separate source areas are confirmed by heavy mineral analysis. The western-derived sand became more important through time, extending its influence progressively further east throughout the Westphalian.

At this locality, the Upper Haslingden Flags occur within the axial part of an elongate sand lobe, around 8 km wide (Fig. 9.6) that has been interpreted as a bar-finger sand in an elongate delta system comparable to the modern Mississippi birdfoot delta (Fig. 4.11). Palaeocurrents are directed towards the east, parallel with the axis of the sandstone lobe (Fig. 9.6). Loose blocks are dominated by ripple cross-lamination,

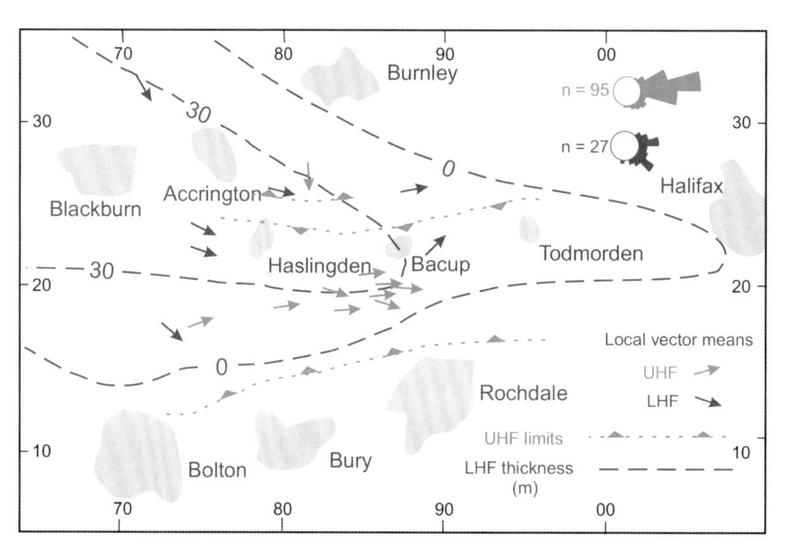

Figure 9.6 Map showing the distribution of sandstone thicknesses and palaeocurrents in the Lower and Upper Haslingden Flags (G_1) in the Rossendale area. Note how the palaeoccurrents are orientated parallel to the elongation of the sandbodies (based on Collinson & Banks, 1975).

although there is also some parallel lamination. Trace fossils are common, particularly *Lockeia (Pelecypodichnus)*, which occur as ellipsoidal bumps (on bases) and hollows (on tops) on bedding surfaces. These are the resting traces of non-marine bivalves that moved up through the sand as it accumulated to maintain their position at the sediment surface. The elongation of these traces shows a preferred orientation that is thought to be parallel with prevailing palaeoflow.

Distributary channels are not present at this locality despite the inferred axial mouth-bar setting. This suggests that the advance of the mouth bar stopped before its feeder channel had advanced to this area. As a result, unusually, the deposits of the upper mouth bar are preserved. Some 12 km upstream (west) distributary channels are present in axial part of the lobe.

Above the mouth-bar sandstones is an interbedded sandstone-siltstone sequence. This is truncated by the erosive base of the Lower Rough Rock, which extends as a uniform sheet over much of the basin. Its relationship with the highest part of the Yeadonian succession is best seen at Scout Moor Quarry. Palaeocurrent directions (mainly to the south) and the feldspathic composition of the Rough Rock demonstrate that it was supplied by an entirely different drainage system to the Haslingden Flags.

Fletcher Bank Quarry, Ramsbottom (SD 802168; 53.6478, -2.3017)

This large working quarry lies just to the east of the A65 and may only be visited by prior arrangement, negotiated either by an initial scoping visit or by a telephone call. It is operated by Marshalls Mono Ltd and is a manufacturing plant as well as a quarry. Any visit should begin at the reception, although the quarry manager's office may be further into the site. Limited disability access (distant view of faces): providing it is possible park at the manager's office.

This large, long-lived quarry now extracts sandstone exclusively for crushing as fine-grained aggregate, which is then used in the manufacture, on site, of concrete flags and other products. It forms a major feature of the landscape, especially as viewed from the western side of the valley.

The quarry provides a magnificent exposure of the Fletcher Bank Grit

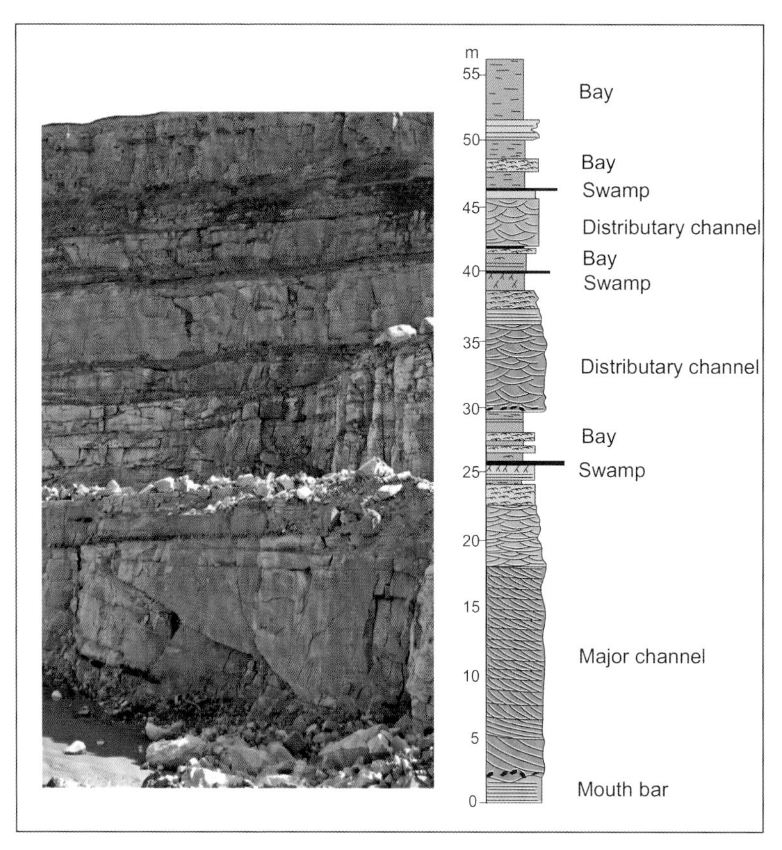

Figure 9.7 Fletcher Bank Quarry, near Ramsbottom, showing the Fletcher Bank Grit (Marsdenian, R_{2b}) and associated sediments overlain by the Helmshore Grit. The main sandstone includes giant cross-bedding and is overlain by an interval that includes thinner channel sandbodies, finer-grained sediments, palaeosols and coals, suggesting a water-logged delta plain with distributaries and crevasse deposits. The Helmshore Grit at the top is a more extensive channel sandbody. The graphic log was measured some time ago and details do not match exactly, as quarry faces have shifted in the meantime (log based on Okolo, 1983).

(Marsdenian, R_{2b}), a major deltaic distributary channel complex, and its overlying succession, in all some 55 m thick (Fig. 9.7). A borehole drilled in the quarry penetrated a thick coarsening-upward sequence of mainly muddy and silty sediments above thin sandstones and mudstones of early Marsdenian age. This underlying succession compares closely with the sequence seen west of Blackburn, where thick, early Marsdenian turbidites (Alum Crag Grit) occur at the same level. This succession appears to be the main basin fill for this area, showing that the Kinderscoutian basin-filling progradation did not extend this far west, but stopped somewhere between here and Todmorden (see Tower Quarries).

As with most actively working quarries, it is not possible to predict exactly what will be visible or accessible at any particular time. The channel sandstone complex, which forms the main quarry face, comprises at least three channel units. The lowest one is the thickest, around 20 m thick and, at one time, around 4 m of erosional relief was visible at its base. It comprises mainly medium-scale cross-bedding with sets up to around 2 m. However, excavations towards the northern end of the quarry exposed giant sets (*c.*10 m) with mudstone apparently infilling and draping the final position of the slip face. The overlying channels are smaller and are eroded into one another. Inclined bedding, which infills the highest of these channels, suggests a gradual abandonment and progressive reduction of its active cross-sectional area as discharge waned.

Above the channel complex is a succession of mudstones, siltstones and sheet sandstones along with a seat-earth and thin coal. This interval represents a shallow floodplain or delta plain setting that at times was emergent. Packages of sheet sandstones, inferred to be crevasse-splay deposits, within a dominantly fine-grained interval show quite marked fluctuations in thickness. The uppermost sandstone in the quarry face is the Helmshore Grit, above which is the *B. bilinguis* (late form) (R_{2b3}) Marine Band.

Scout Moor Quarry, Ramsbottom (SD 813190; 53.6655, -2.3017)

This quarry is run by Marshalls Mono Ltd and is managed from the Fletcher Bank Quarry offices, although there is also a manager on site. It may only be visited by prior arrangement with the management at Fletcher Bank Quarry. Depending on quarrying operations, it may only be possible to view the exposed section from a distance. Distant view is disability friendly. As in any active quarry, it is impossible to predict exactly what might be visible at the time of a visit.

The exposed section in the quarry face is of Yeadonian (Namurian, G_1) age and comprises the Upper Haslingden Flags below and the Rough Rock above, stratigraphically equivalent to the section exposed at Heys Britannia Quarry, some 5 km to the east (Fig. 9.8). The hillside above the quarry face is underlain by the lowest part of Westphalian Coal Measures (Langsettian), above the *G. subcrenatum* (G_{1a}) Marine Band.

The Haslingden Flags have no genetic relationship with the Rough Rock and were supplied by a very different source area. This contrasts with localities further north and east. For example, at Elland and Hallas Rough Quarry, the Rough Rock Flags and the Rough Rock have a superficially similar stratigraphic relationship, but these were both supplied by the same source area.

The main sandstone of the Upper Haslingden Flags occurs gradationally above a coarsening-upward unit some 20 m thick (only known from boreholes) and is interpreted as a mouth-bar deposit that built out to the east (Fig. 9.6). The sandstones are medium- to fine-grained and micaceous with ripple cross-lamination, some parallel lamination and some small-scale cross-bedding. Palaeocurrents are directed strongly towards the east. At Scout Moor, the Upper Haslingden Flags sandstone is overlain by horizontally bedded sandstones and

Figure 9.8 Scout Moor Quarry showing the Yeadonian interval between the Upper Haslingden Flags at the foot of the face and the Upper Rough Rock at the top. Here the Haslingden Flags are separated from the Lower Rough Rock by a shale interval and the Upper and Lower Rough Rock are also separated by a thin unit of fine-grained sediment that includes a bivalve horizon and a thin coal, the Sand Rock Mine.

interbedded finer-grained sediments. These are thought to reflect a delta-plain setting that was eventually drowned at a weak flooding surface, above which is mudstone with non-marine bivalves.

Above is the erosively based, coarse-grained, feldspathic Rough Rock which comprises two distinct channel sandbodies, the Lower and Upper Rough Rock, separated by an interval of finer-grained sediment that includes a thin coal seam, the Sand Rock Mine. This was exploited locally in shallow pits. The Rough Rock sandstones, particularly the Lower Rough Rock, have sheet-like geometries that can be traced from Airedale to North Staffordshire. Both units are interpreted as deposits of wide fluvial braid plains.

The total Yeadonian interval in the Rossendale area is much thicker than elsewhere in the Pennine Basin with two major deltaic intervals (Lower and Upper Haslingden Flags) and an expanded Rough Rock succession. Together, these differences suggest particularly high accommodation in the area, probably driven by movements on deep faults that bounded sub-basins within the Pennine Basin Complex, in this area probably the Heywood Fault.

The hillside above the quarry has sediments of the lowest Coal Measure (Langsettian) but they are not well exposed. The hill is capped by the Scout Moor Wind Farm whose turbines contribute the most recent anthropogenic addition to the landscape.

Ladcastle and Den Quarries, Uppermill (SK 995060; 53.5511, –2.0082)

It is usually possible to park a small number of vehicles on the public car park next to the museum and canal in the centre of Uppermill. If parking there, walk past the museum and take the first left, Moorgate Street, crossing the canal. At the top of the slope, where the road turns to the right, take the path diagonally to the left, to the railway line and ***cross the line with greatest caution****. Beyond the railway, take the first footpath on the right and follow it for around 200 m when old quarries will appear on the left.*

Alternatively, to avoid the railway crossing, access the quarries through a pedestrian underpass off Den Lane, around 400 m from the corner with Moorgate Street (see GPS co-ordinates). This takes you into the northern end of the quarries. ***This should always be the exit route*** *as the curvature* *of the railway track makes the foot crossing particularly dangerous in the return direction.*

These quarries extend along the hillside for around 300 m with good, slightly discontinuous exposure. The faces seem quite stable, and they are commonly used by climbers. However, particular care is needed along the path at the very foot of the faces, which is narrow and sometimes slippery.

These old quarries are in the Lower Kinderscout Grit (Namurian, R_{1c}) and they show particularly well the channelized nature of the giant cross-bedding that characterizes this unit over a wide area. It also clearly demonstrates the multi-storey nature of the unit (Fig. 9.9). At least three major channel sandbodies are present within the exposure, contributing to a multi-storey, multi-lateral complex. The earliest/deepest example is seen at the north end of the complex. Individual channel sandbodies are up to around 20 m thick and probably hundreds of metres wide. The channel facies include massive sandstones in the deeper part of the channels, giant cross-bedding in the middle part of the fill and medium-scale cross-bedding in the upper part. The giant cross-beds, which achieve set thicknesses over 20 m in places, are a characteristic feature

Figure 9.9 Lower Kinderscout Grit (R_{1c}) at Ladcastle Quarries, Uppermill, showing giant cross-bedding in two of the major channel units present there. The channel units make up a widespread multi-storey, multi-lateral complex and are separated by intervals of finer-grained sediments that may be remnants of overbank deposits or the fills of abandoned channels.

of these channel bodies. They seem to be the products of large bars or in-channel deltas, as discussed in Section 4.3. Medium-scale cross-beds, which overlie the giant sets, show similar dips to the giant sets, and are thought to be the products of dunes that migrated over the large bars. Overlying the channel sandstones, and preserved as erosional remnants between them, are intervals of interbedded thin sandstones and siltstone, which may be overbank sediments or the infills of abandoned channels. Channel complexes such as the Lower Kinderscout Grit form extensive sheets, mappable as continuous features with few apparent channel margins. Within the sheets it is likely that large- to medium-scale heterogeneity shown by these quarries is widespread.

Winscar Reservoir (SE 156030; 53.5239, −1.7673) (▶)

The locality is the main public car park on the east side of the reservoir. This is located in the headwaters of the River Don, for which it provides compensation flows. Approaching from the south, the car park lies beyond the hamlet of Dunsop Bridge and is sign-posted. The car park can be busy at warm weekends. It gives clear views across the reservoir to the exposed succession on the opposite side. Ideal for disability access. Unlike most views, this one illustrates detailed stratigraphic information rather than more widespread topographic aspects. The section of interest, on the opposite side of the reservoir and just above water level, is best studied with binoculars. It is sometimes possible to obtain permission from Yorkshire Water to visit the exposure itself, but the view across the water is very informative.

The interval exposed over some 300 m along the west side of the reservoir is in the upper part of the Marsdenian (R_{2c}) cyclothem. The succession dips gently to the north (right). A prominent sandstone interval is overlain by a predominantly fine-grained interval where the tectonic dip increases towards the righthand end of the exposure (Fig. 9.10A). The sandstone, which makes up most of the exposure, is the Huddersfield White Rock. It appears to have a clear erosion surface at its base close to the left-hand end of the exposure, and there is medium-scale cross-bedding within the sandstone. This suggests that the sandstone is a channel sandbody, probably a distributary channel unit, resting on mouth-bar sands. The shoreline cliff, which is about 5 m

Figure 9.10 A mature palaeosol profile in the top of the Huddersfield White Rock cyclothem (Marsdenian, R_{2c}) at Winscar reservoir. **A**) A distant view of the depositional context from the top of the underlying cyclothem, through the bleached palaeosol and its siderite cemented top into the sharply based mudstone overlying a flooding surface. **B**) A close-up of the palaeosol and flooding surface.

high, exposes around 8 m of channel sandstone when traced laterally down dip.

Resting directly on the channel sandstone is an interbedded interval of grey siltstones and thin sandstone beds around 7 m thick, which dips more steeply to the north. A thin coal seam with an underlying seat earth occurs near the base of this section, not visible across the water. The top of this grey interval passes up rapidly into a very pale, highly bleached unit, around 2 m thick. Close examination shows this to have rootlets in its upper part and is overlain sharply by a well-cemented, red/brown unit around 50 cm thick (Fig. 9.10B). Together, these constitute a highly mature, strongly leached ganister palaeosol. This suggests a long period of subaerial emergence under humid conditions that has been correlated with the low stand of sea level that led to the deep incision of the Chatsworth Grit palaeovalley further to the south (cf. Figs 4.14; 10.19). The dark grey mudstone that sharply overlies the palaeosol records the establishment of deeper-water conditions following a flooding event, but no fossiliferous marine band has been recorded in the exposed interval.

Mouselow Quarry, Glossop (SK 023953; 53.4526, −1.9669)

This is a working quarry, operated by Wienerberger Ltd, exploiting both sandstone and brick clay. It should be visited only by prior appointment on working days. Permission to visit should be sought via the Process Manager at the Wienerberger office at Windmill Lane, Denton, Manchester. The quarry operates to strict HSE rules and full hi-vis clothing (i.e. jacket and trousers), helmet, boots and goggles are mandatory. The quarry may be able to help with full hi-vis outfits for a small number of visitors, but this should not be relied upon.

From Glossop railway station, take the Woodhead road to the north (B6105) and, after about 500 m, turn left (west) into Trinity Road. Straight across at the crossroad is Dinting Road, which should be followed for about 1 km to the quarry entrance (Wienerberger sign) on the righthand (north) side of the road. Follow the quarry drive to the car park at the end and check in at the quarry manager's office.

As with any working quarry, it is not possible to predict exactly what might be visible on any visit, and the description here is couched in general terms. The quarry exposes a section in the lower part of the Huddersfield White Rock (Marsdenian, R_{2c}) cyclothem. Note that the BGS mapping of Sheet 86 (Glossop) shows this as a stratigraphically lower cyclothem (Readycon Dean Series; R_{2a}), but the younger age is now accepted based on marine bands below the correlative sandstone just south of Glossop.

The section in the quarry has three main components, a lower mudstone unit, a middle sandstone-rich interval and an upper mudstone (Fig. 9.11). Neither of the mudstone units contains a marine band, but the lower mudstone has a high sulphur content, suggesting that the *B. superbilinguis* Marine Band may not be far below the base of the section.

The sandstone-rich interval, which is up to around 6 m thick, comprises well-cemented laterally extensive sandstones of rather variable thickness, interbedded with thinner mudstone beds. There is erosional relief of around 2 m near the base of the unit where thick massive sandstones overlie thinner, interbedded, parallel-sided sandstones. The thicker sandstones also show some erosional lenticularity between beds, and many contain large mud clasts. The bases of loose blocks on the quarry floor commonly show erosional sole marks, particularly large

Figure 9.11 Mouselow Quarry, Glossop, showing the top of the sandstone-rich interval overlain by mudstones that lie at the base of an upwards-coarsening unit. The turbidite sandstones are mainly rather massive and show considerable lenticularity due to erosion between successive beds. The section is a deep-water equivalent of the Huddersfield White Rock (Marsdenian, R_{2c}) deltaic cyclothem that is present further east.

flutes. The thicker sandstones are mainly structureless, but there are units with both parallel- and ripple cross-lamination. At the top of the sandstone unit is an upwards-thinning interval that passes gradationally upwards into the upper mudstone unit.

This sandstone interval is thought to be a turbidite succession associated with a lobe of the Huddersfield White Rock delta that prograded from the east. Deposition of the turbidites probably took place in a base-of-slope setting and may have been associated with a minor fall in sea level. Mid-cyclothem turbidite sandstones occur within this cyclothem in an area extending from west of Glossop at least to Holme Moss to the northeast. Such turbidites are unusual in Namurian sheet delta cyclothems, which typically show a simple upwards coarsening above the basal marine band. In this area, however, there are no significant progradational cyclothems between the *B. gracilis* and the *B. superbilinguis* Marine Bands (R_{2a}–R_{2b}), in contrast to the successions in Yorkshire and in Lancashire. The lack of significant sediment supply during this interval is thought to have allowed deeper water to develop into which a thicker, more complex R_{2c} progradation took place, with its turbidite component.

The upper, low-sulphur, mudstone unit is the lower part of the main coarsening-upward delta slope succession that probably ends with mouth-bar sandstones of the Huddersfield White Rock. However, this part of the cyclothem has been eroded by the present-day topography.

Suggested Itineraries

This area does not lend itself to compact itineraries. Rather, it is an area where certain themes can be pursued, in some cases including localities in adjacent areas/chapters.

Many itineraries within this area will include a drive along or across the M62. The motorway traverses the Pennine Anticline between Huddersfield and Rochdale, crossing the Yorkshire–Lancashire boundary at its highest point which coincides with the anticline axis, exposing the Kinderscout Grit.

Chapter 10

Kinder Scout – North Derwent Valley

This area displays the topography and limestone facies at the northern margin of the Derbyshire Massif, the clastic fill of the Edale Basin, the control on clastic deposition exerted by end-Dinantian marine topography and the major palaeovalley of the Chatsworth Grit (Fig. 10.1).

The steep margin of the limestone platform probably relates to underlying fault control and to its essentially vertical growth in response to subsidence, giving rise to an extensive narrow Asbian 'fringing reef' belt that extends along much of the northern and western margins of the platform. In this it contrasts with the more gently inclined margins that characterized the eastern and southern flanks of the platform.

The area is particularly important in showing the clastic basin-filling succession deposited during the late Kinderscoutian (R_{1c}). Between deposition of the limestones and the basin filling there was an extended period of Namurian time during which deep-water mudstones were deposited in the Edale Basin. The mudstones also draped the flanks and top of the limestone platform. The basin-fill of this area contrasts with Pendleian basin fill at the northern margin of the basin (Chapter 7) where the basin-fill was at the up-current (proximal) end of the sediment supply system, whilst the fill in Derbyshire reflects down-current (distal) conditions. The influence of basin-floor topography was somewhat different between the two areas. In the north, the rugged basin floor topography favoured channelized flow, whilst in the south, deposition of thick basin mudstones had largely smoothed out initial basin-floor relief before coarse clastic sediments arrived. However, basin-floor topography still controlled Kinderscoutian deposition at a larger scale. Just north of the Derbyshire Massif, the Kinderscoutian succession from the Mam Tor Sandstone to the Lower Kinderscout Grit displays a basin-filling sequence some 400 m thick. This thins spectacularly down current to

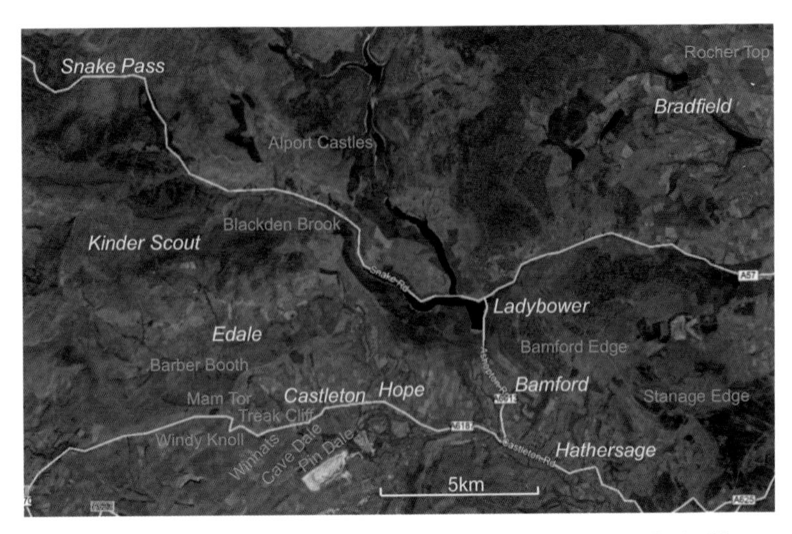

Figure 10.1 Satellite image of the area around the northern edge of the Derbyshire Massif, showing the localities described in this chapter. Birchen Edge lies beyond the southern limit of the image. Google Earth image.

zero over a distance of a few kilometres, against the topography of the limestone platform (Fig. 10.11).

Along the east side of the Derwent Valley, the Lower Kinderscout Grit forms the main escarpment, and various crags and tors display giant cross-bedding directed to the south. The overlying cyclothem (Marsdenian, R_{2c}) demonstrates the scale and nature of the major Chatsworth Grit palaeo-valley and the marked contrast in facies between the valley-fill sandstones and the deltaic sediments into which the palaeovalley is eroded. With the exception of Hope Cement Works, this area is free of heavy industry, although the lower-lying areas are inevitably influenced by agriculture. In the Derwent Valley, the Howden, Derwent, and Ladybower reservoirs led to the drowning of some agricultural land, to large conifer plantations and to changes in the hydrology of the river.

Pin Dale Quarry (SK 160823; 53.3385, -1.7597)

In the village of Hope, take Pindale Road towards the cement works. Continue past the cement works to a sharp bend to the right in the paved road and park at the roadside. The road directly ahead deteriorates into a rough track. Walk up the track for around 450 m and then turn left into the floor of the

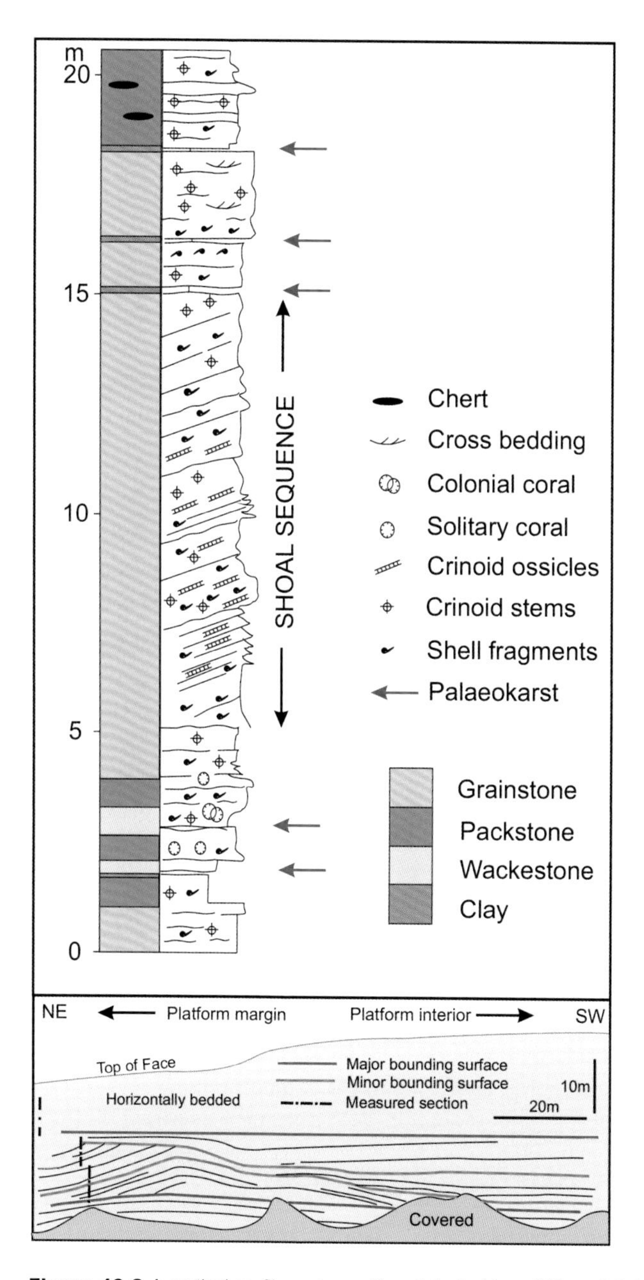

Figure 10.2 A vertical profile and a sedimentological log of Monsal Dale Limestones at the north-eastern part of Pin Dale Quarry, showing the large-scale inclined bedding directed towards the platform margin (based on Gawthorpe & Gutteridge, (1990). Note that the Asbian 'fringing reef' that defines the platform margin to the west lies below this succession.

old quarry. The quarry face is steep and high but seems stable. Take care navigating large blocks at the foot of the main face.

This large disused quarry exposes a succession near the base of the Monsal Dale Limestones (early Brigantian) of platform-margin bioclastic grainstones facies which formed close to the edge of the carbonate platform. The quarry extends for some 300 m laterally and the face trends perpendicular to the platform margin, showing a facies transition towards the margin. Towards the north-eastern end, several units of large-scale (up to 20 m) basinwards-dipping foresets are present, with units separated by inclined bounding surfaces (Fig. 10.2). The lowest bounding surface of the clinoform units shows evidence of emergence as palaeokarst, and calcrete textures. The structures that gave rise to the cross-bedding therefore formed during or after a sea-level rise that flooded the carbonate platform following a period of lower sea level. The clinoform unit appears to have had a multi-stage growth, with periods of advance separated by intervals of erosion that gave minor bounding surfaces. The cross-bedding is thought to result from the basin-ward advance of large bars or shoals that reworked lime sands produced on the platform, possibly by tidal or storm-induced currents. Similar shoal complexes extend further to the east, along the margin of the carbonate platform.

This phase of high-energy sediment transport and deposition was ended by a fall of sea level when the sediments became emergent as a palaeosol developed. The overlying limestones are horizontally bedded and punctuated by palaeosol horizons, suggesting small-scale aggradational cyclothems moderated by fluctuating sea level.

Seen from the northeast end of the quarry, beyond the active quarry works, is a further quarry in the lowest part of the Mam Tor Sandstones, where turbidite sandstones are interbedded with basin mudstones. These mudstones are quarried as feedstock to the cement-making.

Cave Dale, Castleton (SK 150826; 53.3414, –1.7752)

Parking for vehicles is typically difficult within Castleton village, and it is often best to use one of the public car parks at the western side of the village and then walk back through the village. This locality is accessed by a signed footpath that goes from the western end of the Pindale Road at the southeast corner of Castleton village. Initially the footpath is narrowly confined in trees, but opens out into meadow where useful exposure of limestones first occurs.

Cave Dale provides a section, around 400 m long, perpendicular to the Asbian platform margin, the limestones falling within the Bee Low Limestones. The exposures show the transition from the interior platform carbonates in the south into limestones deposited on the upper part of the platform-margin slope. The locality is best studied by walking up the valley to the exposures of the Lower Miller's Dale Lava, a dark amygdaloidal basalt around 8 m thick (Fig. 10.3). The lavas lack pillow structures that would clearly indicate that they were erupted under water, but neither do they show evidence of weathering at the top, which would suggest subaerial emplacement. This volcanic unit defines the boundary between the Chee Tor Rock below and the Miller's Dale Beds above, local and informal subdivisions of the Bee Low Limestone Formation. Miller's Dale Beds continue up into the higher part of the hillsides.

Below the lavas, the valley exposes thick, sub-horizontal limestone beds deposited on a flat-topped platform, close to the break of slope at its edge. The transition from platform to slope occurs abruptly over a 10 m distance and is marked by the loss of bedding and the development of hardgrounds. The hardgrounds are developed on wackestone with depositional cavities lined by marine cement, the cement often dominating the limestone. Whole crinoid stems and brachiopod shells are present. The

Figure 10.3 A unit of basalt lavas that are underlain and overlain by limestones of the Bee Low Limestones (Asbian), in Cave Dale.

hardgrounds are often at high angles to the depositional surface and are encrusted by bryozoans and buried by reworked bioclastic sediments.

The platform-edge facies pass northwards into thin- to medium-bedded limestones deposited on the marginal platform slope. These dip northwards (basinwards) at up to 30°. The presence of occasional geopetal infills shows that these dips are depositional. The slope carbonates consist of wackestone with depositional cavities lined by marine cements, interbedded with beds of re-worked crinoidal limestone. The transition zone between shelf and slope facies in Cave Dale has a vertical upwards trend, showing that the platform margin grew without significant basinward progradation.

Treak Cliff Hillside and Cavern (SK 135831; 53.3444, –1.7958) (▶)

It is usually possible to park along the road heading towards Mam Tor, beyond the junction with the Winnats Pass road. If visiting Treak Cliff Cavern, it may be possible to park on the small car park at the foot of the path that leads up to the cavern. The path is too steep for wheelchairs but may be possible for walking disabled. It is also possible to park in the car park opposite Speedwell Cavern and take the footpath to Treak Cliff that starts behind the public toilet.

The hillside around the cavern entrance at Treak Cliff is the exhumed Asbian platform margin slope, locally trending roughly north–south (Fig. 10.4). Around 150 m south of the cavern, above the first bend in the path up to the cavern, the hillside exposures comprise large blocks of limestone, many metres in size, which form a submarine scree or talus wedge, banked against the steep platform margin. The whole succession is within the Asbian Bee Low Limestones, although the emplacement of the blocks may have been related to a mid-Brigantian episode of uplift and erosion that may have locally removed Monsal Dale Limestones. A talus deposit of such large blocks may have been favoured by a cliff line that developed during the mid-Brigantian. Pockets of Bowland (Edale) Shales occur between some blocks support the idea that the present-day topography closely reflects the end-Dinantian sea-floor topography.

Treak Cliff Cavern provides a unique transect through the carbonate platform margin and shows the sequence of events associated with the

Figure 10.4 The main face of Mam Tor, showing the turbidites of the Mam Tor Sandstone (Kinderscoutian, R_{1c}). The sandstone beds are broadly parallel-sided and continuous, and are distributed in sand-rich and sand-poor packages. The hillside on the left is the exhumed margin of the limestone massif, and the active landslip below the face of Mam Tor gives the hummocky slope to the right.

demise of the Derbyshire platform between late Asbian and earliest Namurian times. Near the cave entrance are Bowland (Edale) Shales of early Namurian age. Further into the cave the mudstones form the matrix between boulders of limestone that make up the talus wedge of the platform-margin slope. The boulders are made up of platform-top and marginal carbonates, formed during the erosion of the platform margin, possibly in the mid- Brigantian. Cavities in the inner parts of the boulder bed are heavily mineralized by 'Blue John', a local variety of purple banded fluorite extensively used for decorative purposes. The unique colour is thought most likely to result from damage to the crystal lattice induced by radioactive decay of uranium in nearby basin mudstones. The fluorite occurs as a pore-filling and replacive phase, and mineralization is thought to have taken place during maximum burial in late Westphalian times. Fluorite-bearing brines and hydrocarbons were probably both sourced from thick basin mudstone successions to the north. The inner parts of the cave are developed in Bee Low Limestones (Asbian) dipping basin-wards at 30–40° (locally to the east).

The show cave displays a spectacular assemblage of re-precipitated calcite in the form of stalagmites, stalactites and various forms of flowstone.

Winnats Pass (SK 138827; 53.3415, -1.7928) (►)

It is important to drive or walk through the Winnats Pass to fully appreciate its scale and structure. Either before or after this drive, park in the public car park opposite Speedwell Cavern and walk up into the lower part of the gorge. Disabled access from car park. It is sensible, for most visits, to limit observations to the accessible exposures low in the gorge and not to climb to the crags at the top.

The Winnats Pass is a large gorge cut into the margin of the limestone massif. It is around 1 km long, up to 250 m wide and with a maximum depth of around 140 m. (Fig. 10.5). The extensive hill slope facing north, into which the Winnats Pass is eroded, represents the exhumed Asbian–Brigantian carbonate platform margin, which is seen in the landscape from Mam Tor (Fig. 10.6) and in close-up at Treak Cliff. Exposures of the Bee Low Limestones on either side of gorge show the transition from platform to platform-margin limestones. Limestones around the transition are mud-rich wackestones with early depositional cavities lined by marine cement. This constitutes the aggrading mud-mound complex that is characterized as a 'fringing' or 'apron reef'. The geometry of the platform to slope transition zone indicates that the platform was

Figure 10.5 The gorge of the Winnats Pass incised into Bee Low Limestones at the steep margin of the limestone platform. The mud mound belt that makes up the 'fringing' or 'apron reef' occurs below the upper slope breaks. The bedded limestones behind the car park are basinward of the 'fringing reef' mass and dip steeply towards the basin.

growing vertically during the Asbian. The break of slope at the top broadly coincides with the narrow but thick marginal 'reef' belt that fringes much of the northern and western margins of the platform.

In the lower part of the gorge, a series of graded, amalgamated coarse-grained bioclastic limestones are exposed. These so-called 'Beach Beds' onlap and post-date the main Bee Low limestones and are likely latest Asbian or early Brigantian in age. The Beach Beds contain highly rounded, coarse bioclasts including fragments of crinoids and brachiopod shells, and their origin has been the subject of much discussion. The name implies an early shallow-water interpretation, but they have also been thought of as a local wedge of proximal turbidites, resedimented from the platform and draping the lower parts of the slope. More recently a shallower nearshore setting has been favoured, related to an intra-Brigantian erosional episode that involved tectonic uplift and greatly lowered sea level. The deposits may have distal equivalents in the Castleton borehole, where thin fine-grained bioclastic turbidites pass upwards into basin shales. Similar limestone turbidites were also inferred in the Alport borehole some 8 km to the north.

There is also some uncertainty about the origin of the Winnats Pass. It has been variously interpreted as a large collapsed cave system, as a sub-marine canyon cut into the Viséan platform margin, and as the result of sub-glacial or periglacial melt-water run-off, formed during the mid- to late Pleistocene. The last of these seems to be favoured at present, but all have shortcomings and provide a basis for discussion.

Windy Knoll (SK 126829; 53.3439, –1.8093)

This small, disused quarry can be approached either from the main Mam Tor car park or by parking along the road between the top of Winnats Pass and the road leading to the Mam Tor landslip. In both cases enter the pasture via a stile and walk across the field. If walking from Winnats Pass, it is best to go straight ahead at the junction and follow the track from a gate on the righthand side of the road after about 100 m. This path allows disabled access for distant view of quarry although muddy when wet. The quarry is in the centre of the field around 170 m from each stile. Avoid the nearby cave entrance.

This is a National Trust property and an SSSI. ***Very strictly; no hammers.***

This classic locality shows degraded hydrocarbons (elaterite) trapped in vuggy pores in the upper part of the Bee Low Limestones, suggesting a thin oil column that has been exhumed, with the oil becoming highly altered as a result.

The limestones in the quarry lie just below the projected unconformity with the Bowland (Edale) Shale (now eroded away). The juxtaposition of the Bowland Shale directly on the Bee Low Limestones demonstrates a significant unconformity whereby the Monsal Dale Limestones and Eyam Limestones are missing in this area. This may be due to erosion of Monsal Dale Limestones during and following a mid-Brigantian uplift and erosion event and to non-deposition of the Eyam Limestones in this area due to late Brigantian eastward tilting.

The limestones here are horizontally bedded bioclastic grainstones with scoured bedding surfaces. Bioclasts are mainly well-rounded crinoids, brachiopod fragments, peloids and aggregate grains. More complete brachiopods, gastropods and crinoids are also present, and some have oolitic coatings, suggesting significant agitation above wave base, a few hundred metres from the contemporaneous platform edge.

The Bowland Shale, which draped the limestone platform, is inferred to have acted as a top seal to hydrocarbons that migrated into the limestone reservoir from thick shales present in the basin to the north. These partially marine mudstones also form the main source rock in the East Midlands and southern Irish Sea hydrocarbon provinces, where they are thickly developed in the Gainsborough Trough and offshore North Wales. They onlap the Derbyshire carbonate platform and once draped the whole platform with a relatively thin condensed succession.

Within the 'oil zone' three zones have been recognized, from top to bottom:

1. Irregular surface overlain by a breccia comprising clasts of blackened limestone in a matrix of elaterite that has the texture of chewing gum. This texture has probably resulted from more recent weathering following exhumation.

2. Blackened zone of hard limestone up to 0.5 m thick. Identical to zone 3 but impregnated by bitumen. Eroded above by an irregular surface which locally cuts out the blackened zone.

3. Pale limestone with clean matrix. Bitumen occurs in internal cavities of fossils and in fissures that penetrate several metres into the zone. They contain limestone clasts in a bituminous matrix. The fissures are cut by calcite veins and are, hence, pre-burial.

The elaterite is highly degraded oil. In this case, it is so biodegraded that it has, so far, defied attempts at geochemical typing and thus relating it, confidently, to a particular source rock.

The nearby cave, excavated in the nineteenth century, yielded Pleistocene animal remains, including reindeer, bison, wolves and grizzly bear.

Mam Tor (SK 129835; 53.3474, −1.8038) (►)

It can be very busy here but it is usually possible to park along the road leading to the Blue John Mine or in the parking area at the end of the road where a fence prevents further progress. Go through the gate at the side of the fence and climb to the top of the highest knoll on the landslip mass. Access beyond the gate will be challenging for many disabled persons, but the view from the car turning place is still very useful. This locality illustrates several different aspects and can take up to an hour or two, depending on what programme is decided. For visits to the base of the main exposure of the landslip back scar it is advisable to wear hard hats. Loose material can fall at random and can also be disturbed by adventurous sheep and by unthinking walkers.

This is the type locality of the Mam Tor Sandstone and is a truly multi-purpose location. Different aspects are set out separately below, and are based on having the highest landslip knoll as the view point.

1) The basinal Bowland (Edale) Shale, dipping gently to the north, underlies the valley floor. Differential erosion of the shales and the limestones has exhumed the margins of the limestone platform so that present-day topography closely reflects end-Dinantian bathymetry, a relationship that is also present along the western margin of the limestone platform. The limestones forming the platform margin here are the Bee Low Limestones (Late Asbian). The Monsal Dale and Eyam Limestones (both Brigantian), which overlie the Bee Low further east, are missing. Any Monsal Dale

Figure 10.6 View to the east from Mam Tor. The steep hillside facing north is the exhumed margin of the Asbian limestone platform with the narrow 'fringing reef' close to the top of the slope. The Hope Valley floor is underlain by mudstones of the Bowland (Edale) Shale that formerly draped the platform. The distant skyline is the scarp of Marsdenian sandstones that dip eastwards towards Sheffield and the Yorkshire Coalfield.

Limestone that was deposited on this corner of the massif was probably thin and was removed by later Brigantian erosion. The late-Brigantian transgression, which led to deposition of the Eyam Limestones further east, apparently never extended this far west. The present-day topography therefore records erosional modification of a depositional morphology.

The steep Treak Cliff hillside is directly underlain by a massive breccia of very large, tumbled limestone blocks, which appear to have formed a scree apron around the edges of the carbonate platform. These sediments are discussed more fully at Cave Dale, Winnats Pass and Treak Cliff.

Mudstones of the Bowland (Edale) Shale are banked against the margin of the limestone massif and are over 300 m thick 2 km to the north in Edale. The mudstones once extended over the top of the limestone platform but in a much thinner sequence. They are patchily preserved on the top of the platform and are known in boreholes at Wardlow Mires, some 10 km to the southeast.

The most distant skyline to the east is the outcrop of the Chatsworth Grit (Marsdenian, R_{2c}), beyond which the successions dips eastwards into the Yorkshire–Derbyshire Coalfield.

2) Turning around to face the main exposure of the landslip back scar, both structural and sedimentological features in the Namurian succession are apparent (Figs 10.4 & 10.7). During Variscan compression, the rather ductile Namurian sediments in the basin were squashed against the rigid limestone massif to give east–west trending folds. The Bowland Shale at Mam Tor is thrust against the northern margin of the limestone massif so that shale exposures on the west side of the main face are now topographically higher than the top of the limestone and above the base of the Mam Tor Sandstone that stratigraphically overlies it. A smaller-scale example of this thrusting occurs low in the face of Mam Tor where a small reverse fault repeats part of the section (Fig. 10.7). At a larger scale, the ridge extending eastwards from Mam Tor follows a gentle syncline with the sandstones preserved along the axis. The south-dipping limb is also the southern limb of the Edale Anticline. These east–west trending folds fade away to the east as regional north–south trending structures come to dominate.

Figure 10.7 The lower part of the face at Mam Tor showing a reverse fault (thrust) that repeats part of the section. The fault resulted from Variscan folding at the northern edge of the Derbyshire Massif. The organization of the turbidite sandstones in sand-rich and sand-poor packages is clearly shown.

3) The Mam Tor Sandstone (Namurian, Kinderscoutian, R_{1c}), which is exposed in the main cliff face, records the first arrivals of sand from the northern source area into the southern part of the Pennine Basin. Prior to that, the basin had received only basinal muds, the Bowland (Edale) Shale. Viewed from a distance, the cliff face mainly comprises rather parallel-sided and laterally extensive sandstone beds, interbedded with finer-grained sediments. This pattern suggests discrete high-energy events that delivered sand into a setting that was otherwise accumulating fine-grained sediment in quiet and deep water. Such high-energy events are likely to have been turbidity currents. The sandstones show no clear patterns of systematic bed-thickness change, but the succession shows bundles of beds that are more and less sand rich (Fig. 10.7). These may reflect the shifting of lobes in the more proximal part of the turbidite system, of which Mam Tor is a distal expression. These first-arrival distal turbidites in the context of the overall prograding sequence are, in fact, unusually thickly bedded compared with other turbidite systems. This may relate to the proximity of the Derbyshire Platform, whose northern margin would have presented a barrier to turbidity currents and led to rapid deceleration and dumping of sediment load. A barrier was almost certainly present, as no sandy equivalents of the main Kinderscoutian sand-rich sequence, spanning from the Mam Tor Sandstone to the Lower Kinderscout Grit, extends into the Staffordshire Basin some 8 km to the southwest. These abrupt changes of thickness and facies support a 'fill–spill' pattern of basin filling (Fig. 10.9).

The Mam Tor Sandstone sediments may be viewed in more detail by walking along the base of the face, best from right to left and thus ascending the succession. The sandstones are interbedded with mudstones and siltstones, and erosional sole marks are present both *in situ* and on fallen blocks. Palaeocurrents, based on *in-situ* sole marks, are directed to the southwest and west. Graded bedding within the somewhat carbonaceous sandstones is uncommon, and beds are mainly structureless. Thin intervals of ripple cross-lamination on the tops of some beds are the only manifestations of the Bouma sequence. Many thicker sandstone beds are internally more complex than classical turbidites, with a mudflake-rich 'slurry' texture in their upper parts. Beds with such shale-clast debrite components can be regarded as turbidite-debrite couplets, where the two components are genetically linked. These hybrid event beds (HEBs) appear to have involved a transition from fully turbulent turbidity currents to slurry-like flows with higher viscosity. Such behaviour could relate to the rapid flow deceleration caused by the barrier of the limestone massif close to the south. One particularly thick sand unit, rich in mud clasts, at the base of the face, is clearly a multi-event interval associated with significant erosion and may have been deposited downcurrent from a channel with unstable banks.

4) The Mam Tor landslip itself also warrants attention. This unstable complex has been moving intermittently for over 3000 years due to instability of an over-steep, post-glacial hillside. Such landslips are quite common around the area with large examples at Back Tor, Alport Castles and along the western side of Kinder Scout.

Movement at Mam Tor is facilitated by the low shear strength of the Bowland Shales, so that rotational slip surfaces have developed. The present landslip mass has been described in terms of three zones. The less mobile upper zone, mainly above the top road, is a series of large, relatively intact, rotated blocks that give rise to the hummocky topography. The middle part is a transition zone that shows greater rates of movement and fragmentation into smaller blocks, as seen in the collapsing road (Fig. 10.8). The lowest zone is a large debris flow, but with a basal slip surface that spills out over the lower gradient at the foot of the slope.

The road, which was finally abandoned in 1979 after many years of patching, used to carry heavy traffic from the Hope Cement Works, and such loads cannot have helped stability. However, heavy rainfall is the most likely trigger for episodes of failure, and the movements that led to the eventual closure were probably initiated by heavy rains that followed the drought of 1976, when intense drying led to the development of deep fissures.

The broken road near the gate illustrates several aspects of extensional brittle failure, similar to that which occurs, at a much larger scale, in tectonically rifted settings. Fractures in the road surface show anastomosing patterns of small normal faults with relay ramps and transfer zones. The vertical surfaces show the internal stratigraphy of the separate fault blocks picked out by different layers of hardcore

Figure 10.8 Collapse of the road on top of the Mam Tor landslide. The brittle fragmentation results from extensional stretching as the underlying plastic mudstones slid downslope under gravity. The blocks dip towards the up-slope 'fault', suggesting that the slip surfaces are concave upwards. There is a hierarchy of mainly synthetic normal faults of a range of sizes.

and tarmac. Such progressive infilling of newly created space in the hanging wall of a normal fault is somewhat analogous to the development of growth faults like that seen at Horn Crag Quarry (Figs 4.15 & 7.9) and, at a much larger scale, in major deltas such as the Niger and Mississippi.

5) Energetic visitors might like to climb to the summit of Mam Tor by following the left-hand side of the landslip back scar. The top was the site of a hill fort dating to the Bronze Age and Iron Age, and

represents some of the earliest human influence on the landscape. From the summit, the view to the north across Edale shows the southern slopes of Kinder Scout, a view described at Mam Nick, below, and not repeated here.

Mam Nick (SK 125836; 53.3486, -1.8131) (►)

Mam Nick is the narrow pass on the road to Edale, just to the west of Mam Tor. It is usually possible to find roadside parking for a small number of vehicles on the Edale side of the pass. Disabled friendly. Ideally walk up the hill a short way to be clear of passing traffic. This stop is entirely for the view, which duplicates that from the top of Mam Tor.

This viewpoint, above Edale, gives a panorama of the southern side of Kinder Scout. The hillside encompasses the whole Kinderscoutian basin-filling succession near the southern margin of the Edale Basin. When traced to the southwest, this succession thins from around 400 m to zero within a distance of around 5 km against the margin of the limestone massif (Fig. 10.9).

The ridge of Mam Tor/Mam Nick is a gentle syncline, whilst Edale itself reflects erosion deep into the Bowland (Edale) Shale in the core of the Edale Anticline. The Kinder Scout plateau is another gentle east–west syncline.

The Bowland Shale, outcropping on the valley floor, extends from the Pendleian up into the late Kinderscoutian with a thickness of around 300 m. This shows that for most of Namurian time, this part of the basin was starved of coarser clastic sediment, and sand was first delivered by turbidity currents that deposited the Mam Tor Sandstone.

The Mam Tor Sandstone crops out in the lowest part of the slope behind Edale village, and is overlain by thicker-bedded, more proximal turbidite sandstones of the Shale Grit. These are well seen in Grindsbrook, behind Edale village, but better sections are seen in Blackden Brook (Fig. 10.11) and at Alport Castles (Fig. 10.12) to the north of the plateau. Topographic steps or benches that extend along the hillside reflect thicker shale intervals within the Shale Grit (Fig. 10.10). These shales were more easily eroded, compared with the sand-rich intervals that produce steeper slopes. The shale units record periods of reduced sand supply to the pro-deltaic turbidite ramp, possibly resulting

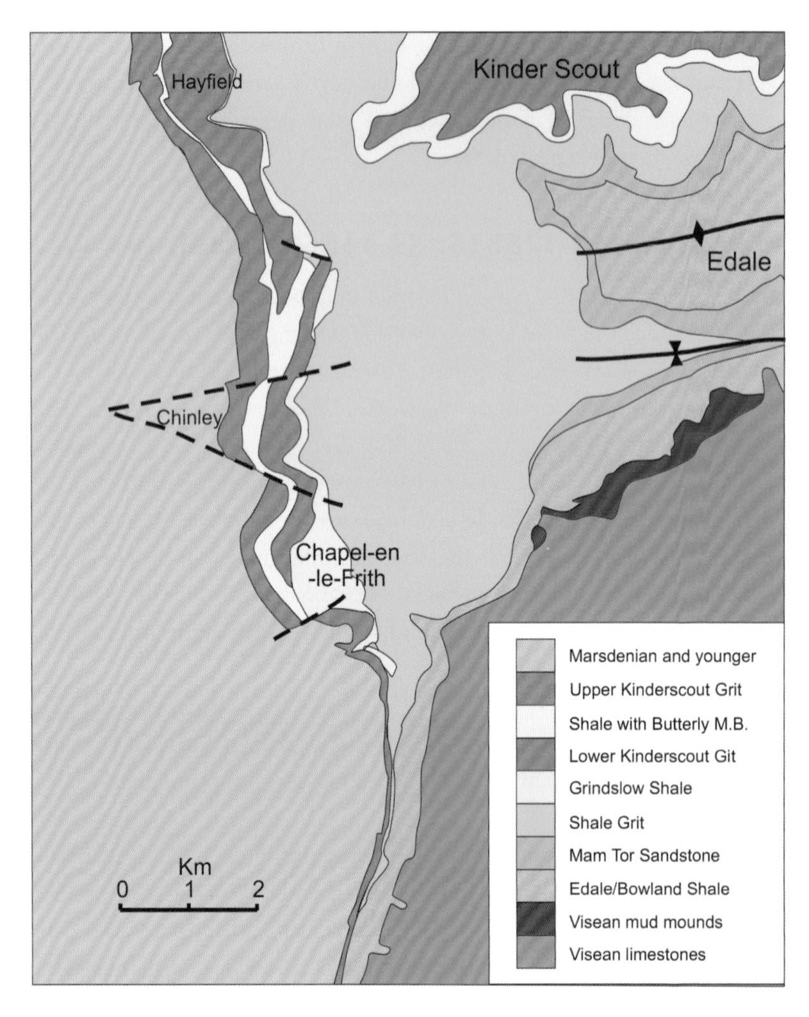

Figure 10.9 Geological map of the northern margin of the Derbyshire limestone massif. The Kinderscoutian succession shows spectacular thinning from the Edale/Kinder Scout area when traced towards the southwest. Only the Upper Kinderscout Grit persists to the south, towards the Staffordshire Basin (adapted from BGS 1:50 000 mapping).

from switching of distributary channels on the delta that lay upstream to the north. The shale benches can be mapped for distances of several kilometres, but they eventually die out, possibly providing an indirect idea of the size of depositional lobes in the turbidite ramp.

Towards the top of the plateau, a mudstone/siltstone unit around 100 m thick, the Grindslow Shale, shows upwards coarsening and

Figure 10.10 The southern side of Kinder Scout above cloud-filled Edale. Bowland (Edale) Shale mudstones underlie the the valley floor whilst the Kinderscoutian basin-fill succession makes up the hillside. Mam Tor Sandstone crops out near the base of the hillside and the Lower Kinderscout Grit forms the top of the plateau. The prominent steps on the hillside are produced by mudstone units within the sandstone-dominated Shale Grit turbidite succession.

contains channel sandbodies up to 15 m thick and hundreds of metres wide. These are thought to be by-pass channels on the delta slope that transferred sand via turbidity currents directly from the river mouth to deeper-water settings. The top of the plateau is made up of the Lower Kinderscout Grit, which forms a widespread sheet made up of multistorey and multilateral channel sandbodies. On the northern side of the plateau, the Lower Kinderscout Grit shows giant cross-bedding in sets up to 25 m thick. The area to the west of the road shows extensive landslipping (Fig. 3.5).

Barber Booth, Edale (SK 112847; 53.3595, -1.8363)

From the Barber Booth junction at the head of Edale, take the minor road to Upper Booth and park at the railway viaduct. If water levels are low enough, climb down into the stream bed a few tens of metres upstream of the footbridge. If water levels are too high, some outcrop of shales can still be seen in the stream bank from the footbridge. Disabled can view from bridge or bank, but climbing to stream bed would probably be unwise.

Exposures of mudstone in the banks and bed of the River Noe just upstream of the railway viaduct are of the Bowland (Edale) Shale. They show typical deep basin mudstone facies with goniatite-bearing marine horizons and siderite concretions. At this exposure, beds are of Pendleian (E_1) age. This mudstone facies extends into the Lower Kinderscoutian (R_{1c}) in this part of the basin. The sequence contrasts with the partly contemporaneous sand-dominated succession in both the Bowland Basin to the north (Pendleian turbidites derived from the north) and the Staffordshire Basin to the south, where Pendleian–Alportian protoquartzite turbidites, derived from the south, occur.

The Edale No 1 Oilwell (1938) was drilled without success in the field across the bridge. It encountered thinly interbedded limestones and mudstones below the Bowland Shale, which would not have formed a viable reservoir. Furthermore, the peak maturity of the Bowland Shale in this area probably took place in Westphalian times, before the formation of the supposed anticlinal trap.

Blackden Brook (SK 130893 to SK 118883; 53.4022, -1.8056) (▶)

Park on the roadside of the A57 (Snake Pass) at a pull-in area created when the road was repaired across an old landslip. This is about 2 km east of the Snake Inn and about 8 km from the Ladybower road junction.

This locality involves a strenuous walk to the top of the Kinder Scout plateau, a climb of about 300 m and a round-trip distance of some 5 km. It is not a trivial undertaking and should only be attempted when water levels in streams are fairly low. It is probably best avoided if heavy rain is forecast. Allow at least 3 hours for this locality. Remember that temperatures can fall significantly over the height involved in the walk.

Cross the wall at the stile and walk down the steep path to the bridge over Ashop Clough. From the bridge it is possible to see outcrops of Mam Tor Sandstone in the bank of the clough (vegetation permitting). From the bridge, continue straight up the hillside, crossing a high wall at a stile. Turn right along the footpath and follow the wall until it ends at a small tributary stream. Cross this and follow path around the corner before climbing down to stream level in Blackden Brook itself which is the base of the measured section (Fig. 10.11). Follow the succession upstream, crossing the stream from one bank to the other as conditions dictate. In the upper part of the stream, there are a few quite steep sections where particular care is needed. For the return trip from the top of the plateau, it is generally best to walk down the hillside to the west of the valley, only dropping down to Blackden Brook from above the base of the stream section. The steep slope back to the valley floor can be hazardous, especially when bracken is high but, with appropriate care, this route is less hazardous than retracing the route down the stream bed.

This long stream section gives a near-continuous section through the main Kinderscoutian (R_{1c}) basin-filling succession from Mam Tor Sandstone through the Shale Grit and the Grindslow Shale to the Lower Kinderscout Grit. The Mam Tor Sandstone is exposed in the banks of Ashop Clough at the bridge and in the bank upstream of the junction with Blackden Brook below the section in Figure 10.11. The logged section begins close to the base of the Shale Grit and shows the typical interbedding of thick, massive, commonly amalgamated beds and units of interbedded thin sandstones and mudstones. Some thick sandstones show rather irregular sub-horizontal cracks whose relationship to bedding is uncertain. Some upper bedding surfaces show current ripples, whilst erosional sole marks can be found on the bases of thinner sandstone beds, usually as loose blocks. The thicker intervals of finer-grained sediment weather back to give steps or benches on the adjacent hillsides, and these can be mapped over distances of several kilometres (see Fig. 10.10), perhaps giving some indication of the dimensions of the sand-rich lobes that are thought to have built the turbidite ramp that gave rise to the Shale Grit.

A thick mudstone unit towards the top of the Shale Grit is mappable regionally and has yielded bivalve fossils and has a high uranium gamma count. This suggests that marine influence corresponded with condensation, a likely response to a rise in sea level. It is possible that this unit correlates with a better-developed marine band elsewhere.

The upward increase in the incidence of channels in both the Shale Grit and in the overlying Grindslow Shale is an important feature of the succession. The Grindslow Shale interval, the deposit of the inferred prograding slope that advanced over the turbidite ramp, begins at a sharp upwards change from thick-bedded sandstones to silty mudstones. Above, these mudstones and siltstones are interbedded with turbidite sands with some small-scale channelling. A larger-scale channel sand

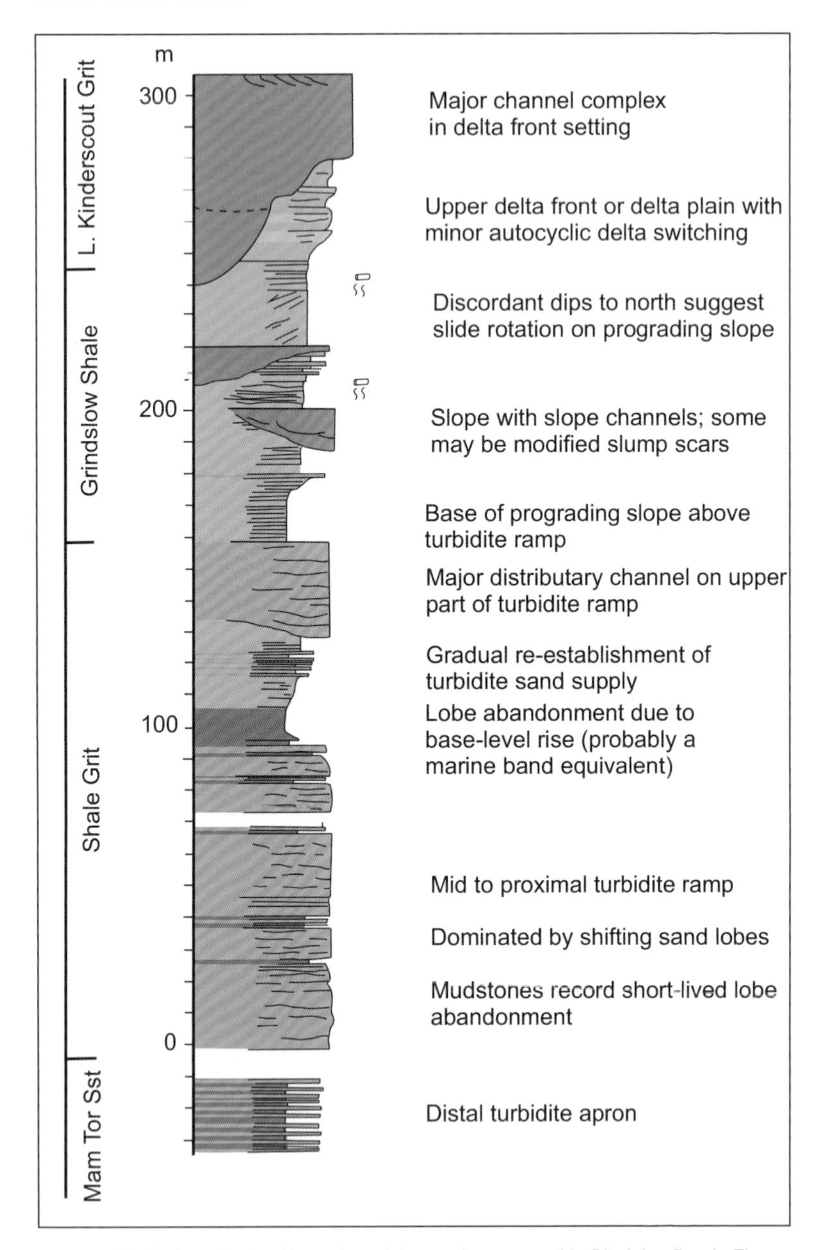

Figure 10.11 Simplified sediment log of the section exposed in Blackden Brook. The lowest part, the Mam Tor Sandstone, can be viewed in the river bank from the bridge. The section exposed in Blackden Brook itself begins at the '0' level of the log. A downloadable pdf file is available for this figure.

body high on the southern side of the valley shows a complete channel cross-section and has large flutes on its base. Higher up the stream bed section, the siltstones show distinctively striped thin, gradational interbedding, reflecting fluctuating supply of suspended sediment. These sediments become increasingly bioturbated upwards.

The tectonic dip throughout the stream section is at less than 5° to the south, and so dips that diverge from this are the result of syn-depositional or early post-depositional disturbance. Complex cross-cutting relationships are thought to reflect slope instability. Such disturbed bedding may be much more prevalent within these inferred slope deposits than is apparent in isolated stream exposures.

The uppermost part of the Grindslow Shale consists of a series of small coarsening-upward units with ripple cross-lamination and small-scale cross-bedding in the sandstones that cap them. This facies pattern suggests quite shallow water depths, possibly approaching a delta plain. However, there is no direct evidence of emergence in the form of palaeosols. One notable feature of the overall succession is the absence of any major mouth-bar sandstones. This may be due to most sand having by-passed the river mouths through channels that transferred it directly to the turbidite ramp. An alternative explanation is that any mouth-bar deposits were erosively removed by the large channel systems of the Kinderscout Grit.

The relief on the base of the Lower Kinderscout Grit, at the top of the Blackden Brook section, is of the order of 60 m, one of the largest seen in the Pennines. It must involve the successive cut and fill of several superimposed channels to give the sheet-like complex of the Lower Kinderscout Grit. The massive sandstones seen at the top of the stream can be traced upwards into giant cross-bedding further along the plateau edge to the west.

The overall succession has been interpreted as the progradation of a linked turbidite-slope-fluvial system, with active sand by-pass during the progradation. More recent attempts to interpret the succession in sequence stratigraphic terms have emphasised the possibility that the major channels were incised during sea-level low stands when sand was more readily by-passed to deeper water.

Alport Castles (SK 143913; 53.4045, -1.7442)

Access to this locality can be restricted in spring, as it is sometimes a protected nesting site. There will probably be warning signs or wardens when this applies. It is a good idea to check the status before committing to a long walk.

This locality is more remote than most, and a visit requires several hours, including some of fairly energetic walking. It is possible to access Alport Castles via Alport Dale, by walking in from the A57, parking at the roadside near the bridge. Walk up the valley to Alport Castle Farm, through the yard and follow the path down to the stream and then up the hillside from the bridge, climbing up the land-slipped hillside.

An alternative route is to drive up the western side of the Ladybower Reservoir and park at the Derwent Dam Visitor Centre (see GPS co-ordinates). From there, it is a 3 km scenic walk to Alport Castles. Walk down the road from the car park and take a path marked by a sign through the plantation. Where the path splits, go to the right and follow up the hill for around 500 m. Take a left on to a farm track and then a path with signpost to Lackerbrook.

Figure 10.12 Uppermost part of the Shale Grit (Kinderscoutian, R_{1c}) at Alport Castles. The exposure, which is about 30 m thick, is dominated by thick-bedded turbidites, commonly in amalgamated beds. The base of a prominent package of thick sandstone beds (arrowed) has clear erosional relief so that the unit is a channel fill. The top of the hillside roughly coincides with the base of the Grindslow Shale, which has been eroded away here.

Follow this to Lackerbrook Farm, cross the field past the farm and turn right at the edge of the wood. Cross a stile and follow the ridge for about 2 km with the Alport Valley to the left and the Derwent Valley to the right. Once at the Alport Castles landslip, climb down to a viewpoint close to the large-slipped pinnacle that gives a good view of the main back-scar exposure.

Whilst it is possible to walk along the base of the main face to observe the sediments closely, it is not sensible to climb or scramble on the face, as it is manifestly unstable.

A large landslip back-scar gives extensive exposures in the uppermost Shale Grit, just below the Grindslow Shale, which is missing here. This is, therefore, the most proximal part of the Kinderscoutian turbidite system. The exposed interval, which is around 30 m thick, is dominated by thickly bedded massive sandstones with thin intervals of thinner sandstones interbedded with siltstones. The thicker-bedded turbidites are commonly amalgamated and structureless, with deposition lamination confined to the tops of some beds. A few thick beds show dish-structures caused by rapid escape of pore waters shortly after deposition. At the larger scale, the outcrop shows several large channel units with thick-bedded sandstones near their base. They are separated by intervals of thinner-bedded sandstones and siltstones (Fig. 10.12). The massive sandstone beds that make up the channel fill become thinner upwards.

The basal erosion surface of a channel unit in the middle of the face cuts rather gently down to the left. Closer examination of the erosional base shows that the surface is locally near-vertical with small steps related to bedding in the eroded thinner beds below. There are scattered mud clasts close to the erosion surface, and erosional sole marks indicate palaeoflow direction. These turbidites were deposited in channels that are inferred to have fed sand to lobes lower on the pro-deltaic ramp, and to have connected back to feeder channels on the actively prograding slope that deposited the overlying (but here eroded) Grindslow Shale.

The Alport Castles is one of the largest landslips in the Pennines and involves both rotation of very large sliding blocks with discrete slip surfaces towards the top, and a zone of increasingly disaggregated masses directly downslope of the main backscar. The lower parts of the hillside remain very hummocky, suggesting that flowage of debris continued almost to the floor of the valley. The landslip is not currently active, and the main phase of movement probably occurred under periglacial conditions at the end of the last Ice Age. Water penetrated fractures in the sandstones and lubricated mud-rich units within the succession, creating instability that was probably accentuated by freeze-thaw processes.

The valley floor, close to Alport Castles Farm, was the site of the Alport exploration well, drilled in the 1940s on the crest of the Alport Dome, a broad, low-relief structure. The well proved the presence of interbedded limestones and mudstones in the Viséan section below the Bowland Shale. The limestones are interpreted as turbidites, probably derived from the Derbyshire Massif carbonate platform to the south.

Viewed from Alport Castles, the northern side of Kinder Scout to the southwest shows a series of topographic benches that reflect the occurrence of mudstone-rich intervals in the Shale Grit turbidite sequence, similar to those seen on the southern side of the plateau from Mam Nick (Fig. 10.10). These landscape features can be related to the succession present in Blackden Brook (Fig. 10.11). The steeper steps, produced by sandstone-rich intervals, are thought to reflect periods of active deposition on sandy lobes, whilst the mud-rich units record periods of lobe abandonment. Mapping of these features suggests that lobes may have had lateral extents of several kilometres.

Viewed from the ridge, the eastern side of the Derwent Valley is made up of Shale Grit in the lower part where it is exposed in stream sections. The Grindslow Shale above is seldom exposed, but the Lower Kinderscout Grit, which forms the skyline, crops out as a series of crags and tors that show giant cross-bedding dipping to the south, similar to that seen at Uppermill, Darby Delph and Bamford Edge.

Rocher Top (SK 262935; 53.4369, -1.6009)

From High Bradfield village, take the road to the north, bear left at the junction about 650 m from the village centre and park on the roadside at a gate and stile near trees on the left about 450 m from the road junction. Secure vehicles, as cars parked on an isolated roadside might be a temptation to thieves. Cross the stile and follow the path alongside the wood. One route is to turn off the main path on the righthand side after about 200 m and follow a path along the top of the escarpment, parallel with the wall. Climb down a gully at a point where the crag takes a kink to the right, after about

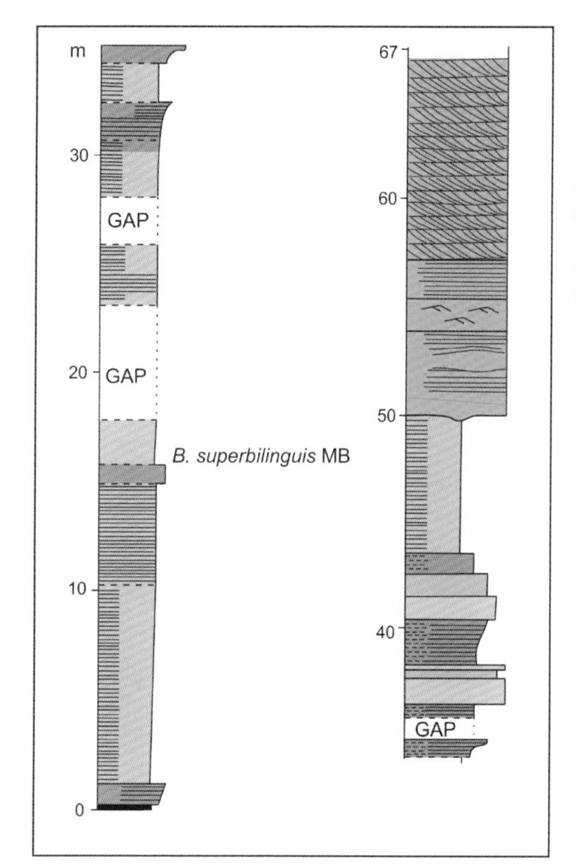

Figure 10.13 Sedimentary log through the Marsdenian (R_{2c}) succession at Rocher Top. The crag itself is represented by the uppermost 20 m of the log. The lower part is based on intermittent exposures in the hillside below.

400 m. This route is not suitable for a group of mixed ability, as the climb down needs some agility and care. The alternative is to continue down the main path for a further 200 m before heading off to the right and following poorly defined paths over rough ground at the base of the crag. The best sections are close to and south of the gully mentioned above. There is no real need to go north of the gully.

This natural crag exposes the upper part of the Huddersfield White Rock/Chatsworth Grit (Marsdenian, R_{2c}) cyclothem. The lower part of the cyclothem, the coarsening-upward, fine-grained interval, occurs in the hillside below, but is not well exposed here. The lower part of the crag shows mainly fine-to-medium grained sandstones with ripple cross-lamination and some small-scale cross-bedding (Fig. 10.13). These sandstones are thought to be mouth-bar sediments at the culmination of the coarsening-upward unit. Some of the sands show soft-sediment deformation.

These mouth-bar sandstones are sharply overlain, with some erosional relief, by channel bodies comprising medium-grained sandstones with medium-scale cross-bedding. There appears to be some repetition of these two major facies components in the inaccessible upper part of the cliff. The channel sandstones are interpreted as normal distributary deposits eroded into associated mouth bars as a result of an overall progradation of a delta lobe, one of several such lobes that characterize this cyclothem (Figs 4.14 & 10.16). Palaeocurrents were directed towards the west and northwest.

The Chatsworth Grit, seen to the south at Stanage Edge (see below), is thicker, more coarsely grained and with larger cross-beds and is interpreted as the fill of a wide palaeovalley eroded into this Huddersfield White Rock deltaic succession in response to a major fall in sea level (Fig. 10.16).

Bamford Edge (SK 205845; 53.3520, -1.6779)

From the A6103, take the lane (New Road) about 200 m south of the Yorkshire Bridge Inn. This lane climbs steeply up the hillside with Bamford Edge on the skyline above. After about 1.5 km, where the gradient starts

Figure 10.14 Giant cross-bedding in the Lower Kinderscout Grit (Kinderscoutian, R_{1c}) at Bamford Edge. The giant set is overlain by cosets of medium-scale trough cross-bedding; (BRR for scale).

to reduce, park on the left-hand side of the road. There is parking space for several cars. Cross the wall at the gate on to the moorland and take the footpath that heads diagonally up to the left after a few metres. Follow this to the crags on the edge of the moor. The exposures stretch along the escarpment for around 500 m, but the best sections are towards the southern end. This gritstone edge is quite popular with climbers, but it is seldom as busy as the more popular edges such as Stanage. It is possible to scramble around from top to bottom without the need for rock climbing. This is not suitable for disabled access, but a more distant view from the road, about 500 m from the road junction, shows the large-scale structures. Park with care.

Bamford Edge is in the Lower Kinderscout Grit (Kinderscoutian R_{1c}), which here shows giant tabular cross-bedding dipping to the south in sets up to about 10 m thick (Fig. 10.14). The base of the sandstone is not seen but it can be safely inferred to be an erosive channel base, several metres below the base of the exposure. Above the large sets is an interval of medium-scale trough cross-bedding, the two units separated by a near-horizontal erosion surface. The large sets are arranged in an en-echelon, off-lapping pattern such that some smaller sets in the upper part of the section expand downcurrent into large sets. The medium-scale cross-bedding above the large sets displays good, large-scale 'rib and furrow' patterns on the top bedding surfaces. These show that palaeocurrents were the same as the dip direction of the underlying large-scale foresets.

Giant cross-bedding, in some cases much thicker than this example, is a common feature of the Lower Kinderscout Grit across much of the area, as far north as Earl Crag, Airedale. Bear in mind, however, that the age of the 'Lower Kinderscout Grit', as recorded on BGS maps, becomes progressively younger from north to south, and so the giant cross-bedding is a feature of several separate progradational phases.

Stanage Edge (SK 243832; 53.3394, –1.6194) (►)

Stanage Edge can be approached either from the Bamford Edge locality, from Hathersage village or from the Ringinglow Road. There is a large car park near the road junction at the southern end of the edge. As Stanage Edge is a classic and popular climbing locality, the car park can be busy at weekends and holidays. The car park has a certain notoriety for thefts from cars and so great care should be taken to secure valuable items. From the car park, the footpath to the exposures is well defined. It is around 400 m to the first worthwhile crag. No disabled access to the exposed faces but view from car park is informative.

This very extensive crag exposes the Chatsworth Grit, a major channel sandbody in the upper part of the Huddersfield White Rock/Chatsworth Grit cyclothem (Marsdenian R_{2c}). The palaeovalley sandbody is up to around 50 m thick and can be traced along the outcrop (normal to palaeoflow) for around 25 km. Its northern margin, between Stanage and Rocher Top, is not precisely located but is probably close to the A57.

Figure 10.15 Large- and medium-scale cross-bedding in the Chatsworth Grit (Marsdenian, R_{2c}) at Stanage Edge. These coarse, pebbly sandstones fill the major palaeovalley that runs east–west through the area.

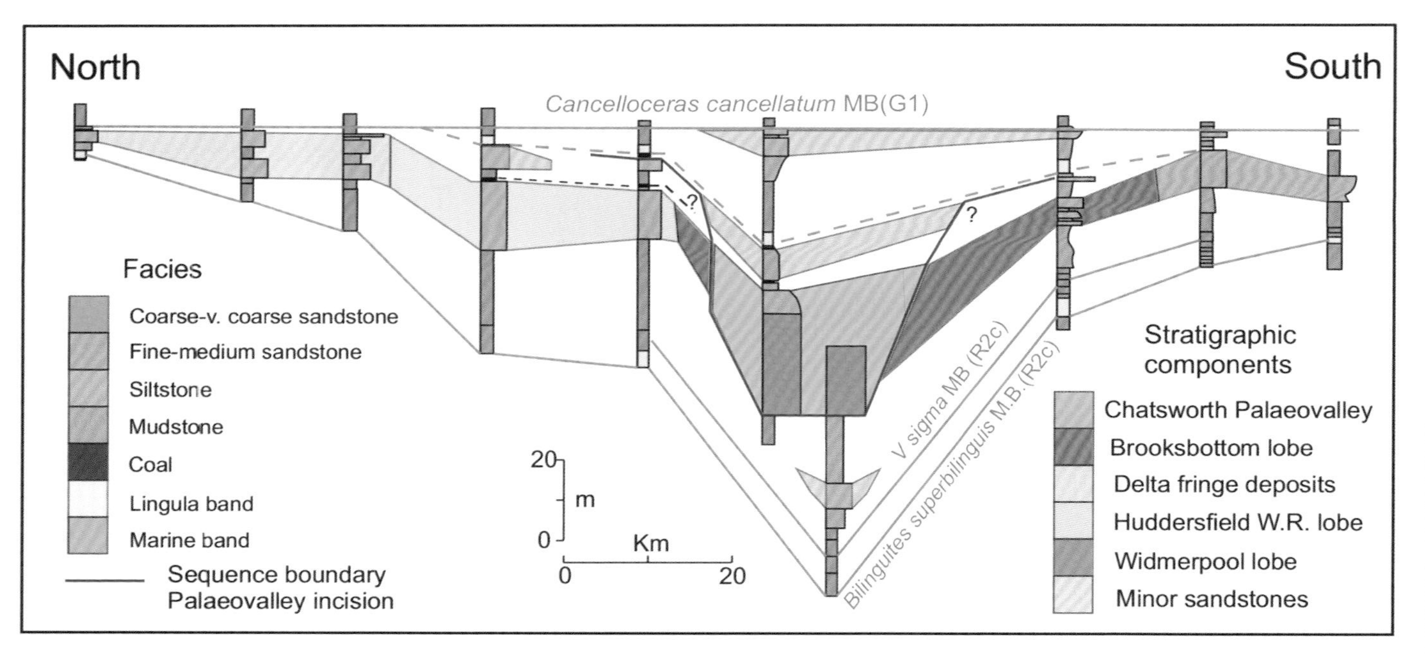

Figure 10.16 Correlation panel of the R_{2c} cyclothem between a series of boreholes to the east of the outcrop. They range from Farnham, near Knaresborough in the north to Melbourne, near Derby, in the south. Subsidence and accommodation space were highly variable across the profile and the Chatsworth Grit palaeovalley is sited in the zone of highest subsidence. Later minor sandstones in the cyclothem also coincide with the zone of maximum subsidence (after Waters *et al.*, 2008).

This margin is more precisely located at Chinley Head (see Chapter 12), some 20 km to the west, in a down-current direction.

Finer-grained sandstone is seen directly below the coarse channel sandstones in a few places. The main channel sandstones are very coarse-grained and pebbly, significantly coarser than the channel sandstones at Rocher Top. They are cross bedded in medium and large-scale sets that indicate a palaeoflow broadly to the west (Fig. 10.15). Some of the cross-bedding is complex, with patterns of descending sets, suggesting large compound bar forms in a wide river channel. The downstream equivalents of this succession are seen at Windgather Rocks, 20 km to the west (see Chapter 12).

The deep incision and the coarse grain size suggest that the Chatsworth Grit is the fill of a palaeovalley that was eroded in response to a period of falling sea level, with the fill being deposited during the low stand and during the early stages of the subsequent rise in sea level (Figs 4.14; 10.16). The lateral equivalent section seen at Rocher Top (Fig. 10.13) is thought to be part of the delta lobe that was incised by the palaeovalley. Tracing the Chatsworth Grit southwards along the Derwent escarpment, through Froggatt and Curbar edges, to beyond Chatsworth suggests that the palaeovalley is around 25 km wide. Beyond the limits of the palaeovalley to the north, the mature palaeosol seen at Winscar Reservoir is thought to record extensive and prolonged subaerial exposure of an interfluve plain during the low stand that led to the incision of the palaeovalley (Fig. 9.10).

Birchen Edge (SK 280725; 53.2455; −1.5817)

From Baslow village, take the A619 towards Chesterfield and turn left at the Robin Hood. Park in the Pay and Display car park to the right of the pub (NB not the pub car park). This car park is very popular and commonly fills up quickly. From the car park, walk along the road to the left for around 100 m and then turn left through a gate. Follow the path to the point where the path splits. The left-hand path (straight ahead) takes a relatively gentle route through the trees up to the edge. The right-hand path involves a very steep and somewhat rough route, directly up the hill. It is not for the faint-hearted, but it rewards with a pleasant walk along the top of the ridge. Either way, it is about 800 m from the car park to the monument that marks the crest of the edge. Birchen Edge faces west below the monument, with good exposure of sandstone around 200 m long. The face is around 10 m high. Climb around the edge with care. This is a popular spot for climbers and for walkers, particularly with children, and it can be quite busy at times. If you are low on the crag, be aware of any activity above you.

Birchen Edge is something of a stratigraphical outlier compared with other localities in this book. However, it is most relevant to understanding controls on late Namurian deposition. It is an exposure of the Crawshaw Sandstone, of early Westphalian (Langsettian, G_2) age, and is the only locality in this Guide involving sediments of this age. The sandstone is stratigraphically equivalent to the Woodhead Hill Rock to the west of the Pennine Anticline, as seen in the hillside above the Cracken Edge quarries at Chinley, and throughout the axial zone of the Goyt Syncline, described in Chapter 12.

Whilst the sandstone falls within the Lower Coal Measures, it does not show a major change in depositional style compared with the underlying Millstone Grit. At this locality, the Crawshaw Sandstone occurs quite closely above the Chatsworth Grit (Marsdenian, R_{2c}), which forms a prominent escarpment along the eastern side of the Derwent Valley, and which crops out lower down the hillside. In most places across the Pennines, the Chatsworth Grit (or its equivalents) is separated from Langsettian sediments by an interval of Yeadonian strata (e.g. Rough Rock Flags, Rough Rock). Here, however, sandstones of this age are

Figure 10.17 The 'Three Ships' tors on top of Birchen Edge, an exposure of the Crawshaw Sandstone showing medium-scale cross-bedding in coarse, pebbly sandstones with palaeoflow to the west.

missing or abnormally thin, suggesting that the major river system that supplied the Rough Rock avoided this area. One reason for this may be that the Chatsworth Grit in this area is a very thick palaeovalley fill, and that differential compaction between this thick, massive sandbody and laterally equivalent finer-grained sediments generated subtle topography that deflected the Rough Rock river. This effect was not suppressed until the Langsettian, when the Crawshaw Sandstone fluvio-deltaic system extended across the area. The Crawshaw Sandstone is a significant reservoir in the small oilfields around the Gainsborough Trough to the east.

The Crawshaw Sandstone at Birchen Edge is very coarse and somewhat pebbly. The base of the sandstone is not seen but is thought likely to be a sharp, laterally extensive erosion surface. Internally, the sandstone is dominated by medium-scale cross-bedding in both trough and tabular sets that show clear palaeoflow directions towards the west. Top bedding surfaces show good examples of large-scale 'rib and furrow' with trough axes that give the most reliable direction of palaeoflow. The three isolated tors, the 'Three Ships', on the moor behind the edge show exceptional three-dimensional exposure of cross-bedding, allowing its fully geometry to be explored in detail (Fig. 10.17). These sandstones are thought to be the channel deposits of an extensive braided river complex and are the youngest of such coarse sandstones derived from the north. Younger sandstones in the Westphalian Coal Measure succession are rather finer-grained, and some are derived from a different source in the west.

Suggested itinerary: Castleton, Mam Tor and Edale

This itinerary demonstrates the nature of the northern margin of the Derbyshire limestone massif and its relationship with the overlying Namurian, including the Kinderscoutian basin-filling succession. It is suggested as a walking itinerary, starting and finishing in Castleton. It could be comfortably done in a day, even with the optional extension to Barber Booth in Edale. The round trip would be around 15 km, including Barber Booth or around 10 km without it. The walk would involve around 650 m of climbing, or 400 m without the trip to Barber Booth. The route includes walking on both roads and on open country, mostly on footpaths.

Castleton village (main car park) – Cave Dale – Treak Cliff – Winnats Pass – Windy Knoll – Mam Tor – Mam Nick (or Mam Tor summit) – Barber Booth (optional) – Castleton village.

Chapter 11

Derbyshire Massif and Southern Derwent Valley

This chapter deals with the interior of the Derbyshire limestone platform and its south-eastern margin. The margin contrasts in depositional style with the northern and western margins described in Chapters 10 and 12. In the interior of the platform, shallow-water settings led to the development of extensive horizontally bedded limestones, locally with small mud mounds. In addition, a small, tectonically controlled sub-basin gave rise to deeper-water facies, some recording reduced oxygen conditions. The south-eastern margin has been termed a ramp, across which the platform margin prograded and retreated throughout Asbian and Brigantian times, with the development of fringing mud mounds and unstable slopes. Platform limestones are locally dolomitized. Interpretations of these limestones have evolved rapidly in recent years and so, unlike other chapters, key references are cited to help document these changes.

The eastern side of the platform is draped by Namurian Bowland Shales that extend up to mid-Marsdenian when the Ashover Grit (R_{2b}) delta advanced into the area. This progradation led to local instability of the delta front giving rise to large gravity-driven slides.

The area is extensively mineralized, and veins of lead/zinc ore were widely exploited with deep and shallow mines. The town of Wirksworth was a centre of both mining and quarrying of limestones. A cluster of disused former quarries has now been incorporated into the National Stone Centre. Because of the concentration of exposures along the eastern margin, and because there are local and important lateral changes in facies in these limestones, some sections of this chapter have a different structure compared to others. Closely adjacent exposures are combined into geo-walks that try to capture local complexity and include examples of the local industrial heritage that depends on it. The important traverse of the Intra-Platform Basin along the Monsal Trail is suggested as a 'geobike' itinerary, suitable for cyclists of modest ability. One Namurian locality at Black Rocks is included here as it is within walking range of the National Stone Centre.

Wirksworth: a platform margin setting

Much of Wirksworth lies on the south-eastern fringe of the Derbyshire Platform, on a peninsula-like outcrop of Viséan limestones. The outcrop reflects a broadly horst-like triangular structure, marked on its northern side by the important Gulf Fault, which downthrows by as much as 100 m to the northeast (Fig. 11.1). The fault limited most of the quarrying in and around Wirksworth and Middleton-by-Wirksworth to the same upfaulted block as the town. The limestones in Wirksworth are surrounded on two sides by softer Namurian rocks, with the limestone creating higher relief at 200–300 m a.s.l. The old centre of Wirksworth is essentially a hill-top town crammed between former quarries on several sides. Quarry faces drop precipitously behind some houses, and the steep narrow lanes comprise randomly clustered houses of different ages and styles.

The town descends on its eastern side to the Ecclesbourne valley, which lies on Bowland Shales (early Namurian). Across the valley, from the higher central parts of the town as well as from the main viewpoint on the geo-walk, the ground rises to the partly wooded Millstone Grit scarp to the east, controlled, in particular, by leaves of Ashover Grit (Marsdenian R_{2b}). The alignment and topography of this valley allowed close rail access to local quarries despite their being located at much higher levels. To the west, the town is bordered by the twentieth-century development of Warmbrook, beyond which the remains of old lead workings can be seen in the open fields around Dream Cave.

Wirksworth was not always so picturesque. For many years, up to the mid-twentieth century, it was one of the most intensively exploited limestone zones in the country, for aggregates and numerous industrial products (Thomas, 2019). During the peak quarrying era, the town was extremely noisy and dusty, with air pollution from quarry blasting and from transportation of quarry products, even through the town centre. Products were carried both by road, and by a network of quarry

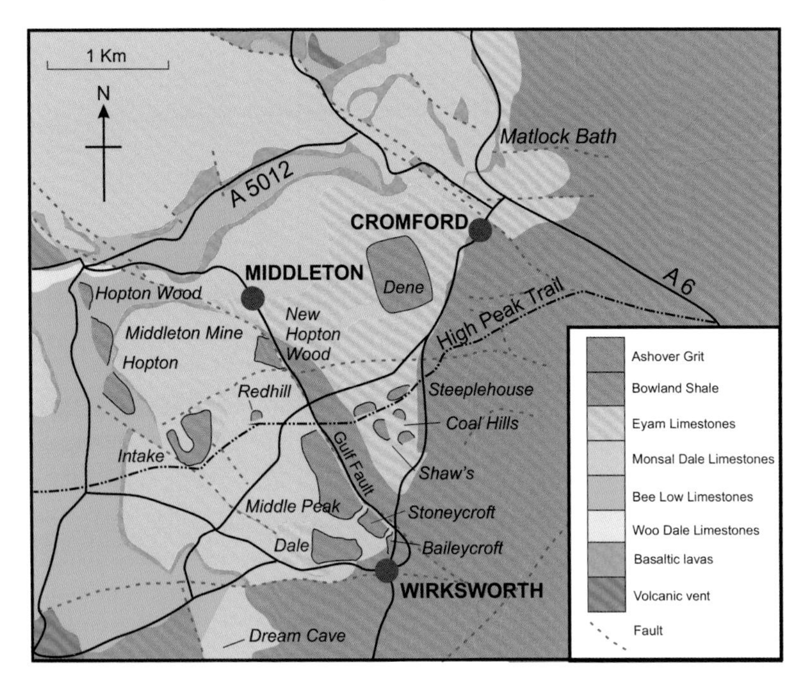

Figure 11.1 Simplified geological map of the area around Wirksworth and Cromford, showing the positions of major quarries (grey) mentioned in this Guide. The names of quarries (italics) are somewhat flexible and different names can be found in other sources. Geology adapted from BGS mapping.

railways leading either to the former Cromford & High Peak Railway or to the Wirksworth–Duffield branch line (now the Ecclesbourne Valley Railway). Although even bigger working quarries now exist elsewhere in Derbyshire, including nearby Dene Quarry, the scale of Wirksworth's quarries, taken together, gives a remarkable visualization of anthropogenic alteration of the landscape. It is an open question whether the dramatic cliffs and occasional pools of abandoned quarries should now be left to develop naturally as an attractive 'neo-landscape' for biological and geological conservation, or reclaimed for leisure use, housing, further industrial purposes or landfill. Examples of all of these are present around Wirksworth and Matlock. From an immediate geological standpoint, it is unfortunate that most of the disused quarries are currently off-limits, having become heavily guarded and strongly fenced without any obvious current usage. Several important geological features are therefore no longer accessible although access may sometimes be negotiable.

Wirksworth Geo-walk: up to the stars and back

Wirksworth is on several regular bus routes and there are car parks around the town centre. All sites on both geo-walk options have full public access, although there are fences and railings in front of some faces on the **National Stone Centre** *Geotrail. In the town itself, walking is on paved surfaces or good gravel paths, but beyond the town, some of the route is on loose gravel or earth footpaths. Walking wear is recommended. Much of the route is steep. The town stops only are suitable for disabled access. Around the town, there are public toilets, shops, restaurants and pubs. Some of the stops of Option 1, but not of Option 2, also have facilities.*

The geo-walks suggested here are based on a guide leaflet, *Wirksworth. Limestone Wunderground Trail* currently obtainable from the Wirksworth Heritage Centre (31 St John Street, Wirksworth DE4 4DS, tel: 01629 707000, www.wirksworthheritagecentre.org). The information in the leaflet is brief and non-technical with little geological detail, but it introduces geologically relevant background and some social and historical history. Its adaptation here provides more geological information for some of the stops but omits others as they are covered as separate localities in this Guide. Information boards at several points along the walk include geological aspects.

The exposures described variously show sections through Bee Low, Monsal Dale and Eyam Limestones close to the southern margin of the Derbyshire Platform. Although the walk does not include examination of specific outcrops, the stops along the route, in conjunction with **Dale** and **Baileycroft** quarries demonstrate the striking southeast-wards thinning of the Monsal Dale Limestones beneath the Eyam Limestones, between **Middle Peak** and the town centre. (Fig. 11.4) With sufficient time, **Baileycroft** and **Dale** quarries can easily be added to these walks to fill out the details of these changes.

Two geo-walk options are suggested:

OPTION 1. Follow the walk set out in the *Limestone Wunderground* leaflet trail. No further details are given here as they are mostly covered by Option 2, which uses the same stop numbers, or by localities covered elsewhere in this Guide. The whole route is 5.5 km and at least 3 hours should be allowed – more if full visits to the Heritage Centre, Wirksworth Church, the National Stone Centre and its Geotrail are included.

If the guide leaflet cannot be obtained, follow Option 2, without necessarily attending to the extra geological details or the extra stops (those with letter suffixes). *Limestone Wunderground* trail **Stops 6 (High Peak Trail)** and **7 (Middleton Incline)** are mainly of heritage interest and **8 (National Stone Centre)** is covered separately.

OPTION 2. This uses the same stop numbers as Option 1, with some extra stops denoted by letter suffixes (e.g. '4a' is between Stops 4 and 5). This route is about half the length of Option 1, but time should be allowed for geological details.

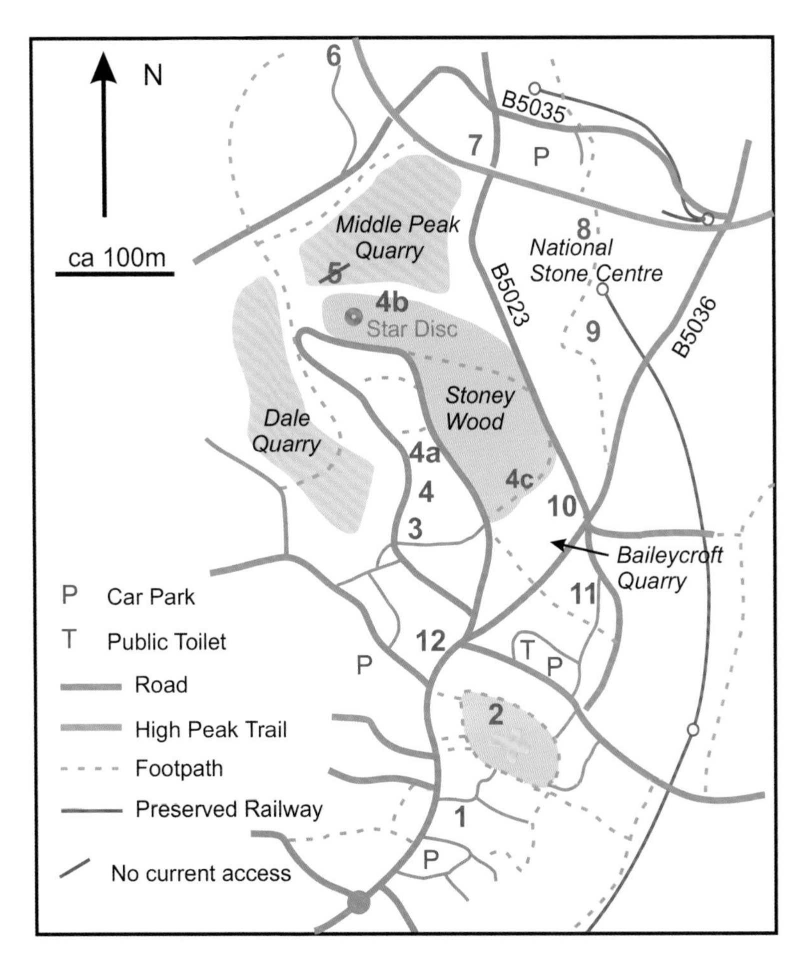

Figure 11.2 The area of the geo-walk around Wirksworth, showing the positions of the stops on the walks. Adapted from a Wirksworth Heritage Centre leaflet.

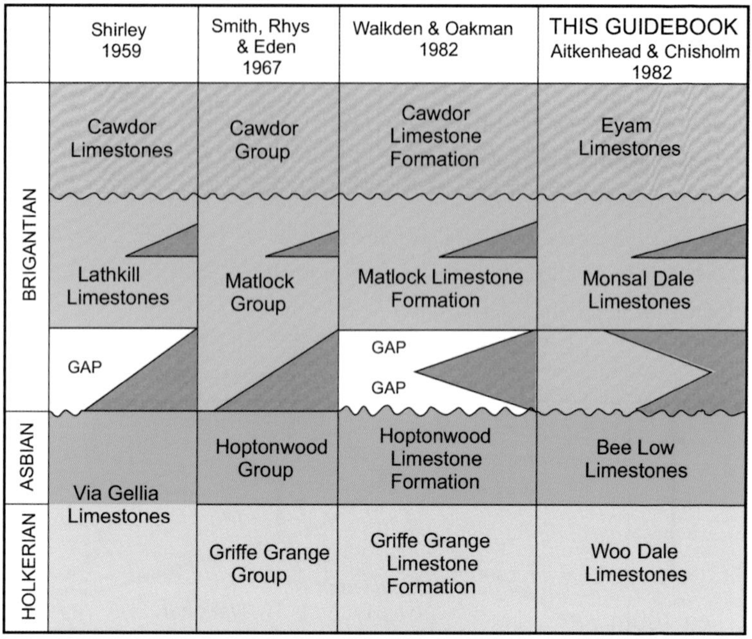

	Shirley 1959	Smith, Rhys & Eden 1967	Walkden & Oakman 1982	THIS GUIDEBOOK Aitkenhead & Chisholm 1982
BRIGANTIAN	Cawdor Limestones	Cawdor Group	Cawdor Limestone Formation	Eyam Limestones
BRIGANTIAN	Lathkill Limestones	Matlock Group	Matlock Limestone Formation	Monsal Dale Limestones
BRIGANTIAN	GAP		GAP GAP	
ASBIAN	Via Gellia Limestones	Hoptonwood Group	Hoptonwood Limestone Formation	Bee Low Limestones
HOLKERIAN		Griffe Grange Group	Griffe Grange Limestone Formation	Woo Dale Limestones

Figure 11.3 The changing stratigraphic nomenclature of the Dinantian limestones of the Derbyshire Massif, particularly in the south-eastern part of the massif. Being aware of these changes is important when reading older literature (modified from Walkden and Oakman, 1982).

Option 2: Geo-walk

The geo-walk is in and around the centre of Wirksworth, starting at the Wirksworth Heritage Centre close to the Old Market Place and ending nearby (Fig. 11.2).

The exposures described are mostly in now inaccessible disused quarries, and variously show exposures of Bee Low (Asbian), Monsal Dale and Eyam (both Brigantian) Limestones whose earlier names are set out in Figure 11.3. During the Asbian, the area lay on the southern margin of the Derbyshire Platform (Fig. 11.4). Following a marked disconformable break in sedimentation, the platform margin stepped back northwards, leading to slightly deeper water on a carbonate ramp. The beds also thin south-eastwards from Middleton-by-Wirksworth towards the centre of Wirksworth and show features suggesting slope instability. Although the walk does not include detailed examination of outcrops, views along the route, in conjunction with Baileycroft

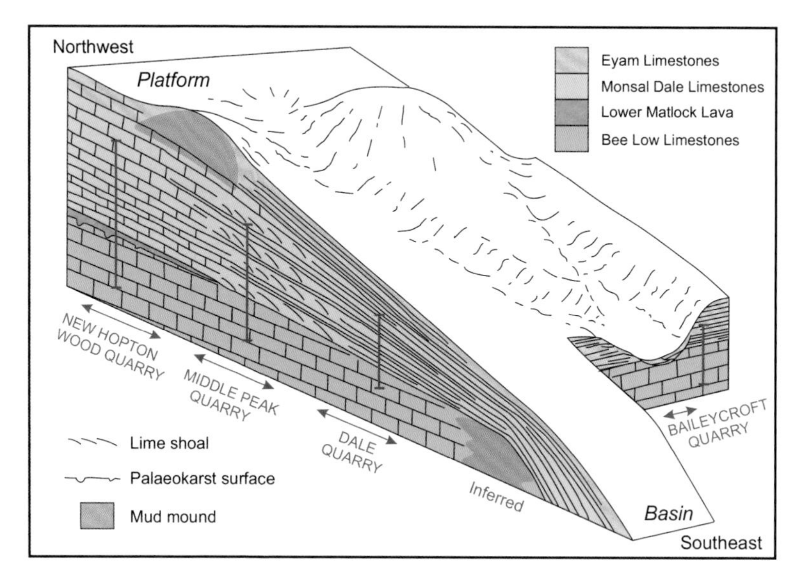

Figure 11.4 A schematic model of the facies distribution within major stratigraphic units near the platform margin in the Wirksworth area. The main section is a projection onto a plane roughly parallel with the Gulf Fault. The geology up to the Monsal Dale Limestones is largely based on quarries immediately southwest of the fault but the 'inferred' mud mounds of the Bee Low Limestones are projected from the Dream Cave area. The Eyam Limestone facies are projected from the exposures around the National Stone Centre, where the mud mounds, although shown here within the Eyam Limestones, are now assigned to the uppermost Monsal Dale Limestones. The main faces at Baileycroft Quarry trend north–south, and its geology is here projected onto a NE–SW pseudo-face. The upper surface of the model is thought to be an imagined sea-floor topography in mid-late Brigantian times. There is no evidence that the 'valley' was directly draped by Bowland Shales. Based on Walkden & Oakman (1982).

Quarry and Dale Quarry (easily included in this walk) illustrate this sedimentary history (Fig. 11.4).

Stop 1: Wirksworth Heritage Centre (53.081359, –1.572976)

The Heritage Centre introduces geological and other background aspects of the area, not just those relevant to this walk. The Centre's museum gives an excellent introduction to many aspects of Wirksworth's heritage including its mining and quarrying, and the social impacts. A section is devoted to an important and unusual fossil discovery made in the **Dream Cave** or **Dream Mine**, *c.* 1.5 km southwest of the town near Godfrey Hole – a Pleistocene Woolly Rhinoceros (covered separately).

This is associated with Dean William Buckland (1784–1856), a key figure in the early history of geology and palaeontology in Britain.

*Directions to **Stop 2**: Walking northwards from the Heritage Centre along St John St towards the Old Market Place in the town centre, take any of the little lanes immediately off to the right to enter the churchyard and the imposing parish church of St. Mary.*

Stop 2: St. Mary's Church (53.081982, −1.572271) (▶)

St. Mary's Church building dates mostly from the thirteenth to fifteenth centuries on a site that goes back at least to the eighth century AD. It is built of local Namurian gritstones, the pink hue of much of which suggests Ashover Grit. The scale of the church reflects investment of local wealth derived from lead mining. The main point of geological interest is the locally renowned relief carving in the south transept of a figure carrying a chisel-headed hammer and a basket or bag (?kibble) (Fig. 11.5). He has long been known in the local vernacular as **'T'Owd Man'** (The Old Man) and is widely thought to be an Anglo-Saxon lead miner (or perhaps an early geologist!). There is a full audio-visual

Figure 11.5 T'Owd Man: a relief carving of an early lead miner in St. Mary's church, moved to Wirksworth from Bonsall, and now a symbol of local heritage.

explanation next to the figure, which was originally located in the nearby lead-mining village of Bonsall. Both there and around Wirksworth, he is a local logo and mascot, reproduced on souvenirs. There are other notable ancient carvings elsewhere in the church.

Directions to ***Stops 3, 4, and 4a****. From the church, return to the main road, St John St., and use the pedestrian crossing by the Old Market Place. Turn right to continue a few more metres along St John St. to reach a steep narrow lane, Green Hill. In a few metres, turn left off this into another steep lane, The Dale, which continues almost all uphill to* ***Stops 4b and 5****, running closely parallel to the northern perimeter of Dale Quarry for much of the way.*

Stops 3, 4 and 4a along The Dale (▶)

Originally, houses along The Dale were homes of lead miners. Lead mining in the area goes back into prehistoric time but declined in the 1800s. The lowest cottages along The Dale, at **Stop 3, Dale End (53.0830, −1.5742),** date from later in that time. As quarrying expanded, it displaced lead mining as the main industry. The houses of The Dale were increasingly encroached upon by **Dale Quarry**, and some were eventually demolished to make room for quarry expansion, or were otherwise left poised high above vertical quarry faces. The quarry depths (formerly even deeper) can be glimpsed between the houses, whose proximity to quarrying operations often caused damage to windows and roofs.

Until the 1950s, the double cottage at numbers 31–33 on the right (**Stop 4, Walkers Cottage: 53.0834, −1.5755**) was once a grocer's whose main customers were quarrying families. A plaque explains that this property was restored by The Derbyshire Historic Buildings Trust in 1980, evidently in a traditional way. Like much of Wirksworth, its building materials are mixed; in this case, an Ashover Grit frontage with mixed local limestones at the side.

A cottage (**Stop 4a, 53.0845, −1.5764**) about 150 m further up the Dale on the right is located in a small former quarry showing thinly (*c.*30 cm) and irregularly bedded limestones (Fig. 11.6). The BGS 6-inch map shows these as Monsal Dale Limestone. The face is on private property and so cannot be examined without permission, but from the lane it offers a glimpse of one of the main facies of the largely inaccessible Dale Quarry. It shows the same 'thinly bedded (*c.*30 cm

Figure 11.6 Thinly bedded limestones in the upper part of of the Monsal Dale Limestones at Stop 4a. The thinner bedding is characteristic of the ramp slope setting.

thick) fine-grained bioclastic wackestones with shale partings' which dominate much of the Monsal Dale and Eyam Limestones in the quarry. Palaeokarstic surfaces are absent or inconspicuous. This facies differs from typical platform-top Monsal Dale Limestones that occur further north. It has been suggested that the facies here represent a change from platform margin conditions to a ramp whose 'unstable marginal slope ... is dominated by fine-grained carbonate sediment that was winnowed and transported off the shelf [platform] by storm and tidal currents and deposited below wave-base.' (Cossey *et al.*, 2004).

*Directions to **Stop 4b**: Continue along The Dale to a sharp hairpin bend to the right, with steps leading up to a wooden gate (**53.0856, –1.5815**). (The Option 1 route continues straight ahead up the steps and on to a gravel track with fences on both sides preventing access to **Dale Quarry** on the left and **Middle Peak Quarry** on the right. **Stop 5** on the* Wonderground *Trail is omitted here but is incorporated as a view across Middle Peak Quarry at **Stop 4b**. Views of the quarries from the gravel track are mostly insufficient to see good geological details. The Wunderground Trail continues beyond the quarries for about 1 km across open fields to reach the **High Peak Trail (Stop 6)**, the **Middleton Incline (Stop 7)**, and the **National Stone Centre***

(Stop 8), *which are treated separately. From* ***Stop 8*** *the Wunderground Trail returns to the town via an old trackway,* ***Old Lane (Stop 9)****.*

Around the hairpin, the lane name changes to Green Hill and climbs to a summit in about 200 m. Bear left through an ornamental arched gateway bearing the name 'Stoney Wood' and 'Gateway to the Stars'.

A visit to Dale Quarry can be added as an optional 'there and back' detour to the main geo-walk, a distance of some 250 m each way. A few metres below the hairpin on the lane, a footpath to the south goes gently down to the quarry floor and through a tunnel to reach the viewpoint.

Stop 4b: Star Disc and Stoney Wood, (53.0857, –1.5782) with view of Middle Peak Quarry

The Star Disc is a twenty-first century stone circle and celestial amphitheatre created by local artist, Aidan Shingler. It is constructed from black and silver granite (not local) with carvings of a star chart that mirrors the northern hemisphere's night sky. The site is intended for star gazing and for staging events, and is an important community focal point, educational resource and performance space. There is also an orientation table (topograph) with compass bearings to places in and beyond Wirksworth. The Star Disc is located at the highest point in the former Stoneycroft Quarry, which was given by Tarmac, its former owners, to Wirksworth. It had been largely filled in by Middle Peak overburden, obscuring most of Stoneycroft's quarry faces, and the quarry has become overgrown. Since 1994, it has been planted by local children as woodland and become a community park, renamed Stoney Wood.

From the Star Disc area, the prominent view is to the east and south. The eastward view is dominated by the broad Ecclesbourne valley that runs to the east of Wirksworth. This is underlain by early Namurian Bowland Shales, partly downfaulted along the Gulf Fault. Beyond the valley, the ground rises to a prominent N–S ridge dominated by leaves of Ashover Grit (Marsdenian, R_{2b}), and marked by masts at each end – Bole Hill (323 m) to the left and Alport Height (314 m) to the right. To the south, the land gradually descends to overlie younger, mostly lower-lying, rocks of the Midlands including Coal Measures. This view emphasizes Wirksworth's position on higher, older Carboniferous ground at the southern limit of the Dinantian carbonate platform.

Figure 11.7 The Bee Low and Monsal Dale limestones of Middle Peak Quarry, showing the scale and number of the cyclothems that make up the succession. 'Palaeosols', which define the tops of many cyclothems, are units or conspicuous bedding surfaces that show evidence of subaerial emergence.

A flight of wooden steps leads up from the Star Disc to a northward viewing point, dominated by the disused Middle Peak Quarry and its impressive north face. This exposes 110–120 m of strata, about 800 m distant from the viewing point (Fig. 11.7). Before quarrying commenced in the 1850s, the fields in the area were used for cultivating barley for brewing, in contrast with their now evident dramatic anthropogenic transformation. Since the quarry faces can only be seen at a distance, binoculars can be useful. Middle Peak Quarry is **Stop 5** on the *Wunderground Trail*, but as it is no longer accessible, this view must suffice. The following basic details, mainly based on Walkden & Oakman's (1982) description, can be observed, and are best lit in the middle of the day.

The entire face is dominated by highly visible sedimentary cyclothems, generally separated by one or other of palaeosols, wayboards and palaeokarstic surfaces. Darker layers can sometimes be discerned and represent facies changes within the cyclic sequence (Fig. 11.8). The very prominent massive limestone in the middle horizons of the face is about 20 m thick and overlies the relatively thin basal cyclothem of the Monsal Dale Limestones whose lower contact is picked out by a noticeable row of shrubs and vegetation, beneath which are Bee Low

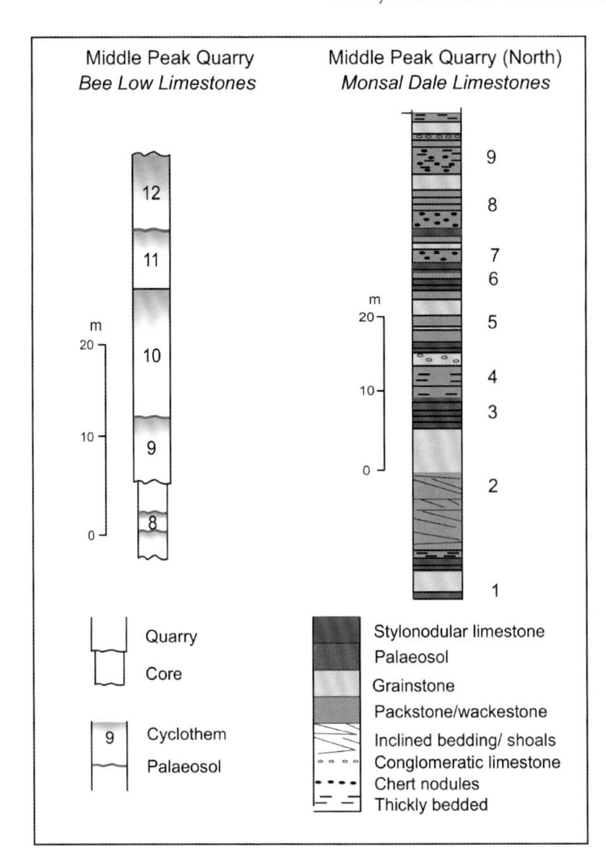

Figure 11.8 Simplified profiles through the limestones in Middle Peak Quarry, showing the thickness and distribution of shallowing-up cyclothems. Note the contrasting scale of thicknesses between the two formations (after Walkden & Oakman, 1982).

Limestones. These consist of about 45 m of very uniform, massive bedded algal and crinoidal packstones, often cross-bedded, although this is not readily seen from the viewing point. Four major cyclothems can be seen between the surface of the pool at the quarry base and the contact with the Monsal Dale Limestones. They are separated by palaeokarstic surfaces, generally associated with palaeosols.

The massive limestone in the lower part of the Monsal Dale Limestones consists of large, coalesced lobes of wackestone, packstone and grainstone, with gigantoproductid brachiopods. The lobes represent generally southerly-directed prograding mud-banks and sandbanks that are strongly cross-bedded similar to those in Dale Quarry (Fig. 4.5). Above this, the Monsal Dale Limestones consist of a further seven visible cyclothems separated by palaeokarstic surfaces. The total thickness of the Monsal Dale Limestones is more than 60 m, a thickness that reduces drastically to the southeast, to just a few metres on the eastern face of Stoneycroft Quarry in the lower part of Stoney Wood (**Stop 4c**) and along the western faces of Baileycroft Quarry only *c*.1 km from the north face of Middle Peak Quarry (Fig. 11.4).

Returning to the Star Disc area, the immediate foreground looking eastwards consists of a vegetated ridge separating the southern faces of Middle Peak Quarry from the northern face of former Stoneycroft Quarry. A few outcrops of the latter that project through the vegetation are difficult to access but appear to show 8 m of Monsal Dale Limestones. Bearing in mind that the summit area here is about 25% lower in elevation than the summit of Middle Peak as seen from the upper viewing point, this outcrop must correspond to a middle interval within the Monsal Dale Limestones seen on the main Middle Peak face. Much of the outcrop seems, from a distance, to be strongly bioturbated, which is typical of a mid-Monsal Dale Limestone cyclothem (Fig. 4.3A). The beds undulate slightly, change in thickness, and dip towards the Star Disc. Published descriptions of the adjacent Middle Peak and Dale Quarries, suggest that this may be slump-related (cf. Fig. 11.11) or possibly correspond to the undulating cross-bedding mentioned above.

Directions to ***Stop 4c:*** *From the Star Disc, several routes can be taken through Stoney Wood to reach its lower entrance* (***53.0863, −1.5743***)*. However, to see the main remaining face of the former quarry* (**Stop 4c**)*, descend via the more gentle route (c.500 m) of the main footpath, approximately south-eastwards from the Star Disc towards the southeast corner* (***53.0846, −1.5743***) *of Stoney Wood, passing various local artists' installations on the way. At the corner, the path enters a small area of woodland, and close by the Community Orchard, there is a choice of paths. Take the sharp left path northwards to turn back on yourself to stay within Stoney Wood, passing an information board about its history in relation to Stoneycroft Quarry. Descend from here and cross a footbridge to reach the lower entrance.*

Stop 4c: Stoneycroft Quarry face (e.g. from 53.0851, −1.5743)

There are few accessible exposures left in Stoney Wood but, on the way down the path to the Community Orchard turn, the eastern face can be seen between and over the trees. There are further views beyond this turn and the information board. The face now has a dense growth of trees right in front of it, so it is not really accessible for geological study. It is located 'back-to-back' with the Lime Kiln faces and Co-op Faces of **Baileycroft Quarry**. Depending on viewpoint, three formations can be distinguished in good light: (1) Bee Low Limestones (grey and massive), (2) Monsal Dale Limestones (dark irregular units), and (3) Eyam Limestones (pale weathering, very irregular) resting unconformably on an undulating surface of Monsal Dale Limestones. Note the greatly reduced thickness (? 6–8 m) of the Monsal Dale Limestones here, compared with *c*.60 m only *c*.1 km to the north on the main face of Middle Peak Quarry (Fig.11.4).

The left-hand panel of the information board near the lower gate shows a view of the north face shortly after Stoneycroft Quarry was abandoned in the 1980s and when large-scale infilling had already progressed. This and three similar views, showing stages of Stoneycroft's history from active quarrying to its transformation to Stoney Wood, are shown in Thomas (2019) and emphasize the sheer scale of these anthropogenic changes to the landscape.

*Directions to **Stop 10:** Leave Stoney Wood by the ornamental lower gate and turn right into Middleton Road (B5023). This road runs directly along the Gulf Fault to reach **Stop 10** at the junction with the B5036.*

Stop 10: Lime Kiln pub (SK 288543; 53.0855, –1.5722) (▶)

The sign of this Grade II listed inn shows an old lime kiln. These were important features of this and many other limestone regions, where lime burning was a vital industry. Limestone pieces were laid in alternating layers with charcoal, later coal, and in more recent kilns a single mass of coal was set only at the base. The kiln was heated to over 900°C to make building materials like lime mortar and quicklime fertilizer for farmers. There are no longer immediately adjacent lime kilns, but there are some near Access Point M of the National Stone Centre. About eight kilns operated at various times between 1880 and 1925 in Stoneycroft and in the area north of Baileycroft Quarry.

The Lime Kiln Faces in the pub garden currently provide the best exposures of the west side of the **Baileycroft Quarry** with Bee Low, Monsal Dale and Eyam Limestones, clearly separated by erosion surfaces.

*Directions to **Stop 11:** Join the B5036 for c.50 m and walk to the south passing Wirksworth Infants School and crossing Cemetery Lane, to enter*

Figure 11.9 The Moot Hall, Wirksworth, the historic meeting place of courts established to apply rules and resolve disputes within the lead-mining industry.

another lane, North End. From here, you can see the Co-op Supermarket, Fire Station and Wirksworth Cars which occupy the former Baileycroft Quarry on each side of the main road. In North End, turn right immediately into Chapel Lane. Stop at the Moot Hall after about 85 m on the right.

Stop 11: Moot Hall (SK 288542; 53.0841, −1.5718) (▶)

Remains of former lead workings can be seen in all the limestone areas around Wirksworth, indicating their earlier importance to the local economy and for Wirksworth's history. Lead mining was controlled by local laws overseen by the Barmaster at Barmote Court held in the Moot Hall (Fig. 11.9). The court registered claims, measured amounts of mined lead ore, settled disputes and collected taxes. The Court has met in Wirksworth since at least 1288, and the present building dates from 1814. In fact, there are two barmote courts, one for Monyash, covering the High Peak, and one for Wirksworth, covering the Low Peak. In 1814, the Monyash court moved its sessions to the Wirksworth Moot Hall, and since 1994 the two have met together once a year, in April. According to tradition, bread, cheese, clay pipes and tobacco are provided at the meetings, and a representative of the monarch, who is the *Lord of the Field,* attends.

Directions to Stop 12: Continue along Chapel Lane and turn right at the T junction with Coldwell St. to reach the centre of Wirksworth at the Old Market Place at St John's St.

Stop 12: Old Market Place (53.0826, –1.5737)

Separate guides exist for the historic buildings of Wirksworth. St. John's St. now continues northwards as Harrison Drive, passing through a walled cutting and the two sides of **Baileycroft Quarry**. This cutting was made after the quarry closed in 1960 and entailed demolition of several historical buildings adjacent to the Red Lion pub on Old Market Place, which had previously been closed off at its northern side.

Dale Quarry Wirksworth: slumping down the slope (SK 282541; 53.0838, –1.5799)

General notes on transport, facilities and under-foot conditions, in the introduction to the Geo-walk, apply equally here.

This disused and partially infilled quarry lies some 500 m west north-northwest of Wirksworth town centre, to the north of Brassington Lane. All the nearby lanes are steep and narrow and while there is a marginal chance of a parking space for a single small vehicle in the upper part of West End, parking is recommended in the town itself, e.g. the Market Place car park. The locality, which is a viewpoint, is reached by walking up West End and turning right at the fork into Brassington Lane, close to the last house. After about 50 m a gap in the wall gives access to the public footpath through Dale Quarry. This public right of way passes under a low face of the quarry in a short tunnel. The path is fenced on both sides restricting public access to the rest of the quarry. The quarry is less than 50 m wide at this point, but this section of the path affords two relatively close views of the faces on each side before crossing the quarry and entering the tunnel. As this locality is only a viewpoint, binoculars might prove useful.

An alternative access route is as a detour from the Wirksworth Geo-walk by leaving that route between ***Stop 4a*** *and* ***Stop 4b*** *close to where The Dale enters the hairpin bend into Green Hill. Take the footpath that meets The Dale at a sharp reverse angle on the left, through the small wood, and pass through the tunnel in Dale Quarry to reach the viewpoint.*

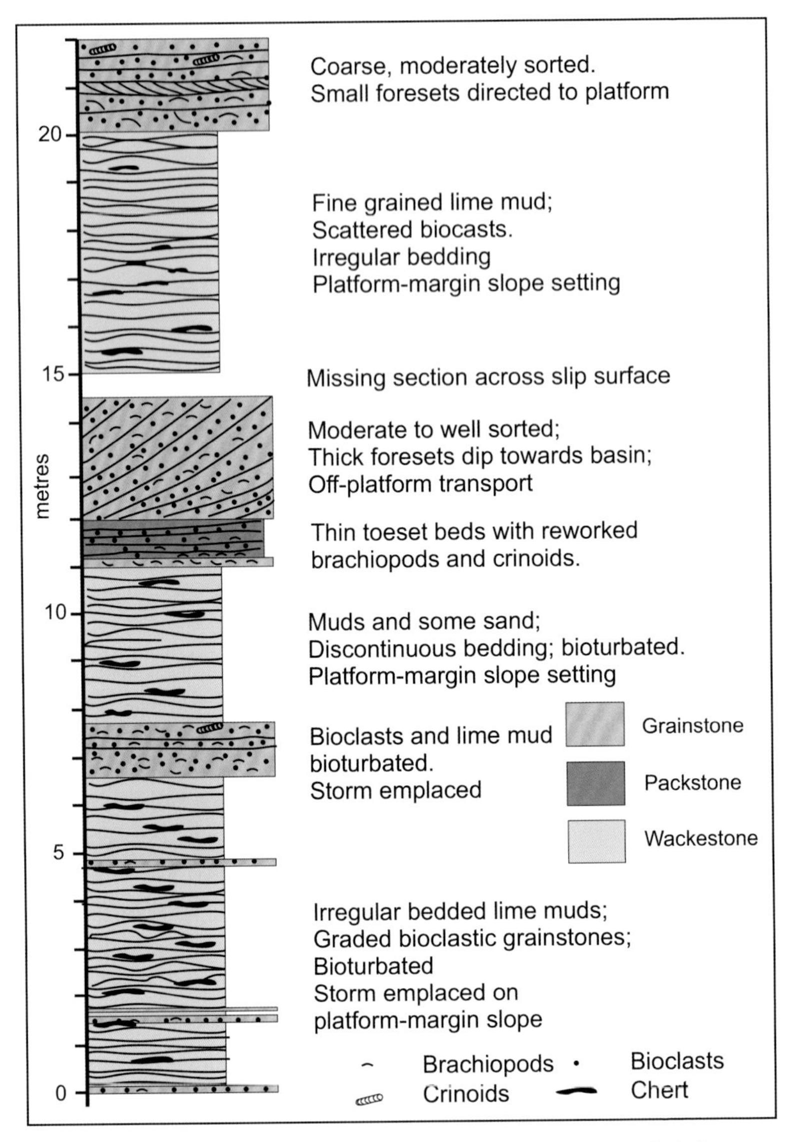

Figure 11.10 Measured section through the Monsal Dale Limestones at Dale Quarry. Most of the measured exposure is now buried through tipping. The stratigraphic positions of the exposed faces are indicated. These are illustrated in Figure 11.11 and Figure 4.6 (after Gutteridge, 2003).

Dale Quarry provides excellent views of Monsal Dale Limestones in a platform margin to slope setting, showing soft sediment deformation structures and large-scale cross-bedding. It is now one of the only

Figure 11.11 Soft-sediment deformation in the upper part of the Monsal Dale Limestones at Dale Quarry.

places left in the Wirksworth area that shows clearly this assemblage of sedimentary features. As the two quarry faces cannot be closely accessed, observations are limited to views from the viewpoint. The underlying Bee Low Limestones and overlying Eyam Limestones are not visible due to quarry infill or to the faces not extending high enough.

With Brassington Lane behind the viewer and the tunnel entrance in front, the long quarry face on the left (looking roughly northward) exposes *c.*20 m of beds, although *c.*30 m of Monsal Dale Limestones are recorded for Dale Quarry as a whole (Fig. 11.10). The section seems to compare with that recorded elsewhere in the quarry before it was infilled. The face shows relatively thin beds (*c.*30 cm) with thin shaly partings, described as fine-grained bioclastic wackestones. In the upper half of the face (Fig. 11.11) the bedding is substantially disturbed along two inclined surfaces that merge to the right. A marked anticlinal fold, whose left limb rises from the lower surface, intersects a higher surface. Its right limb slopes down onto the upper plane, as well as the lower one. Aside from the fold, the dip of the disturbed bedding is predominantly to the left, broadly opposed to the local structural dip (BGS map shows 12° to southeast). This suggests rotational sliding and slumping directed more or less to the right (south-eastwards), consistent with other examples of slumping reported elsewhere from this quarry.

The quarry face further away to the right (more or less eastwards) is the southern face of the exposed top of a largely buried, island-like massif known as Burton's Vein Pillar. It was not quarried away as it contains shafts and chambers associated with the mining of Burton's Lead Vein, which left it in an unstable state. About 10 m of beds are seen, the lower 5 m or so being thinly and irregularly bedded with shaly partings. These probably correlate with the fine-grained bioclastic wackestones seen in the left-hand quarry face. The upper 5 m or so is a single set of cross-bedding whose foresets dip to the right (southeast) (Fig. 4.5). Its stratigraphic relationship with the section in the left-hand face is not clear. Either the cross-bedded interval is impersistent and/ or it is displaced by the cross-faulting shown by BGS 6-inch mapping, or both.

Overall, the facies of the Monsal Dale Limestones in this quarry (see also **Stop 4a** of the Wirksworth Geo-walk), are strikingly different from their equivalents further north (e.g. **Redhill Quarry** and **Middle Peak Quarry**). There, the interval is typically in units 1–2 m thick, stacked in cyclothems with palaeokarstic surfaces (Fig. 11.8), characteristic of a platform-top setting. In Dale Quarry the limestones are darker and more thinly bedded without palaeokarstic surfaces.

Interpretations are largely based on earlier workers who enjoyed better access to the quarry. Gutteridge (2003) described the lower beds on the left-hand face as 'discontinuous bedding in bioturbated peri-platform carbonate muds and occasional carbonate sand'. The upper cross-bedded unit comprised 'thickly bedded foresets dipping basinward; moderate to well sorted crinoidal grainstone with comminuted brachiopods lying parallel to foresets; foresets show off-shelf transport.' Cossey *et al.* (2004) regarded this feature as 'isolated mega-ripples of crinoidal grainstone' which 'migrated down the slope until they were starved of coarse, bioclastic sediment and ... buried by fine-grained slope carbonates.' Similar features elsewhere are described as 'shoals', which could be large ebb-tidal bedforms or possibly ebb-tidal deltas.

Cossey *et al.* (2004) infer that, during deposition of the Monsal Dale Limestones of Dale Quarry, material was 'winnowed and transported off the shelf by storm and tidal currents and deposited below wave-base ... thin bioclastic beds represent occasional storm-deposited influxes of coarser material derived from higher energy environments [on] the platform margin. Some storm beds were mixed with the surrounding fine-grained slope carbonates by bioturbation. The platform margin slope was unstable and was affected by down-slope creep and major slumping.'

In this immediate area, the underlying Bee Low Limestones are of platform facies whilst the facies of the Monsal Dale Limestones in Dale Quarry suggest deposition in deeper water on a slope, contrasting with their platform facies to the north. The platform margin evidently retreated northwards prior to and during deposition of the Monsal Dale Limestones, probably in response to a combination of sea-level changes and tectonic tilting. The sedimentary environment in the upper part of a slope was evidently dynamic and unstable.

National Stone Centre Area: on the edge (SK 286552; 53.0935, –1.5732)

This account of the NSC area is dedicated to the memory of the late Ian Thomas (1945–2024) who was instrumental in its establishment, and to the late Gordon Walkden (1944–2022) who contributed notably to a radical change in our understanding of the geology of the Derbyshire Platform limestones. His graphical reconstructions and synthesis of the geology of the NSC area continue to be particularly influential.

*The National Stone Centre (NSC) lies 1.2 km north of the centre of Wirksworth, and 1.9 km south-southwest of the centre of Cromford. The quarry-complex consists of six disused quarries, centred around the Visitor Centre (**53.0935, –1.5735**), and is generally referred to as the 'Steeplehouse Quarries Complex' or the 'Coal Hills Quarry Complex'.*

*This locality corresponds to **Stop 8** in the Wirksworth Heritage Centre* Limestone Wunderground Trail. *It is included in the **Wirksworth Geowalk (Option 1)** but not described there.*

*There are four main access points to the complex, **B, I, J,** and **M** (GPS details below) (Fig. 11.12). All can be used for foot access. From **B** there is paid parking at **C,** and from **J** it is possible to park along the road at **K.** For **M** there is free parking at the edge of Wirksworth at the southern end of Old Lane.*

*Wirksworth is served by several bus routes, which variously stop at **B, I** and **J,** and at the end of Old Lane for access point **M**. Some bus routes also serve the train stations in Matlock and Matlock Bath. **M** is the best point for combining a visit to the NSC area with **Baileycroft Quarry** and/or the*

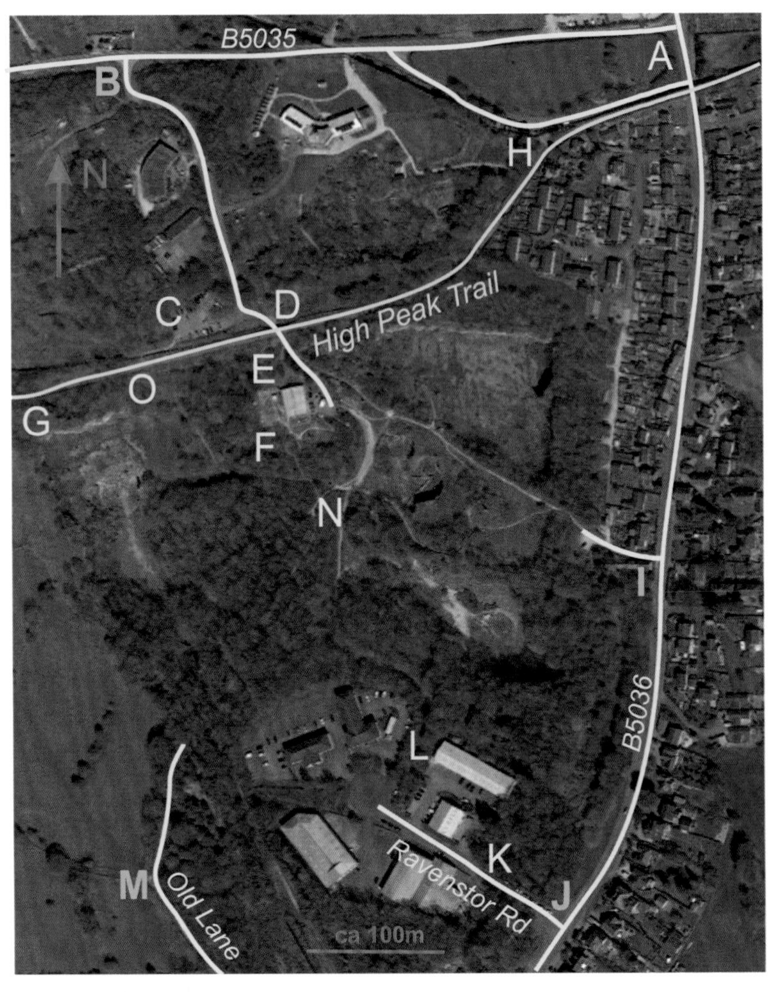

Figure 11.12 Google Earth image of the NSC complex showing main entry points and navigation points within the site.

Wirksworth Geowalk *which they can link up with at the* ***Lime Kiln pub (Geowalk Stop 10)****.*

For Access Point ***B****, take the B5035 towards Ashbourne, by turning westwards off the B5036 at the very top of Cromford Hill* ***(A)*** *close to where the High Peak Trail (HPT) crosses the road in an over-bridge. In less than 500 m, turn left at the numerous signposts (Access Point* ***B: GPS 53.0960, –1.5758)*** *to the National Stone Centre, Derbyshire Eco Centre* *and Mount Cook Adventure Centre. Buses stop at* **B**, *so those on foot can now take the same route as vehicles. The NSC car park (pay) (***C***) is on the right at the lower end of the lane, just north of the High Peak Trail. This gives direct access to the HPT (walking, cycling, horse-riding) which leads to other described geological locations (***Black Rocks, Steeplehouse Quarry (Stop 9)***, ***Intake Quarry, Redhill Quarry*** *and* ***Harboro' Rocks***). *The lane continues under a bridge (***D***) beneath the HPT to reach the NSC Visitor Centre where there are disabled parking spaces. Others should walk from the main car park (c.150 m).*

Access points ***I*** *and* ***J*** *have bus stops nearby on the B5036 and are useful options for those on foot. Access point* ***I*** *(***GPS 53.0927, –1.5695***) is suitable for dropping or meeting people on the main road but not for parking. A short lane leads westwards from the main road and in 50 m ends at a small car park used by local residents. It continues through the NSC gate as a public right of way leading past* ***Coal Hills Quarry (Stop 6)*** *directly to the Visitor Centre in 230 m. From access point* ***J*** *(***GPS 53.0904, –1.5705***) walk westwards along Ravenstor Road to enter a small industrial estate and turn right (north) off the roundabout on to a footpath at* ***L*** *(c.160 m). Walk northwards for 65 m and through the NSC gate. Take the left branch to the footpath hub* ***N*** *(***GPS 53.0930, –1.5730***), marked by one of several turret sculptures by local artist, Denis O'Connor, with signposts nearby to Ravenstor Station and Wirksworth. Continue northwards to the Visitor Centre (***F***); distance from gate c.180 m.*

Access Point ***M*** *(***53.0905, –1.5746***) is on the southwest edge of the NSC area, on Old Lane, that lane forks north-northwest off the B5036, close to bus stops. There is a car park here although it is often full. Proceed on foot along Old Lane, passing old lead mine workings in the fields on the left, to take a right fork into dense woodland at Access Point* ***M***. *The footpath, curves to the right beneath a fine stone archway of an old, but never used, quarry incline. Curve to left by the old lime kilns and pass stone benches to reach the pathway hub* ***N***. *Total distance from the main road is c.800 m.*

Although only a few paths through the area are designated public rights of way, all paths and geological locations in the NSC area are accessible at all times, apart from limited access to Steeplehouse Quarry (explained below). Most quarry faces along NSC's Geotrail are fenced- or walled-off and are not usually accessible for close examination. Permission for such examination should be obtained in advance from the NSC.

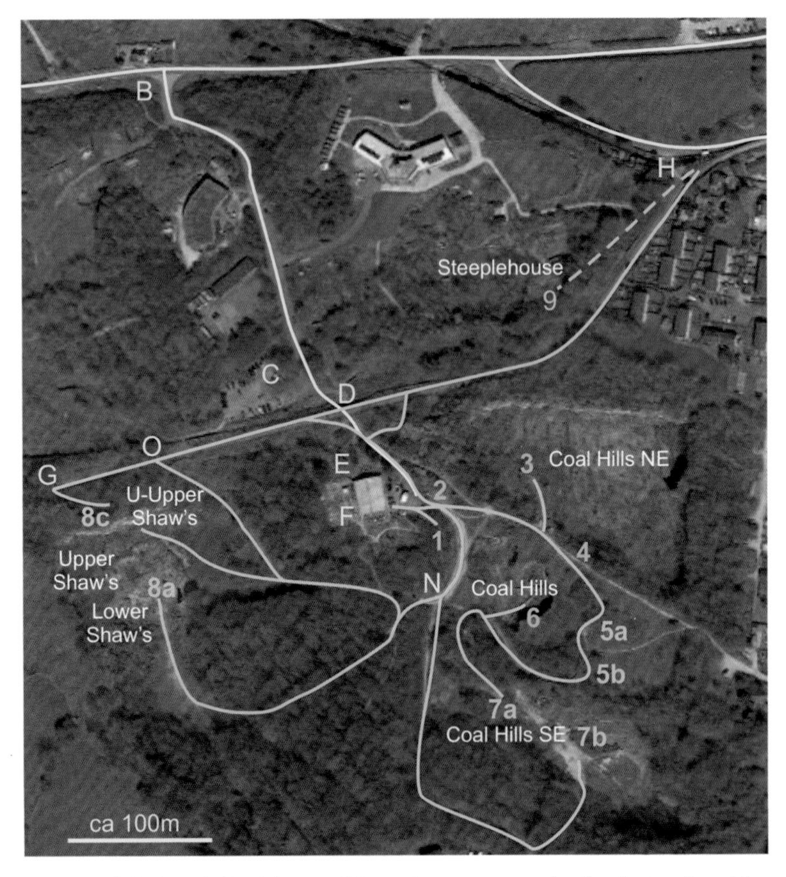

Figure 11.13 Google Earth image of the main quarry area, showing the location of the various quarries and the route of the suggested geo-walk.

The various points of geological interest are linked by criss-crossing paths. First-time visitors might prefer to follow the route suggested here (Fig.11.13) which trends broadly from north to south, from platform facies across the mud mound belt and into the uppermost slope facies, and back again. It begins at the NSC Visitor Centre ***(F)*** *which includes a café, art gallery, souvenir shop and toilets, as well as a small museum area with exhibits explaining the extractive industry, although not directly related to the NSC area. Sculptures by local artist Denis O'Connor are inspired by lead mining shafts in the area. The NSC site also includes the* ***Millennium Wall*** *at* ***Stop 5b****. Various geologically related activities are laid on by the NSC including drystone walling courses and stone carving.*

The total NSC area stretches from the Cromford–Middleton road (B5035) southwards almost to the edge of Wirksworth, although the area of geological interest forms only a part of this. Here, 'NSC site' refers to the area immediately around the Visitor Centre and that covered by its own Geotrail for which a leaflet is available from the Visitor Centre (Fig. 11.13).

The disused quarries of the NSC area expose Monsal Dale and Eyam Limestones and three of its six main quarries (**Coal Hills**, **Coal Hills NE** and **Coal Hills SE**) are covered by the NSC's own Geotrail. The wider area around the site includes the **Shaw's Quarry** complex, and **Steeplehouse Quarry** which is the only quarry not routinely accessible. A sixth quarry, **Coal Hills West** is now very overgrown and therefore not included here, although it could be explored using published information cited here. In addition to the six quarries, the NSC area included some 60 former lead mine shafts, several lime kilns and various smaller outcrops.

The history of the NSC and its surroundings is described by Ian Thomas (2019) who first put forward the idea of a national stone centre. He has been a leading instigator, pioneer and Director of the present facility. The choice of this site emerged from a review in 1984 of some 100 possible sites. The Visitor Centre opened in 1990, and the site is a geological Site of Special Scientific Interest especially on account of the excellent exposures of its mud mound complex. The NSC has recently formed a partnership with the Institute of Quarrying which has co-located on the site. The resulting increase in funding will permit site improvements including a new Visitor Centre with an expanded exhibition area to include local geology. For further information, including permissions for access, activities, events and guided tours, see the website www.nationalstonecentre.org.uk or phone 01629 824833, address: National Stone Centre, Porter Lane, Wirksworth, DE4 4LS.

Thomas's account includes a geological interpretation similar to that presented on NSC's information boards and in its Geotrail guide. It includes a useful aerial view of the NSC quarries labelled with their names, plus the adjacent area to the north. The quarries changed ownership during their working years and consequently their names changed too, resulting in an utter confusion of numerous similar names. As different geological publications have used a variety of names,

Stop number & name This guide	Name NSC Geotrail	Name & locality No. (Walkden, 1970)	Name & locality No. (Walkden & Oakman,1970)	Name Cossey *et al.* (2004)	Selected older names
Stop 1 Viewpoint	Locality 1 Viewpoint	N/A	N/A	N/A	N/A
Stop 2	Locality 2 Rock Shelves	N/A	N/A	N/A	N/A
Stop 3 Coal Hills NE	Locality 3 Lagoon	Locality 70c Coal Hills (North face)	Fig 7 Coal Hills (East) Fig 8 Upper Coalhills	Coal Hills (East)	Coal Hills; Coal Hills NE; M1 North East; Old; Pensend; Penn's End; Steeple Grange Wimpeys
Stop 4	Locality 4 Mineral vein & crinoid beds	N/A	N/A	N/A	N/A
Stop 5a	Locality 5 Top of reef mound (in part)	N/A	N/A	N/A	N/A
Stop 5b Millennium Wall	Locality 5 Top of reef mound (in part)	N/A	N/A	N/A	N/A
Stop 6 Coal Hills	Locality 6 Inside the reef	Locality 70a Coal Hills Complex reef 1	Fig 7 Coal Hills Fig 8 Coalhills Reef	Coal Hills	Cole Hills; Knoll Limestone; Reef; Reef mound
Stop 7a Coal Hills SE viewpoint	Locality 7 Looking into the abyss	N/A	N/A	N/A	Coal Hills SE; Colehill; Cromford Road; Phillips; Shaws; Sugarslide
Stop 7a Coal Hills SE	N/A but see Locality 7 above	Shown but unnamed	Fig 7 Lower Coal Hills Fig 8 Lower Coalhills	Lower Coal Hills	Coal Hills SE; Colehill; Cromford Road; Phillips; Shaws; Sugarslide
Stop 8a Lower Shaw's		Shown but unnamed	Fig 7 Shaw's (lower) Fig 8 Lower Shaw's	Shaw's (lower)	Coal Hills West; Old Lane; West; Shaw
Stop 8b Upper Shaw's	N/A	Locality 70b,Coal Hills Complex, reef 2 (upper)	Fig 7 Shaw's (upper) Fig 8 Upper Shaw's	Shaw's (upper)	Coal Hills West; Old Lane; West; Shaw
Stop 8c Upper-Upper Shaw's	N/A	Shown but unnamed	Shown but unnamed, unnumbered	Shown but unnamed	Coal Hills West; Old Lane; West; Shaw
Stop 9 Steeplehouse	N/A	Locality 71 Steeplehouse	Fig 7 Steeplehouse Fig 8 Steeplehouse	Steeplehouse	Hoptonwood's; Smart's; Steeplehouse; Thistley Close
Coal Hill West	N/A	Shown but unnamed	Fig 7 Coal Hills (west) Fig 8 Ravenstor	Coal Hills (west)	Coal Hills; Coal Hills West; Ravenstor

Figure 11.14 The quarries on the Geotrail with our preferred names in left column and the suggested synonymy of names used in earlier accounts. Stops in red text are additional to those used by the NSC.

a tabulation of suggested synonyms (Fig. 11.14) includes the quarry names used in this Guide and in other publications and should help understanding of the older literature.

This Guide treats the NSC area as a series of Stops which are mostly closely spaced, along easy footpaths starting from the Visitor Centre.

The NSC's Geotrail consists of gravel paths with mostly easy gradients (Fig. 11.13). The site is landscaped around the quarries, and information boards explain the geology with excellent illustrations of the more common fossils on the numbered posts. The boards include reproductions of original paintings by Ian Thomas, showing palaeoenvironmental reconstructions. Copies of some of these paintings are on sale at the Visitor Centre. Guided tours of the Geotrail sites are also available. Additional stops in this Guide are reached by unpaved footpaths that can be muddy and narrow.

Stratigraphical background

Collectively the NSC quarries constitute a 'geologically unique site' (Thomas, 2019) providing 'a spectacular three-dimensional view of the stratigraphical evolution and sedimentology of limestones deposited at the edge of the Derbyshire Platform during Brigantian times' (Cossey *et al.*, 2004), in this case, at the southeast margin. This includes its marginally located mud mounds, which have commonly been referred to as 'apron reefs'. The NSC area allows comparisons with other mud mounds elsewhere on the Derbyshire Platform, particularly along other parts of its margin.

Recent research has raised questions about the more classic view of the 'reefs' of the Derbyshire area, and these quarries are important in that reassessment. These questions are explained along the way through the NSC area, although further research will be necessary before clear answers can be given to some of them. Guide users are encouraged to assess these questions for themselves.

This account depends significantly on the detailed work of several geologists who have each contributed to our understanding. Papers by Cossey *et al.* (2004), Walkden (1970), Walkden & Oakman (1982), Adams (1980), Gutteridge (1991 & 1995) and Cox & Harrison, (1980) are particularly important.

This is an interestingly complicated area geologically, especially regarding facies relationships within the limestones. Fortunately, three-dimensional visualizations (block diagrams) of the NSC exposures show the geology and facies relationships. The pioneer was Walkden (1970) whose original block diagram was updated first by Walkden & Oakman (1982) and then by Cossey *et al.* (2004) (Fig. 11.15). Walkden &

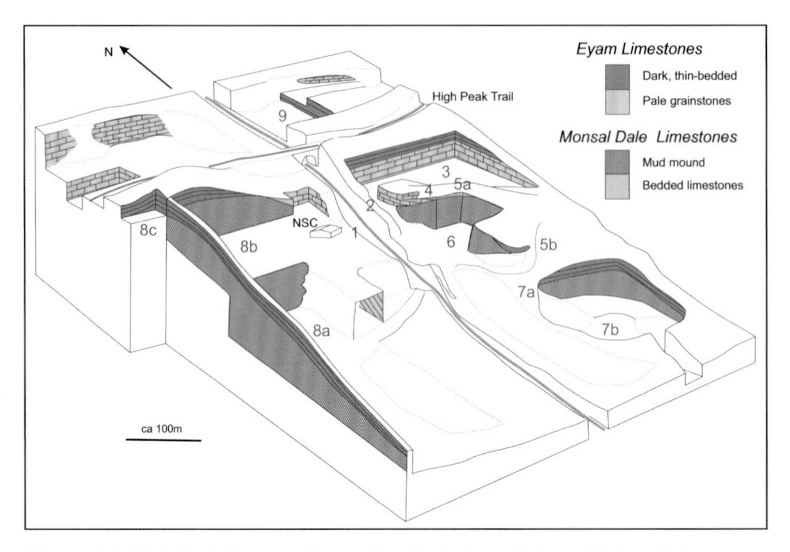

Figure 11.15 A block diagram showing the distribution of Monsal Dale and Eyam Limestones and their facies within the NSC complex. The profile on the left face of the diagram is an imaginary composite section showing the general thickness and bedding changes across the platform margin (adapted from Cossey *et al.*, 2004). A downloadable pdf file is available for this figure.

Oakman also provided two transects through the quarries, giving facies details (Fig. 11.16). Facies and interpretations evolved (not surprisingly) through these different versions and, in general, visitors should refer by default to Figure 11.15.

Walkden (1970) did not include stratigraphic names, but the transects (Fig. 11.16) show almost the entire complex of quarries lying within Eyam Limestones, with only the uppermost Monsal Dale Limestones present at the base of **Cole Hills Quarry SE (Stop 7b)**. BGS mapping also shows the whole NSC quarry area lying on 'Cawdor Limestone [i.e. Eyam Limestones] with knoll-reef'. This interpretation persists on the BGS online interactive map (*Geology Viewer* – https://geologyviewer.bgs.ac.uk).

The present versions of the block diagram and transects (Figs 11.15 & 11.16) show a very different stratigraphy, with all the mud mounds falling within the Monsal Dale Limestones along with the underlying beds. Figure 11.16 is not sufficiently detailed to assess which, if any, beds around the mud mounds also fall within the Monsal Dale Limestones. These issues are discussed at the relevant Stops.

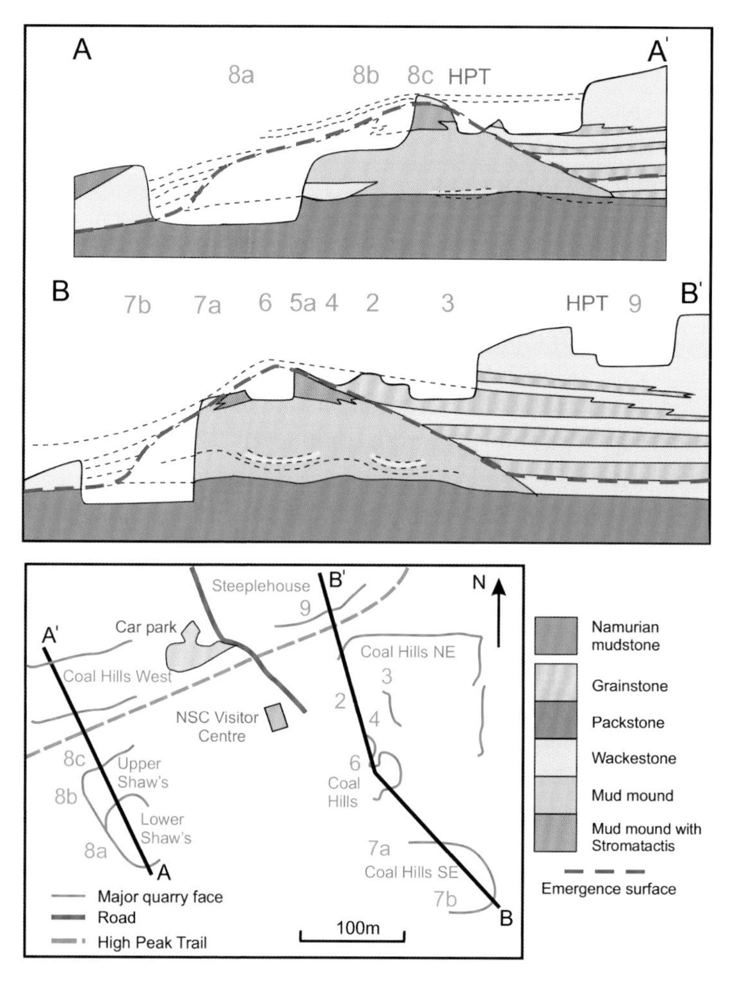

Figure 11.16 Vertical profiles across the mud mounds of the National Stone Centre quarries showing the stratigraphy and facies in the exposed surfaces. The inferred position of the emergence surface, between the Monsal Dale and Eyam Limestones, attempts to apply the Adams-Gutteridge model (Fig. 4.8), but its precise position is not always clear. Further work would be required to locate it fully (adapted from Walkden & Oakman, 1982). Considerable vertical exaggeration.

This stratigraphical revision was explained by Gutteridge, who, following Adams (1980), provided strong evidence (e.g. calcretes) for subaerial emergence of at least the upper 10 m of the older mud mounds, originally mapped in the Eyam Limestones. Such evidence exists for the uppermost layers of the **Coal Hills Quarry (Stop 6)** mud mound and for other mud mounds across the Derbyshire Platform. On this evidence, it was argued that these mud mounds belong not in the Eyam Limestones but at the top of the Monsal Dale Limestones (Fig. 4.8). Their subsequent emergence coincided with a widespread intra-Brigantian fall of sea level.

Reef and mud mound concepts

Prior to the stratigraphic revision mentioned above, the mud mounds and the facies around and above them were regarded as contemporaneous or penecontemporaneous within the Eyam Limestones (Fig. 4.9). The mud mounds have long been regarded as a form of reef, although not obviously similar to any modern coral-algal reef, and older literature generally used terms for off-mound sediments based on modern reef concepts. Inevitably when dealing with the subject of reefs (*sensu lato*), matters relating to 'reef definition' have to be mentioned, where 'reef' embraces almost any kind of buildup including mud mounds. The three most widely accepted criteria for reef recognition are (1) organic framework, (2) relief above the surrounding floor, (3) presence of reef-surface at or close to the water's surface (imagined as a threat to safe navigation). Discussion of these tenets and their application to mud mounds of the Derbyshire Platform is beyond present scope, although the depth of the water in which they occurred continues to be debated.

For a north to south transect from platform across the belt of mud mounds, the sediments have been described as passing from 'lagoonal' and 'back-reef facies', across the mounds and into 'fore-reef facies'. The geology of the NSC site is explained in such a way in its Geotrail leaflet and on its information boards, and with picturesque illustrations which aim to capture the public's imagination. Consistent with this approach, the overall context presented by, for example, Ford (1977), is that the Derbyshire White Peak was an elongated atoll-like structure within which were one or more volcanic islands. It was, thus, supposedly similar to Darwin's (1842) 'barrier reefs' in his well-known scheme for modern coral reefs.

The recent stratigraphical revision calls into question some of this interpretation, since the mud mounds, now in the Monsal Dale Limestones, must have been dead when at least some of the supposedly contemporaneous off-mound sediments were deposited during Eyam Limestones time. Thus, much of the off-mound material must have been deposited around and over dead mud mound relicts (Fig. 4.8).

However, perhaps some sediments were genuinely contemporaneous with the mud mounds in the form of the inter-mound deposits of Gutteridge (1991, 1995). Although these relationships have been worked out for the **Coal Hills Quarry** mud mound (**Stop 6**), further work is needed on other mud mounds and their off-mound sediments around the NSC area to fully elucidate the stratigraphic and facies relationships.

NSC area route

Approaching the NSC Visitor Centre from the car park (**C**) and the HPT, a small Geosteps arena at **E** overlooks the Visitor Centre. This consists of terraced seating whose risers are constructed from different stone types from all over Britain, including Derbyshire, in their order of stratigraphic superimposition. There is also a preserved lead mine shaft near the Visitor Centre entrance.

*The suggested route through the NSC area quarries begins at the **Visitor Centre** (**F**). From the side entrance (Blue Lagoon Café), take a short cul-de-sac (c.30 m to east-southeast) to a small walled area with the viewpoint.*

Stop 1: Viewpoint (53.0934, −1.5729)

This is designated as a viewpoint by the NSC Geotrail, but at the time of writing, vegetation is beginning to obscure the view.

The skyline to the west-northwest is the main ridge of the Marsdenian (R_{2b}) Ashover Grit at Barrel Hill and Bole Hill. To the south, where the skyline dips lowest, the country is underlain by Namurian Bowland Shales and younger Triassic rocks, while the ground to the southeast is predominantly the same as the NSC area, i.e. Dinantian limestones. The closer view shows faces of the **Coal Hills Quarry** (**Stop 6**) mud mound in Monsal Dale Limestones. The gentle southwards slope of the Carboniferous ground away from the viewpoint corresponds, at least in part, to the palaeoslope of the Brigantian limestones at the edge of the Derbyshire platform. An information board explains other details of the view.

*Return towards the Visitor Centre and bear right to a junction of two paths at the edge of the driveway. A small outcrop (**Stop 2**) beside the path, is close to one of the tower sculptures.*

Stop 2 (53.0935, −1.5728)

This is a small outcrop *c.*20 m long, showing 2–3 m of Eyam Limestones in well-bedded units dipping gently into the hillside. The units are pale fossiliferous grainstones, in beds 10–50 cm thick, with crinoidal debris, reworked corals, with a dense layer of gigantoproductid brachiopods mostly in life position (Fig. 11.17). Unlike most of the NSC Geotrail stops, this one can be examined close up.

These beds, including their equivalents at **Coal Hills NE Quarry (Stop 3),** have often been interpreted as 'lagoonal' and the NSC Geotrail information does so too. This is probably because they were laid down on top of the Derbyshire Platform, sometimes perceived as an atoll-like structure, as mentioned above, where these deposits were thought of as 'back-reef', lying 'behind' (i.e. on the platform side of) a contemporaneous rim of the nearby 'reef' mounds as seen at **Stops 6 (Coal Hills Quarry) and 8b (Upper Shaw's Quarry)**. However, lagoonal deposits more typically record calm shallow-water conditions, which is clearly not the case.

Aside from the *in-situ* gigantoproductid facies, these limestones are now better interpreted as having been deposited as carbonate sand bodies in high-energy shoals at the edge of the platform. Crinoidal grainstone shoals likely resulted from recurrent storms regularly breaking up

Figure 11.17 Coarse grainstone with abundant gigantoproductid shells, mainly in a concave-upwards orientation, their *in-situ* life position. Eyam LImestones at Stop 2.

in-place crinoid meadows. Crinoid stem orientations in the uppermost levels of **Cole Hills NE Quarry (Stop 3)** lie in a dominantly north–south direction (Walkden, 1970), and 'southerly-directed foresets' are reported in the same deposits. This is approximately normal to the contemporaneous platform margin indicating off-platform, downslope transport. The predominance of convex-down preservation of giganto-productid shells suggests they grew during interludes of lower energy and/or slightly deeper water (cf. Le Yao *et al.*, 2016; Nolan *et al.*, 2019).

However, as the mud mounds are now thought to belong to the older Monsal Dale Limestones, these crinoidal grainstones must post-date the mud mounds, and their palaeoenvironmental conditions cannot be compared with those of 'back-reef' settings.

Descend from the lower end of the **Stop 2** *face along the path which lies closest to it, keeping to the left. The path curves slightly to the right between two recently built drystone walls. After c.50 m, take the path branching sharply to the left to reach a viewing point into* **Coal Hills Quarry NE (Stop 3)**. *There is another viewing point just where this path branches off.*

Stop 3: Coal Hills Quarry NE (53.0937, –1.5721)

This is the largest quarry in the NSC area, the east face opposite the viewing point being *c.*130 m long. The face to the right is at right angles to it and is a few metres shorter. The main face is *c.*12–14 m high, less at the southern end, and exposes Eyam Limestones. The faces are not accessible without prior arrangement.

Four facies are represented, the darker ones being clearly distinguishable in good light. The succession here is as below:

5. *c.* 4 m of crinoidal grainstones, as below, but becoming more noticeably thinner bedded.

4. *c.*1 m of fine, dark impure limestones, containing fish debris including dermal denticles of *Petrodus* (a shark relative), not visible from the viewpoint of course, but see **Stop 9**.

3. *c.*3 m of crinoidal grainstones.

2. *c.*1 m of fine, dark, impure limestones containing a light-coloured angular grainstone. The bedding from the

viewpoint is chaotic in places and consists of channel-like lenses (Fig. 11.18). These contain randomly orientated shelly debris.

1. 3–4 m of pale grainstones variously crinoidal with productid layers, as in **Stop 2**, and interpreted in the same way.

*Return along this side path and turn left by the other viewing point for **Stop 3** to re-join the main path. In c.10 m take a right fork and short flight of steps into a little hollow with seating and trees (**Stop 4**).*

Figure 11.18 Rather disturbed bedding in calcarenites of the Eyam Limestones, possibly the result of slumping or channeling near the platform margin. Stop 3; Coal Hills NE.

Stop 4 (53.0931, –1.571570)

This outcrop is located in a woody hollow on the lowermost slope of a small knoll which overlies the mud mound at Stop 6. It is mostly accessible, although a fence prevents visitors from accessing the higher levels of the knoll surface. The rock surfaces are heavily blackened. There is a low rock wall *c.*10 m long and 2 m high on the right as you enter the hollow. This is what remains of a lead vein working in Eyam Limestones, and the face still shows some mineralization, mainly galena and barite.

With your back to the rock wall, look at the convex sloping rocky bank a few metres high in Eyam Limestones, flanking a hillock underlain by the mud mound and associated beds of **Stop 6 (Coal Hills Quarry).**

This is partly fenced off, but it can be inspected close up where it comes down to meet the floor of the hollow. It shows numerous crinoid stems (columnals). Crinoids are seldom preserved in this way, being more usually broken up into their smaller constituent disc-like plates (ossicles) usually from their arms, and generally into shorter lengths. This state of preservation suggests calmer conditions than for the finer crinoidal grainstones at **Stops 2 and 3.**

The sloping surface is the uppermost bedding plane of a series of crinoidal grainstones with brachiopods, overlying the mud mound of **Coal Hills Quarry (Stop 6)**. Such crinoidal deposits in the vicinity of mud mounds on the Derbyshire Platform and elsewhere have been interpreted as flanking deposits contemporaneous with the adjacent mud mounds. In some cases, they have been thought to contribute to mound building by trapping sediment (cf. bafflestone). However, while crinoidal beds elsewhere are integral with mud mounds, there is strong evidence that, here, they are post-mound sediments deposited on and around relict mud mounds following prolonged mid-Brigantian emergence (see **Stop 6**). In fact, the crinoidal beds here at **Stop 4** pass into the crinoidal grainstones seen at **Stops 2 and 3**. The mud mound below them, seen at **Stop 6,** is now thought to lie within the Monsal Dale Limestones, necessitating a period of emergence between them and the grainstones.

With the crinoidal bed on your right, take the path out of the hollow and follow it to the right for about 30 m to reach another viewing point, ***Stop 5a,*** *surrounded by a circular stone wall.*

Stop 5a: Viewing point (53.0929, –1.5716)

The NSC Geotrail treats this site and the next one, **Stop 5b**, together as its Locality 5, but they are not on the same spot and are not related geologically. This viewpoint is located on top of the same hillock as **Stop 4** and is similarly underlain by the mud mound and associated beds of **Stop 6** in the Monsal Dale Limestones. The visible surface is continuous with that at **Stop 4**, but here it is almost entirely overgrown and is completely fenced off. The bedrock, as at **Stop 4**, consists of crinoidal facies with brachiopods in Eyam Limestones. However, the present-day topography here more or less follows that of the mud mound and conveys well its scale and form. In the distance to the north, the northern face of **Coal Hills Quarry NE (Stop 3)** is visible. The beds underfoot here are at the same stratigraphical level as the lower ones in that quarry.

Leave this viewpoint by continuing SW-wards to join another path after c.15 m. Alongside this path is a series of short lengths of drystone walling (the ***Millennium Wall****) which start a little way back along this other path, and continue past* ***Coal Hills Quarry (Stop 6)*** *to the right then curving round through 180° towards* ***Stop 7a****.*

Stop 5b: Millennium Wall (53.0928, –1.5714)

Built by members of the Dry Stone Walling Association during the May Day bank holiday in 2000, the Millennium Wall consists of separate segments of walling, each a sample of the stone and building style used in different parts of Britain and each with an information board. The wall continues for about 110 m alongside the path. However, apart from the fact that some segments are built of local Derbyshire Carboniferous stone, including limestone, it does not directly illustrate the local geology, although it does demonstrate the long-standing value of these building stones. As in Derbyshire, drystone walls are a key feature of the human influence on the landscapes especially of much the Pennines.

Follow the path and Millennium Wall downhill from where you joined it, to the same level as the floor of ***Coal Hills Quarry (Stop 6)*** *on your right next to the drystone wall segment built from Milnrow Sandstone from Cheshire and into the quarry (total distance c.100 m).*

Stop 6: Coal Hills Quarry (53.0930, –1.5720)

The faces are not closely accessible without prior arrangement. At the time of writing, the initial fencing barrier is being replaced by drystone walling.

This is the most commonly cited, almost celebrated, highlight of the NSC geology and shows particularly well the relationship of the mud-mound structure to overlying and adjacent beds. The disused quarry faces lie along two irregular embayments, together *c.*60 m across, with a maximum height of *c.*8 m (Fig. 11.19).

The larger face on the right (south) mostly exposes mud mound facies, except at the highest level. This facies consists of largely unbedded, fine-grained carbonate mudstone with irregular depositional cavities

Figure 11.19 Mud mound in Monsal Dale Limestones draped by flanking beds of Eyam Limestone on top and to the left. The irregular top of the mound may record palaeo-karstic dissolution. Coal Hills Quarry, Stop 6.

lined by marine cement and pocket-like accumulations of reworked brachiopods and bivalves (Fig. 4.8). The same facies can also be seen less prominently in the lower part of the face to the left (north). Gutteridge (1995) classes this as a 'vertically accreted mound' and the exposed deposits are assigned to his 'mound-core' facies. If there are true (i.e. internal) inclined 'mound flank' deposits (*sensu* Gutteridge), they are not obvious. There may be at least two mud mounds in this quarry.

Overlying the mud mound facies are Eyam Limestones in 30–50 cm beds. These are crinoidal grainstones with brachiopods, as described at earlier stops. The uppermost beds are continuous with the outcrops at **Stop 4** and the overgrown surface at **Stop 5a**. They correspond to the beds at **Stop 2** and to the lower beds of **Stop 3**. The bedding is less obvious at the top of the main mound on the right where the sediments seem to fill the uneven upper surface rather randomly, but towards the left the beds become more regular, including the very conspicuous overhang.

The contact of the Eyam Limestones on the mud mound facies is irregular and undulating and corresponds with a surface that is marked up on a photo of the quarry face on the NSC information board. Although not visible from the quarry floor, an impersistent calcrete marks much of the contact. This may correspond to the brown veneer that can be seen on a prominent fallen block of the mud mound facies. This can be accessed at the foot of the right-hand quarry face, as it is now

Figure 11.20 Thin beds of Eyam Limestones dipping to the left off the draped margin of the mud mound in the Monsal Dale Limestones. Coal Hills Quarry; Stop 6.

incorporated into the protective stone wall. An alternative possibility is that this apparent calcrete has a more recent origin (?Quaternary). The calcrete, if Brigantian in age, and other related evidence indicate a subaerial break in sedimentation of the mound, favouring the mound belonging to the uppermost Monsal Dale Limestones.

The crinoidal grainstones overlying the mud mound are clearly banked against it, with successive beds onlapping the mound surface. The facies is similar to that seen at **Stops 2, 3** and **4** in the lower part of the Eyam Limestones. The beds encroach upwards and laterally on the sides of the mound and the highest beds drape right over the irregularities of the entire structure. They dip away from the mound and have therefore been regarded by some as contemporaneous 'flank beds'. However, they do not continue into the mound-core facies and are therefore not contemporaneous. Following the subaerial emergence and death of the mound, it was re-submerged in Eyam Limestones time, and initially must have formed a relict feature, elevated some 10 m above the sea floor.

The irregular upper surface of the mound proper suggests its three-dimensional shape might be broadly concave, although this might be partly a trick of perspective. If truly concave, this might suggest that during emergence it was eroded more in the middle, which would fit with Purdy's (1974) model of karstic erosion of carbonate edifices.

Alternatively, was this concavity the final surface shape of the mud mound prior to its intra-Brigantian emergence? Flattish or concave top surfaces of mud mounds generally develop by aggradation to a water surface (Gutteridge, 1995). In comparison, modern Bermuda 'boilers' (algal–vermetid reefs) are aggraded to sea level and have flat to concave tops, and within Indo-Pacific reef complexes, small atoll-shaped structures less than 10 m across, called 'faros', are similar in scale to this Coal Hills mound. Perhaps the uppermost surface of this Coal Hills mud mound resulted from a combination of such erosional and constructional factors.

An unusual feature on the right (southeast) side of the quarry is a large cave-like cavity within the mound. There have been various interpretations of its age and history. It is lined by brown flowstone that has traces of classic Derbyshire mineralization and, together with other evidence (Cossey *et al.*, 2004), this suggests that the cave formed subaerially (vadose zone) when the mound had emerged post-depositionally, and was not a submarine cave formed penecontemporaneously with the mound. While the NSC's Geotrail information gives the latter explanation, it is extremely difficult to explain cave formation in the course of submarine carbonate deposition, as they are in effect contradictory processes. Nearly all supposed 'modern' examples turn out to have formed above sea level in the vadose zone or in the uppermost (phreatic) zone during Quaternary lower sea-level stands.

Return to the path by the ***Millennium Wall*** *and continue to the west. The path and Wall then curve back on themselves through c.180°. Proceed SE-wards to reach another viewing point (****Stop 7a****), total distance c.60 m. This overlooks a deep, narrow quarry.*

Stop 7a: Coal Hills SE viewpoint (53.0925, –1.5723)

This is 'Locality 7' of the NSC Geotrail, as the latter omits a visit to the quarry below. For this Guide, it is designated **Stop 7a** and the quarry proper, below the viewpoint, is **Stop 7b**.

This quarry mostly exposes a large mound complex now taken to be within the Monsal Dale Limestones. The view is mostly of a large face on the left, *c.*85 m long and up to *c.*35 m high, trending approximately NW–SE. The face lacks obvious bedding or other structure, except at the very top where at least 4 m of thin and well-bedded limestones can be made out. The unbedded mass is recognizably part of a large mud mound, which wedges steeply downward south-eastwards away from the viewpoint. Since the regional slope at the edge of the Derbyshire Platform descends from the Wirksworth area in a broadly southerly direction, this mound might either be a downslope continuation of the mound in **Coal Hills Quarry** (**Stop 6**) or a separate one, located vertically en-echelon slightly downslope from it. The thinner far end therefore appears to be the downslope toe of the mound.

The NSC information board suggests that the well-bedded strata on top of the mound are high-energy deposits eroded penecontemporaneously from the mound by storms and strong wave action. However, elsewhere, the upper surfaces of mud mounds like that at **Stop 6** are marked by an emergence surface, and it seems likely that this is also the case here. Beds on top of the mound are therefore most likely Eyam Limestones grainstones.

The information board points out a thin bed of volcanic ash *c.*2 m above the quarry floor. For this and other details, see **Stop 7b**.

*There is no direct path to reach the quarry seen from **Stop 7a**. It is necessary to retrace the path past the **Millennium Wall** and **Stop 6 (Coal Hills Quarry)**, and then bear WNW for c.25 m to reach a hub of footpaths at **N (GPS 53.0930, −1.5730)**, marked by a turret sculpture. Turn sharply to the left onto a path running southwards (towards Wirksworth) and follow this for c.100 m through a densely wooded area. Follow it round to the left, initially ESE but then curving round to the right, for another c.75 m to where a small industrial estate is seen on the right through the trees. At this point, turn sharply left on a short branch path to arrive at the floor **of Coal Hills SE Quarry (Stop 7b)**.*

Stop 7b: Coal Hills SE Quarry (53.0923, −1.5716)

This quarry is publicly accessible, but high, nearly vertical faces make close examination of the higher levels difficult, except by the climbers who frequent it. However, additional to the information given under **Stop 7a**, this stop allows examination of the lowermost faces including those around the entrance area where thick, densely crinoidal grainstones can be seen. Upper facies can be examined with binoculars from the viewing point and the quarry floor. This is arguably the most difficult of the NSC quarries to explain, and probably requires further detailed research.

The most striking feature of the lowest part of the main face is a very even thin (*c*.2 cm) bed of clay, formerly volcanic ash, *c*.3 m above the quarry floor. This is pointed out on the information board at **Stop 7a**, which suggests that it might have originated from known vents at Ible or Bonsall 3–4 km to the north. However, the known products of these vents are near the base of the Monsal Dale Limestones whilst this clay lies nearer their top. The band seems to correspond to a clear discontinuity at about the same elevation around the entrance to the quarry. Below this ash level, the beds around the quarry entrance are very irregular (?slumped) although not on the corresponding face on the opposite side of the quarry, where the limestones seem better bedded, but not very fossiliferous. It is not clear whether these beds on each side of the quarry belong to the same stratigraphical level. Neither this nor the ash are documented in earlier publications. The disturbed beds around the entrance seem to lie in a downslope position from the mud mound, which might account for the slumping.

Above this break, on the left side of the quarry entrance, are densely crinoidal grainstones in a bed or lens at least 2 m thick. This face continues upwards and along into the main part of the quarry in rather blocky thick units. The corresponding horizon on the main face is also very fossiliferous in places with a lens(?) of orientated crinoidal fragments. These lower levels of the quarry face seem not to be part of the mound but to lie below it.

It is not clear whether these crinoidal facies on the western faces of the quarry are an integral part of the mud mound, bound by microbial action, or are shoals in the Monsal Dale Limestones. Gutteridge (1995, his fig.3) indicates that crinoidal grainstones generally underlie laterally accreted mounds. But it is also not clear whether, and how, such pre-mound crinoidal grainstones differ from those that occur within mounds' internal flank areas (both Monsal Dale Limestones), or from the post-mound crinoidal grainstones (Eyam Limestones). All of them locally show current orientation. More than one of these possibilities seems to apply within this quarry but further study is needed. There are similarities between this quarry and Shaw's Quarry, and related questions for **Stops 8a, 8b, 8c** are discussed further there.

Above this facies, and becoming increasingly thick towards the northern end of the quarry, is a pale grey, dense limestone with numerous brachiopod and crinoid fragments. According to Walkden & Oakman (1982) the mound proper begins about halfway up the main face, but in practice it is difficult to identify the contact as the upper part of the quarry face is mostly heavily blackened. Likely blocks of it on the floor of the quarry are very different from the densely crinoidal facies. It is poorly bedded and blocks are very pale grey, dense and fossiliferous with crinoid fragments and brachiopods. From the quarry floor, the upper surface of the mound seems to be highly irregular, perhaps erosive, as at **Stop 6.** Its hollows seem to be filled by relatively unbedded rock that becomes better bedded at the top of the face. It is not clear whether the unbedded material is a further phase of mud mound growth or whether, as at **Stop 6,** it belongs to the Eyam Limestones. The well-bedded strata above are described as dark and brashy calcarenites by Walkden (1970) who referred to these beds as 'forereef' even though they lie above rather than to the side of the mud mound. More likely the mound belongs within the Monsal Dale Limestones, and these crinoidal beds are within the Eyam Limestones that followed flooding of the emergent mound.

Stop 8: Shaw's Quarry

Stops 8a, b and c *are all in a single mound complex within* ***Shaw's Quarry****, but have to be reached separately by rather circuitous routes, returning in each case to the footpath hub* **N** *(Figs 11.12; 11.13). They can therefore be visited in any order. Starting with* **Stop 8a** *(***Lower Shaw's Quarry***), retrace the route from* **Stop 7b** *towards the hub* **N***. Turn sharp left to curve to the southwest for 30 m, and then take the right fork back through the woodland. The path runs more or less straight for 90 m before curving to the right to arrive at the entrance to* **Stop 8a** *after c.35 m. The main quarry face is a further 85 m ahead with flanking faces on each side. The one on the left (west) is longer and clearer of vegetation than that on the right. Lower Shaw's Quarry is separated from its upper counterpart by a substantially high (c.16 m) main face and upper ledge. Although some of this main face looks as if it can be examined by climbing the spoil, it is unstable and there are warning signs about falling rocks.*

The lower two quarries (**Stops 8a, 8b**), combined, show the most striking of the three mud mound complexes in the NSC area, this being the largest. They also show a much thicker cover and more extensive wedged flank of beds above and around the mound. However, unlike at Coal Hills, the onlapping pattern of the post-mound beds here at Shaw's is less clear. It seems more stepped than wedge-like. However, while the exact nature of the contact is not clear and not easily accessible, these beds do not seem to pass from the flank directly into the mound. This and the fact that the highest beds pass right over the mound seems to rule out the possibility of them being intra-mound flank beds, (i.e. the 'mound-flank deposits' of Gutteridge (1995)), but must be post-mound Eyam Limestones.

Cox & Harrison (1980 p.104) give a detailed section for the combined upper and lower quarries. In summary there are 3.9 m of Monsal Dale Limestones, overlain by 23.1 m of 'apron-reef facies' (taken here to be mud mound facies), followed by 2.1 m of Eyam Limestones. They assign the mud mound ('apron-reef') to the Eyam Limestones in accordance with contemporary BGS classification. In this Guide it is placed in the uppermost Monsal Dale Limestones, following the Adams-Gutteridge Model (Fig. 4.8).

Also of interest is the very marked irregularity of the Eyam Limestone cover beds. These show strong channel-like structures, 5 m or more in scale, suggesting either very dynamic sedimentation conditions (?large scale cross-bedding/? channelling) or instability on a palaeoslope giving rise to slumping and channelling. The latter might be due directly to the influence of the mound, if and when it was an upstanding relict during Eyam Limestone times and/or of the position of these deposits downslope from the platform edge, or some combined effect of both. The downslope possibility would fit with the general picture of the Brigantian around Wirksworth, recording a transition from a platform to a ramp-like setting, with overall onlap of deeper-water conditions during this time. The thickness and dips of bedding suggest prograding of Eyam Limestones cover beds down the palaeoslope.

Stop 8a: Lower Shaw's Quarry (53.0930, −1.5754)

If the small thickness of bedded Monsal Dale Limestone recorded by Cox & Harrison was present at the foot of this face, it is now buried by spoil but some can be seen locally in the lowest part of the face. The main part of the face is unbedded and rather chaotic-looking. Although freshly broken surfaces are pale grey to white, most of the faces are blackened, perhaps dating from when steam-powered plant and locomotives were used in the quarry and nearby lime kilns were in action.

The mound here is a crinoid- and brachiopod-rich boundstone that continues round to the lefthand (western) face where its upper surface is seen. It descends from **Stop 8b: Upper Shaw's Quarry** to reach the lower quarry along its western face. Its upper surface is very irregular and uneven and further descends steeply towards the quarry floor near the entrance area. Above it is a substantial interval of bedded facies, which becomes thicker towards the quarry entrance than is recorded by Cox & Harrison. The constituent beds thin in the same direction and are extremely disturbed in places, perhaps through thrust-faulting, slumping, or perhaps as large-scale cross-bedding. This may reflect instability on the flanks of the mud mound and/or on the overall platform palaeoslope (see example in **Dale Quarry**). The Adams-Gutteridge Model requires that this Shaw's mud mound must be assigned to the Monsal Dale Limestones whilst the bedded facies must be Eyam Limestones. Presumably there is an emergence surface at the top of the mud mound, although this needs further investigation. The Eyam Limestones can be examined at Stop 8c.

To visit ***Stop 8b: Upper Shaw's Quarry****, return to the footpath hub at* ***N****. Bear left immediately before the tower sculpture to join an E–W footpath in a few metres. A more or less straight path through a wooded area leads onto the main ledge of Shaw's Quarries in c.90 m. Note that this ledge lies immediately above the high main face of* ***Lower Shaw's Quarry (Site 8a).*** *Warning signs and large stone blocks indicate the potential danger. However, there is no fencing, so keep well to the back of the ledge and pass behind the low ridge of overgrown spoil to reach the main face.*

Stop 8b: Upper Shaw's Quarry (53.0932, –1.5757)

This part of Shaw's Quarry is an upward continuation of the section exposed at **Stop 8a**. However, Figure 11.16 suggests that the mud mound changes upward from being a crinoidal and brachiopod boundstone to a stromatactoid boundstone, which has been interpreted as a 'reef crest' facies, as has also been suggested for most of the mound at **Stop 6**. This implies some kind of ecological succession as the mound grew upwards, presumably into shallower water more influenced by waves and currents.

Figure 11.21 Mud mounds in Monsal Dale Limestones, overlain by thinner bedded Eyam Limestones that appear to abut the side of the mound irregularly and also pass over the mound summit. The Eyam Limestones are irregularly bedded with apparent channelling and evidence of sediment instability, particularly to the left. Upper Shaw's Quarry, Stop 8b.

The overlying Eyam Limestones seen in the lower part of the quarry at **Stop 8a** can also be seen here, starting with just a few metres above the mound and becoming thicker in total to the west (Fig. 11.21). This face merges with the corresponding face of the lower quarry where these beds become more disturbed and the overall interval thicker whilst individual beds become thinner. Closer examination of these beds is possible at the uppermost levels of this quarry (i.e. **Stop 8c: Upper-Upper Shaw's Quarry**).

*To reach **Stop 8c** one must leave the main NSC area to join the High Peak Trail (HPT). The shorter route is to return eastwards along the main quarry ledge to reach a lightly wooded area, where a rough footpath branches left (north) from the main path, skirting the wooded area. The path gradually swings westward, climbing to meet the HPT at point **O (53.0936, −1.5756)** in about 100 m.*

*This path is rather uneven and irregular. An easier but longer route to **O** involves returning to navigation point **N** and then to the **Visitor Centre (F)**. Pass under the HPT bridge (**D**) to enter the car park (**C**). Keep to the left (south) side of the car park alongside the HPT to take a short path that rises from the car park to join the HTP through a gate. The total length from the hub at **N** is c.270 m. From here it is another 20 m to where the shorter route joins the HPT at **O**. From **O**, walk westwards along the HPT for c.100 m, passing a large upstanding outcrop of Eyam Limestones on the south side,* *and an information board about the trail, to reach a gap in the low wall on the left (**G: 53.0936, –1.5769**). Enter the rough, overgrown area to access a series of small quarry faces (**Stop 8c**) which are an upward and lateral continuation of the upper faces of **Stop 8b**, but well fenced off from it.*

Stop 8c: Upper-Upper Shaw's Quarry (53.0933, –1.5762)

This little quarry gives an opportunity to examine close up the beds that are seen at the top of the faces at **Stops 8a and 8b**.

The quite steeply dipping beds here are coarse crinoidal grainstones in Eyam Limestones, with large-scale cross-bedding, sometimes graded. They overlie the core facies of the large mud mound below seen in the main faces of the other parts of Shaw's Quarry and continue along the west sides of the quarries below. The beds dip away from the mound face in a southward direction from the HPT. These beds were evidently laid down over the upstanding mound, perhaps as a relict feature, similar to the overlying Eyam Limestones at **Stop 6 (Coals Hills Quarry)**. They are at least 10 m thick in relation to the top of the mound, so the relict mud mound must have had a diminishing influence during the deposition of these beds. These beds represent a crinoidal sand body, typical of an Eyam Limestone 'facies mosaic' over this part of the platform, characterized by numerous carbonate sand bodies recording southward off-platform transport between and around the mud mounds (Gutteridge, pers. comm.).

*For **Stop 9 (Steeplehouse Quarry)** return to the High Peak Trail and walk directly along it for about 500 m, passing **O** and then the Visitor Centre on the right, to reach a gated entrance on the left that leads to the terminus yard and station area of the **Steeple Grange Light Railway (SGLR) (H)***

Stop 9: Steeplehouse Quarry (53.0948, –1.5721) (with remarks on Middleton Quarry)

*The terminus area of the SGLR is accessible to the public at all times from both **H** and from Old Porter Lane on its north side. From here one can walk along the line when it is not running to see the small **Dark Lane Quarry** and cutting faces in Eyam Limestones: dark mudstones and wackestones. Although popular with local walkers, this is not a formal right of way, and the SGLR store equipment along some of it. Please do not interfere with this or hammer the faces (rock falls have been quite common, and CCTV cameras monitor the equipment areas).*

For Steeplehouse Quarry proper, the only officially correct route into the quarry is to follow alongside the short SGLR branch from the entrance close to the main SGLR terminus area. However, Mount Cook Adventure Centre have taken over the quarry for young people's outdoor activities and placed a gate across the entrance and this is padlocked when the line is not running. The SGLR runs mostly at weekends from late spring to early autumn and in holiday seasons. Those who do the train ride are invited to see the fossils and will be given a brief explanation of them. At all other times, Mount Cook are obliged to provide someone to accompany visitors to safeguard any children who are using their quarry facilities. For this reason, visitors must contact Mount Cook on 01629 8232702 Option 2 or email them (explore@mountcook.uk), ideally in advance. Visitors entering the quarry without permission could be challenged by any Mount Cook staff present, especially if children are present. For most visitors, geological interest focuses on the slabs by the end of the line. Beyond there, there is a sizeable quarry face drop, making further progress into the quarry risky.

The Steeple Grange Light Railway (SGLR: https://www.sglr.co.uk/) is a volunteer-operated narrow gauge, industrial-style line, with the unusual gauge of 18 inches built on the track bed of the former standard gauge Killers' Branch which branched off the Cromford and High Peak Railway (C&HPR) mineral line, now the High Peak Trail. The line was built by the Killer Brothers in 1884 to serve **Middleton Quarry (53.0980, -1.5880)** (and later Middleton Mine) which they then owned. The SGLR operation was started on Killers' disused track bed in 1985. Access to Middleton Quarry may be possible with advance permission from Hopite (https://hopite.co.uk/contact-us/) who run a retail business from the quarry, and/or from the quarry owners, Tarmac. Apart from brief comments here, the quarry is not included as a locality in this Guide. A visit (subject to permission) might be combined with a ride on the SGLR on running days as the line ends in the lower part of Middleton-by-Wirksworth village on the opposite side of the B5023 to the quarry entrance, and adjacent to the Derbyshire Wildlife Trust building, where copies of Thomas (2019) can currently be obtained.

The main Middleton Quarry face shows massively bedded, cyclothemic Hopton Wood facies of the Bee Low Limestones as also seen in the lower part of **Middle Peak Quarry (Wirksworth Geowalk; Stop 4b)** but with at least eight cyclothems. There is a large gated entrance within the quarry into the mine workings, now heavily protected by security devices. Around 40 km of mine gallery workings underlie Middleton Moor to the west and connect with other main entrances in Hopton Wood Quarry on the opposite side of the Moor, 1.5 km away. The location of the Middleton Mine entrances and the network of mining galleries are shown by Thomas (2008, fig.1). Further information on the geology of Middleton Quarry/Mine and its workings is found in Walkden & Oakman (1982) and Thomas (2019).

The SGLR offers a 25-minute round trip to Middleton-by-Wirksworth passing through a series of cuttings in dark Eyam Limestones. In addition, the trip also includes a much shorter ride on a branch which begins at the terminus and runs for around 100 m parallel to the HPT into **Steeplehouse Quarry**. Some SGLR volunteers like to provide basic information about the geology and, in particular, they point out that the bedding planes of some of the large slabs around the quarry terminus show numerous dermal denticles of the shark-relative *Petrodus*. Additionally, Walkden & Oakman (1982) report remains of other fossil fish species that are almost certainly reworked.

An example of the memorial (and decorative) use for which Hopton Wood Stone is famous can be seen by the gate into the SGLR terminus, where there is a replica WWI headstone, like those used for war graves of former battle fronts, along with an explanation. There is further information by The Green **(53.101511, –1.589460)** at the upper end of the Middleton village.

Baileycroft Quarry, Wirksworth (SK 287542; 53.0846, –1.5731) (►)

*The quarry lies 200–300 m north of Wirksworth town centre, on Harrison Drive, (B5036) with exposures on both sides of the road. Harrison Drive was built in 1960 through the original quarry, dividing it into two main 'sub-quarries': (1) a larger western part, much of which lies behind a Co-op supermarket, here termed the 'Co-op Faces' and (2) a smaller eastern part, most of which runs alongside the Wirksworth Cars plot, here termed the 'Wirksworth Cars Faces'. The Co-op Faces extend to the north into the Lime Kiln Faces behind the Lime Kiln pub (**Stop 10** of the **Wirksworth Geowalk**). Two smaller faces extend from the Wirksworth Cars Faces; one, the Pavement*

Face, runs southwards along the public pavement for about 20 m. The other, the North End Face, is located in a small parking lot in the apex between Harrison Drive and North End. The quarry complex is approximately oval shaped with a long axis of about 190 m. The maximum height of its almost entirely vertical faces is c.20 m at the northern end of the western faces, decreasing to less than 10 m at the southern end. On the eastern side, the Pavement Face is 5–6 m maximum height, and the Wirksworth Cars Faces are a little higher and c.85 m long. The Co-op supermarket car park and Wirksworth Cars parking lot can provide almost continuous access along the quarry faces although they are now becoming overgrown in places. **Recent (2024) rockfalls on the Co-op faces has led to close access being denied by fencing. If and when remediation work eventually allows closer access, the following account should still be applicable. The Lime Kiln Faces are currently the most accessible.**

Buses stop right by the quarry. The supermarket, petrol station, fire station and other buildings occupy the western side of the quarry. The supermarket car park is intended for Co-op customers only and is limited to a strictly enforced 1 hour. Geological visitors should check with supermarket staff, especially if more time is needed. Alternatively, there is charged parking in and around the town centre, 5–10 mins walk away. Access to the western faces from the supermarket car park allows the Bee Low Limestones to be easily examined close up, with younger strata approaching ground level at the southern end. The Lime Kiln Faces can be seen at a distance from the grounds behind the pub when it is open. When crossing Harrison Drive note that it often carries fast traffic, and there is no pedestrian crossing.

The Wirksworth Cars Faces lie behind the parked cars in the 'showroom' and permission should be obtained from the office when open. After hours, those entering without permission will set off a loud pre-recorded warning from a PA system controlled by CCTV. To avoid this, use the phone contact on the company's website (https://www.wirksworthcars.com) to obtain permission. The company people are constructively aware of, and cooperative about, the geological importance of the locality. The disordered and unstable upper strata can cause potential rock-falls, so the uppermost parts of the faces are secured by netting, which somewhat obscures them. Take care not to damage the sale cars' bodywork when moving around the area – and no hammering please! The northward and southward pavement extensions of the eastern faces are public and easy to examine.

The Co-op Faces were amongst the best limestone localities for disabled access. The quarry floor is almost entirely level and hard-surfaced, and parking is possible close up. However, at the time of writing, there have been recent rock falls with temporary barriers preventing access, and the car park can be busy. For the Wirksworth Cars Faces, wheelchair access may sometimes be limited by parked cars. Food and other provisions are available from the supermarket. Car fuel and charging, and an ATM, are available in the forecourt. Public toilets, shops, restaurants and pubs are to be found in the town centre.

This quarry, together with the **Wirksworth Geowalk** and **Dale Quarry**, makes a major contribution to understanding the complex development of the south-eastern margin of the limestone platform in Asbian and Brigantian times. It is an important locality whose complexity is the main point of interest. Western quarry faces show Hopton Wood facies of Bee Low Limestones (Asbian) overlain disconformably by Monsal Dale Limestones (earlier Brigantian). These are, in turn, overlain by Eyam Limestones (later Brigantian) on an irregularly undulating unconformity which progressively cuts down through older strata to the south (down dip). Some of the faces are mineralized (e.g. dog-tooth spar calcite and galena). The Wirksworth Cars faces look surprisingly different, being more complex. They show strikingly irregular and slumped strata interpreted as several episodes of channel fill deposits, making it difficult to correlate the stratigraphical relationships between the two sets of faces. They look so different that it is almost as if there were a fault between them, although none is shown on published maps. Whilst unconformities and channel deposits provide the main interest, this is also a good locality to examine the Hopton Wood facies of the Bee Low Limestones. The combination of southerly tectonic dip (22°) and the progressive southerly downcutting by the younger strata in the same general direction allow examination of most of the exposed strata at ground level at one point or another.

Numerous, often conflicting, stratigraphic and thickness details have been published for Baileycroft Quarry. These inconsistencies probably related to where and when the intervals were measured, bearing in mind the southerly dip of 22°, whereby older strata disappear southwards from view. There may also have been differences in what was visible in the quarry when these sections were measured. Published thicknesses seem

Figure 11.22 The limestone succession on the western side of Baileycroft Quarry, showing the relationships of the main formations. SW corner of Co-op Faces. Height of exposure *c.*9 m.

to be based on the western faces. The stratigraphy of the eastern faces (Wirksworth Cars and Pavement) is even more problematic, because there is little consensus in the literature about which formations are present there. Improved dating of the strata in this quarry is needed.

It is best to first examine the quarry complex along the Co-op Faces because the stratigraphy here seems clearer. The lowest beds are paler-weathering, massive-looking units 2–3 m thick, consisting of grainstones and packstones rich in small crinoid fragments, and typical of the Hopton Wood facies, which is much valued as a decorative and building stone. To the north, these beds increasingly make up most of the quarry faces. The units are separated by gently undulating palaeokarstic surfaces and thin wayboards. Towards the top, one bed, 30–50 cm thick, consists of a brachiopod-rich facies between two substantial wayboards. One to two metres above this there is a change to a dark wackestone with much dark clayey material. This passes upwards into massive, well-sorted, crinoidal packstone-grainstones that are eroded out to the south. The lower contact of these darker beds is more uneven than the partings within the Bee Low Limestones and is taken to be the disconformable base of the Monsal Dale Limestones (Fig. 11.22).

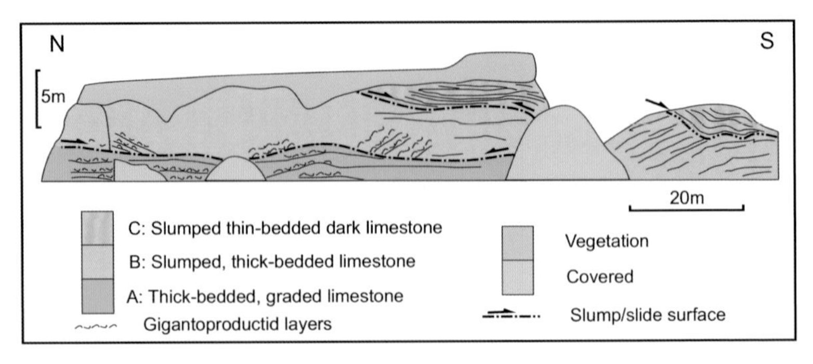

Figure 11.23 A profile of the Wirksworth Cars Face showing its division into three units, A, B & C, whose significance is discussed in the text. Deformed bedding results from slumping in units separated by slip surfaces. This profile pre-dates Wirksworth Cars, whose vehicles now obscure distant views of the lower face. When this profile was produced, the lower part of the face was partially covered by stored road salt and these areas are now uncovered (after Cossey *et al.*, 2004).

Above the Monsal Dale Limestones are dark, more thinly bedded limestones with clayey partings. Their contact with the Monsal Dale Limestones is strongly undulating, and bedding in these highest strata is very irregular, wavy and with sudden thickness changes. These beds contain Brigantian fossils and are taken to be Eyam Limestones. They are often obscured by vegetation and are locally eroded back, leaving a clearly undulating top to the faces below, especially at the northern end of the Co-op Faces. Going southwards the faces progressively decrease in height, and this unconformable contact descends progressively until it appears to have cut out the Monsal Dale Limestones completely in the southwestern end of the quarry. There is therefore a disconformity between Bee Low Limestones and Monsal Dale Limestones, and a much more erosive unconformity between Monsal Dale Limestones and Eyam Limestones.

The loss of intervening outcrop adds to the difficulty of visually correlating beds on the Co-op Faces with those on the Wirksworth Cars Faces. The Wirksworth Cars Faces have a very different and interestingly confused appearance. Figure 11.23 shows a complex range of features. When this section was drawn, the quarry was used for road salt storage and some of its faces were hidden. These are now visible. This figure recognizes three distinct units, A, B and C (upwards). It might seem that these should correspond to the Bee Low, Monsal Dale and Eyam

Limestones respectively, as seen along the Co-op Faces, albeit in a very disturbed state here. However, while the beds in the lowest unit (A) somewhat resemble the Bee Low Limestones, lithologically they differ in being thinner bedded (*c.*1 m) and graded. They contain gigantoproductid brachiopods indicating a Brigantian age, and this rules out Bee Low Limestones. It seems that the combination of the southerly dip and even deeper downcutting by younger strata compared with the western faces, cause the Bee Low Limestones to be absent on the eastern faces.

It is possible that Bee Low Limestones are present to the north along North End Face according to BGS mapping, but if so, to be consistent with the interpretation of Figure 11.23, they must be cut out completely, immediately southwards where the outcrop adjoins the Wirksworth Cars Face (by the little shrub garden). Alternatively, the beds interpreted by BGS as Bee Low Limestones must correspond to Unit A of the three Brigantian units. Aside from the North End Face, it follows that all three units along the Wirksworth Cars Faces must be Brigantian. This leads to three interpretations as to which formations are present – Monsal Dale Limestones only, or Eyam Limestones only, or both together? At least two of the units must belong to the same formation. For want of better dating, it is not possible to be certain.

Unit C (thinly bedded dark argillaceous limestones) is typical of the Eyam Limestones and mostly occurs along the top of the southern half of the Wirksworth Cars Face. Most of this face, however, is dominated by the strikingly undulating and chaotic-looking Unit B. The lithology is similar to that of Unit A but the bedding planes lie in various orientations, indicating substantial slumping. Unit C is also slumped on a smaller scale and tends to drape and infill the uppermost surfaces of Unit B. The best interim interpretation is that the units represent three different phases or events of sedimentation within the Brigantian, above a substantial but local unconformity cut into the Bee Low Limestones.

There is a well-documented, progressive thinning of the Monsal Dale Limestones in a southerly direction in the larger Wirksworth area (e.g. Fig. 11.4) This thinning has been attributed, in part, to the above marked unconformity that progressively cuts down into the Monsal Dale Limestones, as well as to phases of non-deposition within them. Furthermore, these beds have become thinner in this same direction because of a change from shallow-water deposition on the platform top to platform-margin deposition of thinner beds on a palaeoslope. In addition, there was probably greater compaction of these slope deposits compared with platform limestones.

The quarry also lies close to the southern margin of the Derbyshire Platform, with conjectured concentrations of mud mounds fringing the southern margin within the Bee Low Limestones (Asbian) (Fig. 11.4), one of which can be seen near the **Dream Cave**. The Bee Low Limestones in the quarry would therefore have lain a few hundred metres behind the shelf margin and its conjectured mud mounds. In the Brigantian the marginal mud-mound complex stepped back northwards towards the platform. It is exposed locally in disused quarries in and around the **National Stone Centre** in the hanging wall of the Gulf Fault. The Brigantian limestones of Baileycroft, therefore, were deposited basinward of the contemporaneous mud-mound complex.

The general consensus is that the Bee Low Limestones in the Wirksworth–Middleton area were deposited on a flat-topped platform in water depth up to a few tens of metres. Sedimentation was episodic and cyclic, controlled by eustatic changes of sea level. During lowstands, the platform was emergent and subjected to karstic erosion, giving rise, in this quarry at least, to relatively small-scale undulations on the uppermost surfaces of the principal beds. Volcanic material accumulated on these surfaces and weathered into palaeosols seen as thin clayey intercalations (wayboards). A more pronounced emergence with palaeo-karstic features marks the top of the Bee Low Limestones, creating the disconformity of the Monsal Dale Limestones which is taken to be the Asbian–Brigantian boundary.

The facies of the Monsal Dale Limestones on the Co-op Faces reflect their inferred platform margin setting. Syn-sedimentary slumps, poorly developed turbidites and shelf spill-over deposits of grainstones most probably formed on a palaeoslope suggesting deeper water than for the Bee Low Limestones. How much of the Monsal Dale Limestones are present on the Wirksworth Cars Faces is currently not well established. If, however, as some argue, the entire Wirksworth Cars Faces represent a later Brigantian (i.e. Eyam?) channelling event and infill, then Unit A is likely to be of this age too, and the facies of all three units reflect deposition within an Eyam Limestones channel rather than on an open palaeoslope.

The characteristics of the uppermost beds on the Co-op Faces seem similar to the uppermost unit (C) on the Wirksworth Cars Faces and are typical of Eyam Limestones on the platform-margin palaeoslope, although in neither case can they be examined closely. The recent consensus on Baileycroft Quarry is that the unconformity represents a channel (the 'Baileycroft Channel') which was cut down through the Monsal Dale Limestones and perhaps into the uppermost Bee Low Limestones at the southern end of the quarry and below car park level. It has been suggested that the channel was submarine in origin, but possibly followed the course of an earlier subaerial erosional feature that formed when Bee Low Limestone deposition ended. Others favour an entirely karstic origin for the channel. No similar correlative channels have been recorded from the immediate Wirksworth–Middleton area, and the wider unconformity, generally associated with the Baileycroft Channel elsewhere, lacks palaeokarstic relief on a similar scale. However, this may be due to lack of sufficient outcrop. Perhaps its apparently very restricted occurrence reflects local factors that led to more prolonged or karstic erosion in this one area. Two possible scenarios are as follows:

Various authors have suggested possible influence of the underlying mud-mound belt in the Bee Low Limestones. This idea has not been fully explained, but it perhaps implies that during the end-Asbian low stand, when the channel first formed, the former mud mounds had greater upstanding subaerial relief than the emergent platform behind it. The channel was then eroded as a gorge through which platform surface waters drained into the open sea beyond.

Alternatively, or perhaps coevally, the proximity of the channel to the Gulf Fault, which are almost parallel to each other, suggests possible syn-tectonic control of the channel formation. BGS accounts refer to 'the abrupt facies change in the [Eyam Limestones] across the Gulf Fault' and imply a 'penecontemporaneous' origin. This facies change is not specified but might refer to the channel deposits in Baileycroft. When relative sea level rose again, the channel became a conduit for the observed sediment flow off the platform in several episodes within the Brigantian, whilst syn-depositional gradients down and across the channel led to slumps.

Dream Cave Area, near Wirksworth (SK 273531: 53.0743, -1.5943)

This locality comprises a knoll (GPS above) and its immediate surroundings that include a cluster of small sites, 1.6 km southwest of the centre of Wirksworth. It can only be reached by pleasant routes on foot, or perhaps by mountain bike. It can be closely approached by public rights of way and, although the surrounding pastures are criss-crossed by them, the sites described lie at short distances from these paths. Strictly, permission to cross the land to geological sites should be obtained from the farmer (telephone 01629 825842). The Cave site itself is fenced off. Its vertical entrance, just visible from the fence, is an obvious danger to people and animals. The shortest walk starts from Stainsborough Lane, which is a single-track lane with parking limited to a single car or minibus. With multiple vehicles, a walking route from Warmbrook is suggested. Wirksworth is on various bus routes, some of which pass along the B5035. A convenient stopping point near the junction might be negotiable with the driver.

Walk from Stainsborough Lane

The lane runs southward opposite Sycamore Farm at a T-junction on the B5035. About 160 m along the lane, there is a generous passing and turning point on the left **(53.0766, –1.6019)***. Although not strictly a parking place there is room for a single car, whilst leaving space for others to pass or turn. Continue on foot along the lane for a further 100 m and take the public right of way on the left (Summer Lane). This rises gently and, after 250 m, follow it round sharply to the right, then bear left after c.60 m. There are old lead-mine workings in the field on the right. After another 300 m, there is a gate on the right next to a wooden bench* **(53.0761, –1.5938)***. Summer Lane continues sharply to the left and several paths meet here. Ignore these and pass through the gate into the field. Follow the derelict drystone wall southwards for about 200 m. Veer a little to the right to reach the summit of the Dream Cave knoll. Much of the off-road walk has numerous ruts, cobbles and tree roots, and can be muddy in places.*

Walk from Warmbrook, Wirksworth

There are various parking places around Wirksworth, but the quickest routes to the Dream Cave area begin by parking on a road near the western edge of the Warmbrook area, to the west of Wirksworth. From there, several westward footpaths (including Summer Lane) reach the gate on Summer Lane mentioned above. These routes are only c.100–200 m longer than that from Stainsborough Lane.

This locality is significant for several reasons. (1) The cave site is important for the history of palaeontology and the find of a rare Pleistocene woolly rhinoceros. (2) It is the southernmost, partly upfaulted area of Derbyshire Platform Dinantian, with a mud mound in Bee Low Limestones (Asbian), and dark slope facies of Monsal Dale Limestones (Brigantian). These Dinantian beds dip beneath Namurian and younger strata to the south. Together the limestones illustrate the step-back of the platform margin (or onlap of deeper-water facies) from Asbian to Brigantian, complementing other Wirksworth localities. (3) The knoll provides a 360° viewpoint. (4) The area is mineralized with old bell-pit lead-mine workings that provide interesting spoil for minerals and fossils.

This locality is best understood with a short walking itinerary, starting with the Dream Cave, which, being on a knoll, is an obvious starting point to get one's bearings. Small lead mining spoil heaps and bell pits reflect the dense network of mineral veins here (Fig. 5.1). Although the pits seem to be mostly capped, care should be taken. The knoll is an attractive open grassy hill-top with excellent views all round which, especially in good weather, makes the Dream Cave name seem appropriate. In addition to the Dream Cave site itself, there are scattered, somewhat overgrown but accessible, small disused quarries and outcrops of interest. BGS maps indicate that the summit area, and the southern and western sides of the hill including the Dream Cave itself, are underlain by Bee Low Limestones, partly in mud-mound (so-called reef-knoll) facies. Monsal Dale Limestones underlie the eastern and southeastern side. Main points of interest lie within 150 m of the knoll.

Site 1: The Dream Cave (53.0743, −1.5943)

The Dream Cave (**Site 1a**), also known as Dream Mine, Dream Hole, Stafford's Dream and The Bone 'Ole, is located at the summit of the knoll amongst a few ash trees. It is completely surrounded by a fence. It is a listed site, but unfortunately not (yet?) a geological SSSI. The

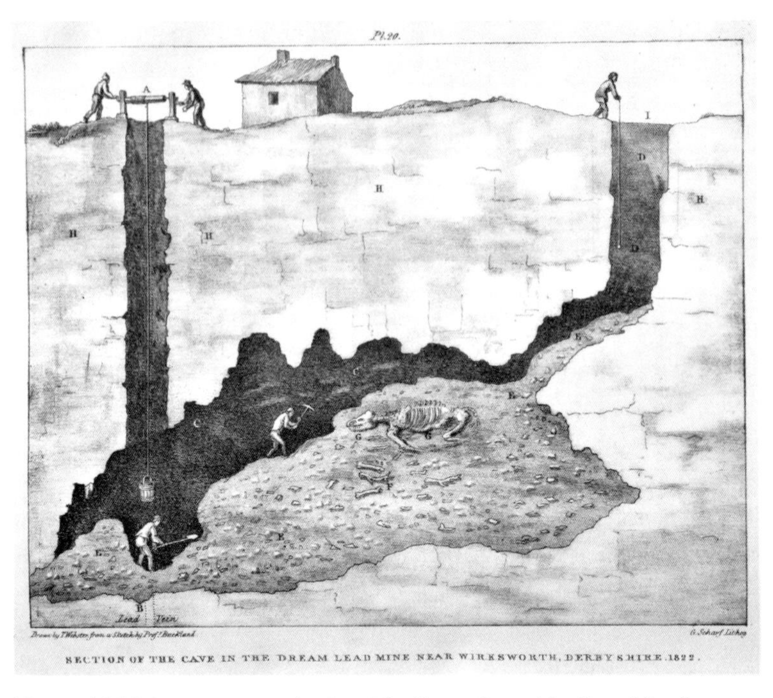

Figure 11.24 A contemporary drawing of the Dream Cave at the time of the discovery of the skeleton of a Woolly Rhinoceras (from Buckland, 1824).

cave entrance proper is within a deep hollow inside the fence. There is an interconnected lead-mine shaft nearby (**Site 1b**), now capped, from which the cave was first penetrated and discovered, along with its Pleistocene woolly rhinoceros. The deep hollow includes the open cave entrance, dissolution pipes and excavations, the cave itself being a karstic fissure. The site is dangerous and those interested in entering should contact local caving organizations (e.g. Peak District Mines Historical Society or the Wirksworth Heritage Centre). The exhibition at the latter includes a good exhibit about the Dream Cave and its history.

The current version of the BGS online app suggests that the cave lies within the Bee Low Limestones, although a recent paper by McFarlane *et al.* (2016) state that it is in Monsal Dale Limestones, citing an older version of this app. The bedrock at the upper part of the shaft is just visible from the fence and shows rubbly facies with poor bedding, probably related to the mud mound. Apart from its original discovery, the Dream Cave had attracted surprisingly little attention until the paper by McFarlane *et al.*, which largely forms the basis of this account.

In December 1822, lead miners sank a shaft about 20 m ESE from the then hidden cave entrance. As they worked, they uncovered a largely complete, well-preserved skeleton of a woolly rhinoceros, subsequently identified as *Coelodonta antiquitatis*, in a mass of unconsolidated earth and gravel of Pleistocene age. As material collapsed into the shaft, the miners had to continually clear it to keep the mine open, revealing a cave fissure, some 14 m deep and the original opening into which the animal must have fallen. The landowner realized the find was unusual and William Buckland, the first Reader of Geology at the University of Oxford and an important figure in the history of geology and palaeontology, was duly informed and visited the cave in 1823. At that time, he was writing his famous work on cave palaeontology, *Reliquiae Diluvianae* [relics of the [Biblical] Flood] and its copper engraving of the cave was based on a sketch he made during his visit (Fig. 11.24). Buckland acquired the rhinoceros bones for his personal collection at Christ Church College Oxford, but after his death in 1856, they were transferred to the Oxford University Museum of Natural History when it opened in 1860.

The many points of interest in this find include:

(1) The completeness and excellent preservation of the specimen, suggesting rapid burial before scavengers could attack and disperse the remains.

(2) Subsequent dating of the skeleton, in the context of current knowledge about Pleistocene climate change and glaciation, shows it be *c.*45 000–48 000 years old, at which time the climate was characterized by large amplitude, millennia cycles of abrupt temperature rises followed by gradual cooling. This was a time when the overall cooling trend had ameliorated somewhat, and conditions were intermediate between present-day warmer conditions and the severely cold glacial conditions of, say, the Last Glacial Maximum (*c.*20 000 BP), with temperatures ranging from –16 to –22°C in winter to 8–11°C in summer.

(3) The contemporaneous fauna included the famous woolly mammoths which, with other herbivores, grazed on the

cold grassland plains ('steppe-tundra') in these latitudes, a biome which has no counterpart today. Further information about the life habits of the woolly rhinoceros is given by Lister (2014, pp.54–56)

(4) Woolly rhinoceros are relatively rare in the British late Pleistocene fauna (fewer than 20 known specimens and locations) and known only from between *c.*56 000 and 28 000 years BP.

(5) A puzzling feature is that the animal's rapid burial was in a sudden flow of sedimentary material initially thought to be caused by a flood. However, if the topography was similar to today, it is difficult to see how its hilltop location could be flooded. Perhaps the area was covered wholly or in part by glacial or periglacial deposits which, when saturated after prolonged rains or rapid thawing, became unstable and collapsed by solifluction into hollows and fissures as mobile debris flows.

(6) In a broader context, the specimen provides evidence of climate change, driven primarily by Milankovitch cycles (obliquity) in the Pleistocene. Similar cycles were probably responsible for the cyclic sedimentation seen in much of the Bee Low and Monsal Dale Limestones of the Wirksworth area, e.g. in Middle Peak Quarry and throughout the Namurian. Both the Pleistocene and Carboniferous are regarded as 'Ice House' times in Earth history.

Site 2 (53.0738, −1.5945)

This site lies about 120 m SW and downslope from the **Dream Cave.** It consists of a small quarry face *c.*35 m long and *c.*8–10 m maximum height. Bedding is poor to non-existent. The lithology is a pale grey, very dense fine carbonate mudstone with a few crystalline fragments probably derived from crinoids. Small, mineralized cavities, including fluorite, are also present. BGS mapping suggests this is part of a mud mound within the Bee Low Limestones, a likely analogue for the 'inferred' mud mounds of Figure 11.4.

Site 3 (53.0735, −1.5947)

This site lies *c.*15 m southward more or less normal to the southern end of the **Site 2** face. The exposure is *c.*8 m long and *c.*3 m high, the lower part consisting of a small cave entrance. Above the entrance, the limestone is better bedded than at **Site 2**, bedding being wavy to lenticular and irregular, *c.*10–30 cm thick. At the western end of the exposure, the lower beds above the entrance pass into a pocket of breccia with fragments *c.*30–40 mm across. The lithology varies from fine grainstone, which is very finely laminated in places, to the same dense mudstone seen at **Site 2**. Fossil fragments are common. Small-scale mineralization in cavities is quite pervasive, fluorite being particularly noticeable. The outcrop is thought to be part of the same mud mound as **Site 2**. The fine grainstone and brecciation suggest this outcrop represents an internal flank facies of the mound, similar, at least in part, to flank facies described by Gutteridge (1995). However, the lithology might also have been affected by mineralization.

Site 4 (53.0736, −1.5950 for centre of this outcrop)

A series of closely located exposures start about 10 m west of **Site 3**, from which they spread along the same contour. They extend *c.*50 m along strike, the highest being *c.*6–8 m high. A cave or mine entrance is located on one of the faces. Bedding is poor to non-existent, and petrographic details have not been published. BGS mapping indicates that this exposure lies within the same Bee Low Limestones mud mound as **Sites 2 and 3**.

Site 5 (53.0738, −1.5925)

This very small exposure, with other possible bedrock blocks in the immediate area, is perhaps part of a former quarry face or mine working. It lies *c.*150 m ENE of **Site 3**, close to a footpath, a broken drystone wall, and the boundary of Sprink Wood to the south. Patches of ochre-coloured soil and mounds of mining spoil are present. The exposure is only 2–3 m long and *c.*1 m high. Bedding is weak and very irregular to absent. The limestone looks very similar to that of **Site 2**, and it might be taken for part of the same Bee Low Limestones mud mound. Although the rock is bituminous and dark compared with that of **Site 2** and the BGS Geology Viewer places the exposure within Monsal Dale

Limestones. However, its lithology and bedding are distinctly different from that of the definite Monsal Dale Limestone outcrop at **Site 6**.

Site 6 (53.0741, –1.5922)

This small exposure, *c.*10 m long and 2 m high, is part of former small semi-circular quarry facing eastwards, now largely overgrown. The exposure proper lies *c.*40 m NNE over a small ridge from **Site 5**, and *c.*14 m downhill, ESE of **Site 1**. The limestone is much more darkly weathering than that at other sites described here. It is well bedded, almost shaly with slightly irregular units from *c.*3 cm to 150 cm thick (Fig. 11.25). The thicker units contain crinoid fragments and may be turbidites. The overall lithology is a dark carbonate mudstone to wackestone containing minute crystalline fragments. BGS maps this outcrop as Monsal Dale Limestones.

There is good evidence elsewhere around Wirksworth that the Derbyshire Platform retreated northwards after a break in sedimentation at the end of the Asbian (Bee Low Limestones) to the area around the **National Stone Centre**, the northern part of **Middle Peak Quarry**, and **Redhill Quarry**, whilst the earlier platform became a carbonate ramp, perhaps due to tectonic tilting. The facies of this site contrast strongly with those of the same formation further north in being much

Figure 11.25 Thinly bedded limestones, with shaly interbeds, within the Monsal Dale Limestones at Site 6 of the Dream Cave locality.

thinner-bedded and darker. Although there is too little exposure to be sure, there appear to be few if any palaeokarstic breaks. The limestones here are therefore taken to have been deposited in a slope environment on the ramp. There is evidently a gradation of facies in the Monsal Dale Limestones via **Dale Quarry** to the more northern localities mentioned above, from deeper slope here to shallow platform.

Lead Mining

Two features indicate the presence of former lead mines in the area– the presence of spoil tips, and the pronounced disturbance of the ground by repeated, closely situated pits with strongly banked ring-shaped rims. Because of the way in which the light often falls on the sloping surface of the hill here, these provide some of the best photographed examples in Derbyshire (Fig. 5.1). They mark the presence of former shallow bell-pits, typically aligned along a vein. This area is one of the most intensely mined around Wirksworth. The spoil tips can yield small specimens of galena, fluorite, barite and flowstone, as well as fossils. These can also be seen in (but not collected from) blocks of the nearby broken drystone wall.

There is a substantial literature and other information on lead mining in and around the Peak District, the focus being the Peak Mining Museum in Matlock Bath and the associated Peak District Mines Historical Society and its publications. Ford & Rieuwerts (2000) provides an overview of lead mining history in the region.

Harboro' Rocks: the 'Brassington Dolomites' (SK 242553; 53.094604, -1.640334) (▸)

Free parking is possible in laybys and on verges near the Hoben works just below the Rocks, or by cycling or walking along the High Peak Trail. From the eastern side of the usually noisy Hoben works, take a small footpath northwards and uphill for a short distance, past a derelict farmhouse on the left, then through a stile and a gate to reach the first terrace.

Harboro' Rocks is an interesting and attractive large outcrop in dolomitized Bee Low Limestones and Monsal Dale Limestones near Brassington, along the north side of the High Peak Trail, with many original details of the macrofacies and bedding styles still recognizable,

Figure 11.26 Dolomitized limestones at Harboro' Rocks. Monsal Dale Limestones overlie Bee Low Limestones with the boundary near the base of the uppermost limestone face.

and a wide range of features to examine. The exposure is a striking rocky massif and landmark several hundreds of metres long and at least 50 m in elevation from the lowest outcrops. Weathering and erosion of the dolomite have made this an impressive, elongated tor, whose serried ranks of pinnacles are like a miniature version of the mightier and more famous dolomitic mountain range of the Dolomites (Southern Calcareous Alps), inviting the 'Brassington Dolomites' nickname. Formation of these dolomite tors is discussed by Dalton *et al.* (1999).

The siting of Hoben works seems unfortunate on scenic considerations, but was originally related to the exploitation of the local silica sand pockets for manufacturing refractory bricks. It now imports materials from elsewhere and supplies customized special sands and other industrial materials processed on the site (Thomas, 2019). The BGS sheet memoir (Aitkenhead *et al.*, 1985) makes little mention of Harboro' Rocks but Manifold *et al.* (2021) included them in their sampling localities, seemingly under 'Carsington', and the Rocks also feature in the content of the paper.

The south-facing (main) outcrops consist of a series of dissected cliffs, more or less at the same level, separated by grassy ledges or narrow terraces. As seen along the cliffs, the bedding appears as more or less horizontal with a slight northward dip into the faces. Most of faces above here are in Bee Low Limestones whose contact with the Monsal Dale Limestones is placed at the easily recognized, even from a distance, double recessive interval about 50 cm thick near the top of the faces, above the third topographic terrace from the base (Fig. 11.26). It consists of a rubbly and noticeably irregular series of finer beds (Fig. 11.27). In the absence of direct dating, the Bee Low Limestone–Monsal Dale Limestone contact in many Derbyshire Platform outcrops is mostly recognized on this kind of lithological–visual evidence. This horizon has been established as a consistent emergent feature across the Derbyshire Platform, denoting the major sequence stratigraphic boundary between the sedimentary packages (Manifold *et al.*, 2021). The same feature is recognized at non-dolomitized localities included in this Guide at (e.g.) **Redhill Quarry** and **Baileycroft Quarry**. At Harboro' Rocks, however, the usual intercalated clays are missing presumably as a result of dolomitization processes (Fig. 11.27).

Figure 11.27 The boundary between dolomitized Bee Low Limestones and Monsal Dale Limestone at Harboro' Rocks. The contact (grassy surface) clearly shows irregular relief typical of palaeokarstic dissolution, but the clay band that typically occurs in undolomitized exposures is missing here, presumably removed by the dolomitizing processes. Trekking stick for scale.

Bee Low Limestones at other localities (e.g. Baileycroft West side, Middle Peak Quarry, Middleton Quarry) consist of very regular cyclothems of massively bedded pale packstones and grainstones deposited at least in part as shoals on the platform proper or platform margin. The cyclothems often show an upward progression from higher-energy shallower conditions through slightly deeper, calmer conditions with bioturbation and back to higher-energy conditions (Fig. 4.3). Each cycle is commonly separated by evidence of emergence such as mammilated palaeokarst and clayey palaeosols of volcanic origin. This pattern seems to apply broadly to the upper level of Bee Low Limestone outcrops of Harboro' Rocks, bioturbated intervals being especially striking and obvious in 1–2 m thick beds weathered back as concave faces. The grassy ledges correspond at least in places to emergent horizons. Manifold *et al.* (2021) analysed the occurrence of these cycles in space and time across the Derbyshire Platform and concluded that they do not persist or correlate laterally for more than a few hundred metres in any given strike direction, making the overall sedimentary succession more like a space-and-time mosaic of lenticular units.

In strong contrast, the Bee Low Limestone outcrops seen along the first terrace are very different from those at higher levels in exhibiting slumps and rudstones containing irregularly rounded cobbles of mud-mound clasts indicating much greater instability than at higher levels. This could have included subaerial emergence in a neighbouring area during a low stand that generated talus, or steepening of the palaeoslope and possible channelling, although there is no obvious exposure which gives support for such events at the Rocks themselves. However, Manifold *et al.* (2021), using broader-based evidence, suggest that such processes probably account for what is seen in the exposures on the first terrace.

Scattered corals including phaceloid forms like *Siphonodendron,* and brachiopods, are moderately well preserved throughout the section although probably not in life position.

Although the faces above the third terrace have the greatest elevation (they are popular with rock-climbers), the Bee Low Limestone–Monsal Dale Limestone contact can be reached fairly safely without climbing skills, by ascending one of the grassy gullies between the faces, leading from the third to the fourth terrace.

Models for the dolomitization of these limestones are discussed in Chapter 4 and illustrated in Figure 5.4.

In addition to the main exposures, the area shows other geological and related features with anthropogenic implications. The broadly southward views are dominated by Carsington Water (reservoir) and by Carsington Pasture Wind Farm (Fig. 1.5). As a source of energy, the latter presents a striking technological contrast with the distant view to the SE of Ratcliffe-on-Soar power station, reflecting the on-going switch in energy production from coal-fired methods to more carbon-neutral technology. Construction of the wind farm encountered difficulties, explained by Jones & Banks (2015), arising from deep karstification of the limestones and complexes of related younger poorly- or unconsolidated deposits associated with dolines. Across the road, but with little visible sign from this distance, is the disused Bees Nest silica sand quarry. Unlike the deposits in other dolines in the area, these are of Miocene age, and in this stratigraphic respect, unique for all of Britain (Dalton *et al.*, 1999, Walsh *et al.*, 2018). The quarry is now largely overgrown and difficult to access.

Middleton Moor, Redhill Quarry, Intake Quarry: just on the platform

Middleton Moor is an attractive area with interesting geology, industrial history, and natural history. It rises gently to a height of 358 m and gives fine views in all directions. It is easy to access and popular with local walkers, and a walk can include stops at Middleton Top Visitor Centre and Redhill and Intake quarries. Cycle hire, café, toilets and picnic tables are all available at Middleton Top.

Middleton Top Visitor Centre (53.0929, –1.5905) (▶)

The High Peak Trail is the track base of the former historically early (1831) and remarkable Cromford & High Peak Railway (C&HPR) which closed in 1967. It was 54 km long with nine steep inclines that carried the line up to the White Peak plateau. Eight of them used cable systems to haul trains up them, as for the Middleton Incline here. The ninth (Hopton), between Intake Quarry and Harboro' Rocks, was the steepest adhesion line in Britain at 1:14 (*c.*7%).

Figure 11.28 Engine house for the winding engine that hauled trucks on the Middleton Incline of the Cromford and High Peak Railway along with a typical truck that was used for the transport of quarry products.

Opposite the café, in the uppermost level of the car park, is a viewing point with topograph. The view is mainly southwards towards Derby and is similar to that in Figure 11.31. The nearer view is mostly across Dinantian limestone country and then over the lower-lying, younger, softer Namurian Bowland Shales and beyond to more low-lying country of younger Carboniferous and Triassic rocks. The high ridge of Ashover Grit (Marsdenian, R_{2b}) can be seen on the left.

From the café, walk eastwards along HPT just a few paces to the top of the Middleton Incline with the Engine House **(53.0932, –1.5894)** (Fig. 11.28), and remains of winding gear. The Engine House can be visited on operating dates, though now with an electrically-driven simulation. In the café there is a scale working model of the Middleton Incline.

Redhill Quarry (53.0935, –1.5909)

From the Visitor Centre, cross over HPT to go through a gate labelled 'Redhill Nature Reserve and Picnic Site' and take the footpath straight ahead into the quarry, which is a nature reserve of unclear ownership or management. It is now rapidly becoming overgrown by a vigorous garden escape (*Cotoneaster*) but shows about 15 m of Monsal Dale Limestone lying disconformably on *c*.6 m of Bee Low Limestones. Of particular interest is the contact between the formations. Manifold *et al.* (2021) describe several volcanic clay horizons weathered into palaeosols alternating with thin, irregular limestones (?calcretes). The uppermost Bee Low Limestone surface is mammilated (*sensu* Walkden 1970) and palaeokarstic. This can be compared with the same but dolomitized contact at Harboro' Rocks and at numerous other outcrops across the Derbyshire Platform.

In almost the lowest exposure of the Bee Low Limestone, there is an irregularly coarse 'laminite' unit about 1 m thick. Each paired layer is less than about 1 cm thick, and under a hand lens, it seems to consist of irregular microbialite alternating with a recessive layer of pale mudstone/wackestone, but this needs petrographic confirmation. Monsal Dale Limestones above the disconformity consist of numerous cyclical units typically around 2 m, mostly separated by thin palaeosols and slightly mammilated palaeokarstic surfaces. When thicknesses of the Monsal Dale Limestones in this quarry and Intake Quarry (below) are compared, the uppermost surface in Redhill seems to correspond approximately with the disconformable top of the Monsal Dale Limestones at Intake Quarry.

Intake Quarry (53.0936, –1.5968)

From Redhill quarry, return to HPT and walk westwards for 570 m to reach a gate across the Trail at **53.0899, –1.5966.** The locked gate on right is the entrance for **Intake Quarry**, which is not accessible to the public but is regularly used by rock climbers. Much of the succession can also be viewed from the northern perimeter fence on Middleton Moor and reached from various public footpaths along the HPT.

Eyam Limestones here consist of thin, strikingly cherty beds (Fig. 11.29), which occur in several broad lenses, perhaps filling shallow channels. The cherts occur as flattened nodules occupying at least 25% of any given bed, often more, becoming tabular sheets in places. Their main orientation is parallel to the thin undulating sedimentary breaks. This suggests that they formed prior to these breaks and are perhaps

Figure 11.29 One of the main faces of Intake Quarry with Monsal Dale Limestones overlain by much finer-bedded Eyam Limestones. The Eyam Limestones are here particularly rich in chert nodules that are pale in colour and flattened parallel with the bedding.

related to early diagenesis. According to numerous publications, Eyam Limestones lie disconformably on Monsal Dale Limestones in this quarry, with a breccia at the base of the Eyam Limestones.

As covered elsewhere in this Guide, the Eyam–Monsal Dale Limestone disconformity can be traced all the way from Intake Quarry through the various quarries and outcrops along the Gulf Fault down to the strikingly channelled and slumped feature in Baileycroft Quarry *c.*2 km to the southeast (Fig. 11.23). This is accompanied by progressive thinning of Monsal Dale Limestones from here to just a few metres (Fig. 11.4), although most of the quarries concerned are no longer accessible, or have been filled in or become overgrown. This suggests that the shallow channels here are orientated broadly northwest–southeast. Some have argued that this thinning is not due to the disconformity, but the presence of channelling here and at Baileycroft suggests that erosional removal is also a factor.

Middleton Moor itself can be explored spontaneously in any direction, as the grassland mostly provides easy walking. There are usually stiles or gates through the few walls and fences. The Moor sits between the major quarries of Middleton Quarry to the east and Hopton Wood Quarry to the west, which are connected beneath the moor by a labyrinth of mines that formerly exploited the Hopton Wood facies of the Bee Low Limestones.

Black Rocks: sliding down the delta front (SK 295557; 53.0977, –1.5684) (▶)

From the B5063, take Oakerthorpe Road alongside the cemetery. Take the signposted track on the left to the Black Rocks car park. Alternatively, add this locality to the NSC Geowalk by walking along the High Peak Trail from Steeplehouse Quarry, a distance of c.750 m. From the car park, with its toilets and café, various footpaths lead up the steep hill towards the main outcrop, some skirting the spoil heaps on the western side. This is popular as a general tourist spot and for rock climbing and easier scrambling. Take care that there may be other people above or below you.

This is the only locality in this chapter that deals with sediments of Namurian age. The locality is one of several known examples of delta-front instability associated with the progradation of the Marsdenian (R_{2b}) (Ashover Grit) delta. Such instability appears to be quite rare in the Pennine Namurian, but it is likely that other examples are hidden by Quaternary landslips, and by solifluction and vegetation. These examples are noticeable because they involve a thick sandstone unit that creates strong topographic features. Other, less conspicuous examples may involve only fine-grained facies that produce little expression in the landscape.

Examples at Birchover and Black Rocks are characterized as 'slides' where the primary deformation is concentrated on discrete slip surfaces with slumps or sandy debris flows occurring in the upper parts of the displaced masses. The slip surfaces are, in effect, syn-depositional faults with strongly concave-upwards (listric) profiles that flatten out in fine-grained mudstones at the base of the progradational (coarsening-upward) interval (cf. Fig. 4.15B). Mapping suggests that the faults are also strongly curved in plan-view, so the slip surfaces are scoop- or spoon-shaped, opening in a down-current direction as shown by cross-bedding in both displaced and undeformed sediments (Fig. 11.30).

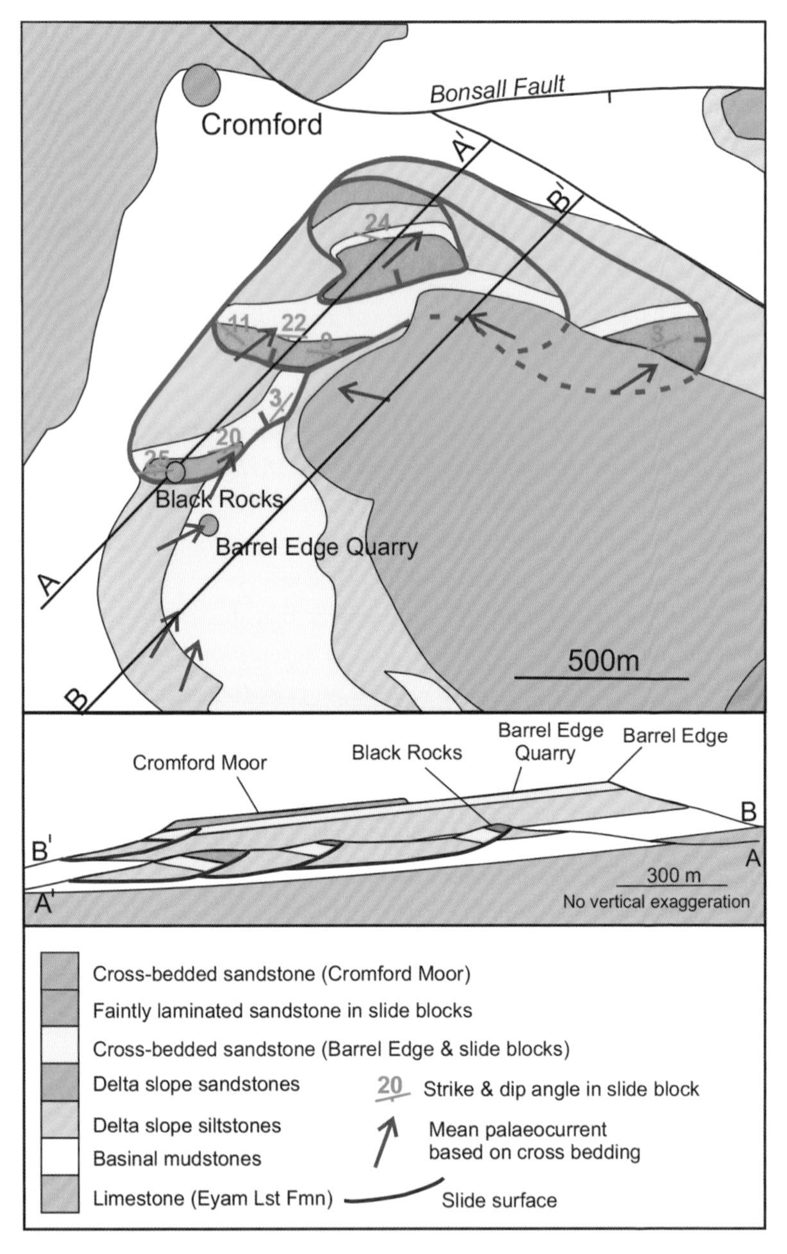

Figure 11.30 Geological map of the area around Black Rocks, along with two vertical profiles that together illustrate the geometry of the slide masses within the collapsed delta front of the prograding Ashover Grit delta. (after Chisholm & Waters,2012)

The sediments that shifted downwards within the fault scoops (hanging wall) were rotated by the shape of the fault so that they dip into the fault (i.e. opposed to the palaeoflow). Published accounts suggest that two distinctive sandstone facies are present within the rotated intervals. Cross-bedded sandstones occur in the lower part of the interval and are overlain by structureless or weakly bedded sandstones. These facies have been related to the development of the structures. Prior to deformation, recently deposited, thick channel sandstones would have been unconsolidated and water-logged. When the sliding began, sands on the downthrown (hanging wall) side moved as a coherent slab, with deformation concentrated at the slip surface and original fabrics and structures preserved. On the upthrown side of the fault (foot wall), the same sands, once exposed in the fault scarp, had little cohesive strength and collapsed to give sandy slumps or debris flows that deposited largely structureless sands on top of the cohesive slab. Weakly defined bedding in these upper sands is inclined in a similar direction to that in the sands below, suggesting that debris flows occurred whilst movement was on-going.

This model is based on extensive detailed mapping (Fig. 11.30), and it is difficult to fully demonstrate its features within the constraints of a typical field trip. For the Black Rocks example, some idea of the scale and complexity can be obtained from a combination of a distant view of the outcrop within the surrounding landscape and a visit to the outcrop itself.

Useful views of Black Rocks from the west can be had from Middleton Moor (Fig. 11.31). The Ashover Grit forms positive features in the landscape and these can be related, to some extent, with the mapping shown in Figure 11.30. The regional tectonic dip here is gently to the east, so the steeper dips in the outcrop are syn-depositonal, related to the sliding.

At the main crag, about 15 m of the Ashover Grit are well exposed. The coarse, pebbly sandstone shows rather poorly defined bedding with dips of around 20°. Towards the top there is medium-scale cross-bedding with small trough forms on some bedding surfaces. It is quite difficult to identify any systematic trends in bedding style within the exposure. White mine spoil that lies on the footpath side of Black Rocks contrasts with the colour of Black Rocks itself. It consists of fragments

Figure 11.31 View to the east from the north-east corner of Middleton Moor. Above Bowland Shales, low in the valley, are the steeply dipping beds of the Ashover Grit (R_{2b}) at Black Rocks, rotated by sliding on the delta front. The tree-covered Barrel Edge ridge behind is also formed of the Ashover Grit, whilst the distant ridge, beyond the Derwent Valley, shows the extensive quarrying of Dinantian limestones within the Crich Anticline. The visible quarry face is on the eastern limb of the fold. The furthest skyline is formed by Westphalian Coal Measures and possibly the Magnesian Limestone (Permian) east of Alfreton.

of underlying limestone, not scree or spoil from Black Rocks. Building ruins of the Cromford Moor Mine can be seen by the café. Its sough (Fig. 11.36), which discharges in Cromford village, is the oldest known in Derbyshire.

There is an extensive view from the top of the crag across the Derwent Valley to the north and across to Middleton Moor to the northwest, with the large active Dene Quarry in Dinantian limestones.

Matlock Bath: tales of tufa and the Matlock Gorge (▶)

Several short walks are possible along and above the Derwent Valley through the Matlock Gorge, and these can also be combined in various ways. Some involve pedestrian use of the busy A6 trunk road, although it is usually possible to use the relatively narrow pavements. There is good public transport access along Matlock Dale from Cromford to Matlock, and Matlock Bath has its own train station and a large parking area next to it. There are hourly trains between Cromford, Matlock Bath and Matlock. Paths at the lower levels have good wheelchair access.

1. Tufa outcrops and the Peak Mining Museum

Parking (some paid) in Cromford, Cromford Mill, and in Matlock Bath (use the Peak Mining Museum car park if possible ***(53.1194, –1.5625)****). Unfortunately for this walk, it is impossible to avoid the A6. It can be started at either end, but at the main Cromford crossroads on the A6 with the traffic lights behind you, walk north-westwards, using the pavement on the right overlooking the River Derwent where it swings downstream through a large meander. Proceed as far as the historic red-brick Masson Mill* ***(53.1128, –1.5615)****. Cross the A6 using the pedestrian crossing and continue to the New Bath Hotel service roads.*

Here, to the right, there are views of the high limestone cliffs above the Derwent with two small mud mounds apparently within Monsal Dale Limestones. These are smaller versions of the big mound which underlies High Tor north of Matlock Bath (see below). Although BGS publications map these mounds as Eyam Limestones, there is now a consensus that they lie within uppermost Monsal Dale Limestones (Fig. 4.8).

Along this side of the road, especially beneath the New Bath Hotel **(53.1159, –1.5604)** are outcrops of calcareous tufa, or more strictly, travertine. Pentecost (1999) summarizes how the tufa here forms. The abundance of limestone charges groundwaters with calcium carbonate and tufa is precipitated 'when a CO_2-rich groundwater discharges and loses excess gas to the atmosphere and aquatic plant photosynthesis. Re-equilibration of the groundwater usually results in the precipitation of calcium carbonate downstream from the spring.' Additional factors are discussed by Banks *et al.* (2012). In the Matlock Bath area, moss is the main photosynthetic plant, as seen where tufa is forming today, but there is also a rich algal flora present including diatoms (Fig. 11.32). The tufa accumulates by precipitation over the moss, and the moss then recolonizes the tufa surface, presumably seasonally. Precipitation is aided by warm temperatures. The springs of Matlock Bath are the second warmest in the country (*c.*20 °**C**) after the famous Bath waters (*c.*45 °**C**).

Figure 11.32 Natural spring water (presumably piped beneath the A6) actively depositing tufa over moss and building a delta-like structure into the River Derwent, close to the Peak Mining Museum. Five springs are currently depositing tufa, four on the western side of the valley and one on the east (although supply for the latter seems to be partially piped from a western source). This is partly because the geological structure dips east across the valley, and the topography is also asymmetric (Fig. 11.33), with most of the cliffs on the east side and steep, less rocky slopes on the west. The combined effects of the easterly dip and the conformable impermeable Matlock Lavas cause more springs to emerge from the west side.

Tufa (and other mineral products) have played an important part in the history of Matlock Bath (Charlton & Buxton, 2019). In the nineteenth century in particular, tufa was quarried for use in gardens, and it was popular for making souvenirs and ornaments by arranging for precipitating spring waters to trickle over objects placed beneath them in 'petrifying wells'. The thickness of the tufa deposits around Matlock Bath suggests that they have been accumulating for a long time. Pentecost (1999) used molluscan evidence to estimate that they go back about 8000 years. Banks *et al.* (2012) state that deposition rates have varied, however, and reaching a maximum between 5000 and 2500 BP.

After passing Holy Trinity Church on the left, recross the A6 at the junction with Temple Road to reach the distinctive pink and cream Pavilion building of the Peak Mining Museum on the right (53.1194, −1.5625), where the Matlock Bath shops begin. In a pond by the car park a large mass of mossy tufa is forming around a fountain. The Museum includes an excellent 'life-size' reconstruction of a small Derbyshire lead mine, as well as the very fine mineral collection of the late Professor R.A. Howie, formerly Lyell Professor of Geology at Royal Holloway, University of London. The Museum is currently open only at weekends, when it is also possible to book a guided tour to Temple Mine (lead) located up Temple Road.

2. Derwent Gardens, Lovers' Walk, and views over the gorge

Those passing quickly through busy Matlock Baths can sometimes be critical of its seaside resort atmosphere and will leave unaware of its quieter attractions, like its various gardens and green spaces along the river between here and Hall Leys Park in Matlock.

The preceding walk can be continued to see the best of these gardens by exploring the historical ornamental Derwent Gardens, well worth a visit in their own right. Enter through a gate on the south side of the museum. Geologically it is notable for the many small buildings, pavilions and other structures made of the local tufa, some with live springs depositing tufa around moss.

Cross to the opposite bank over a gracefully arched modern footbridge to reach the Lovers' Walk, which provides a scenic circular route along the riverbank and up to the crest of the ridge above the river. At the far southern end of the riverbank, the view across the river includes a tributary cascade close to the museum, which is actively depositing a substantial delta of mossy tufa (Fig. 11.32). At each end of the riverbank path, steep irregular steps lead up through woodland and the side of the gorge, passing outcrops of Monsal Dale and Eyam Limestones. Evidence of the Upper Matlock Lava can be seen in basaltic blocks at various points along the path. From the crest of the cliffs, there are fine views over the gorge on one side, northwards along the flanks of the cliffs towards High Tor (Fig.11.33)

Figure 11.33 Matlock Dale gorge from the top level of Lovers' Walk in Derwent Gardens, looking northwards (upstream) towards Matlock. Matlock Dale, on the left, is asymmetric with long steep slopes to the west and craggy sides to east. Their highest point is High Tor with its Monsal Dale Limestone mud mound (Fig. 11.34). The overall dip is to the east and land surfaces more or less parallel this. Monsal Dale and Eyam Limestones together dip beneath the village of Starkholmes, which lies on on Namurian Bowland Shales. The ground behind forms the scarp of the Ashover Grit. Starkholmes lies in a 'false valley' and it has been argued that the Derwent originally flowed along this feature before being captured or diverted along the incised Matlock Dale.

and eastwards across a gentle, broad topographic ledge with the village of Starkholmes **(53.1222, –1.5517)** below the ridge of Namurian Ashover Grit. This provides a surprise element of pastoral contrast with Matlock Bath itself.

3. Views of High Tor mud mound

It is useful to take binoculars, and for photographers a telephoto lens. In any case, choose good clear weather for viewing the High Tor face. A copy of Gutteridge's (2003), a short field excursion, especially its Figure 2, would be ideal. The High Tor face is an iconic landmark of the area, much painted and photographed. A range of views can be had by walking up Saint John's Road ***(53.1324, –1.5576)*** *from the A6, and at the gate into Cliff House Estate* ***(53.1298, –1.5623)****, take the narrow path to the right to ascend to the perimeter fence of the Heights of Abraham. It is not necessary to enter the grounds of the latter (which requires payment) to see good views of High Tor from the upper levels of this walk. On the other hand, one can go to the Heights of Abraham by cable car from the base station* ***(53.1243, –1.5573)*** *and take in the views from the grounds.*

Figure 11.34 High Tor from Heights of Abraham, with Riber Castle behind. The face is made up mainly of a large mud mound (Monsal Dale Limestones) draped by well-bedded Eyam Limestones. The core of the mud mound is massive but it passes laterally into dipping beds that are clearly within the mound complex, and separate from the draping Eyam Limestones. The face is around 50m high.

The face of High Tor demonstrates the typical features of the mud mounds of the eastern side of the Derbyshire Platform with underlying Monsal Dale Limestones and overlying Eyam Limestones (Fig. 11.34; Fig. 4.8). The mound shows the transition from massive, unbedded mound core facies to inclined internal flank beds. These should not be confused with the so-called 'flanking beds' seen, for example, in Coal Hills Quarry (NSC trail; Stop 6), which are post-mound Eyam Limestones deposited above an emergence surface. Analogous post-mound beds can also be seen at the top of the High Tor face.

Once-a-Week Quarry (SK 158681; 53.2093, –1.7650)

This small quarry is the only one in Derbyshire that still works limestone for building and decorative stone. It lies about 1.5 km NNE of Monyash and is approached from the village green by taking the lane to the north and then taking the branch to the right after 300 m. Follow this lane for 2 km and turn left at the 3-lane junction. After about 1 km, the entrance to the quarry track is on the left. Sensible parking close to the gateway should not impede access. Follow the track, which goes through a copse, for around 400 m to reach the quarry.

This is a working quarry and prior permission to enter should be obtained from the quarry manager, at the time of writing Mr Chris Cooling; 07845 828055. If this fails, contact the head office: Natural Stone Sales Ltd; 01629 735508; enquiries@naturalstonesalesltd.co.uk. Follow directions from the management and confine sampling to waste material. The faces are somewhat unstable and are best avoided.

The quarry exposes around 10 m of the Eyam Limestones whose base is probably only a few metres below the quarry floor. Bedding in the limestones is broadly horizontal, although slightly undulating. The limestones are mainly highly fossiliferous grainstones, with interbedded units of fossiliferous packstone at the base and in the middle of the exposed section. The dominant fossils are crinoid columnals, in various stages of disarticulation. Subsidiary fossil components are bryozoans and brachiopods, which are mainly thick-shelled *Gigantoproductus*, which occur as scattered bioclasts and in discrete layers where the shells

Figure 11.35 A vertical section through one of the crinoidal beds within the Eyam Limestones at Once-a-Week Quarry with fossils naturally etched out. The high-density fossil assemblage is dominated by crinoid columnals and ossicles and gigantoproductid brachiopods that have been moved by currents after death.

are highly concentrated and in orientations that suggest reworking. A minority are in life position, suggesting that conditions were, at times, quiet enough for larvae to settle, and more relevantly, they must have been quiet enough for long enough for the brachiopods to stay in place and survive, to grow to their large mature size. Systematic petrography and palaeontology in Nolan *et al.* (2017) includes excellent photomicrographs.

The crinoidal limestones record vigorous growth of crinoids and their post-mortem reworking by currents and waves (Fig. 11.35). Long columnals suggest that currents were not too vigorous, but the absence of the intact calices and the low levels of micritic matrix indicate that conditions were not entirely quiet. It has been suggested that the hydrodynamic conditions favouring this reworking might have been influenced by the proximity of relict mud mounds, but the evidence is tenuous. Many of the 'knoll reefs' shown on BGS maps of this area are hard to demonstrate either at outcrop or in the landscape.

The limestones extracted here are dressed for building stone, whilst larger slabs are polished into decorative surfaces and tabletops at the company's workshops in Rowsley.

Magpie Mine, near Sheldon (SK1727; 53.201, –1.7434) (▶)

This site, about 2.5 km to the northeast of Monyash, is a Scheduled Monument that consists of well-preserved heritage mine workings that can be visited at any time. Some of the buildings are open, mainly at weekends, on Heritage Open Days, when volunteers from the Peak District Mines Historical Society are on hand to explain the site and its historical workings. The site is also an SSSI for its botany. From Monyash village green, take the road to the north and take the right branch after 300 m. Follow this lane for about 2 km to a three-lane junction and take the right-hand lane (Grin Low) for about 300 m. A gate on the left accesses a track leading to the mine, about 400 m away (Fig. 1.4). There is parking space for several cars near the gate, but do not block access to the track. Alternatively, the mine can be approached from Sheldon village, about 600 m away on footpaths that cross the open-access land. This locality and Once-a-week Quarry are within easy walking distance.

Magpie Mine was the last working lead mine in the country, finally closing in 1954 after a colourful history spanning more than 200 years.

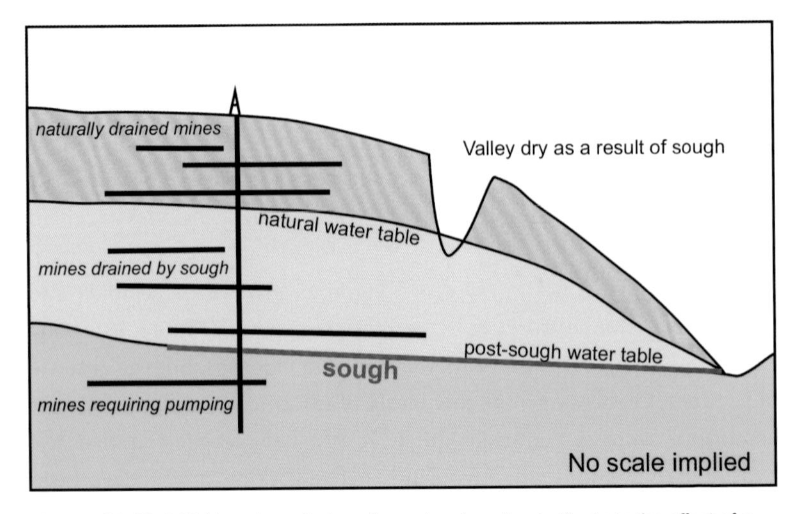

Figure 11.36 A highly schematic two-dimensional section to illustrate the effect of a drainage sough on the water table and its influence on the drainage conditions in affected mines. A secondary consequence of the lowered water table is to change the hydrology of streams within the landscape such that previously flowing streams cease to flow in normal circumstances.

There are five mines on the site, the deepest of which is 208 m deep, and the surrounding fields are pock-marked with bell pits. As with many mines, progress was inhibited by groundwater and the initial solution was to install a Newcomen pumping engine in 1824, which gave a period of high production, in excess of 800 tonnes/year. The pump was eventually not up to the task, nor was a 40' (12 m) Cornish pumping engine, installed a few years later. The eventual solution was to drive a drainage tunnel, a sough, some 2 km long, which continues, to the present day, to discharge large volumes of mine water to the River Wye near Ashford-in-the Water (see Fig. 11.36). Production died out the early twentieth century. A short-lived phase of activity in the 1920s made little headway and the most recent activity, using more modern technology, ended in 1954, partly due to the falling price of lead.

The site preserves the Agent's House and buildings relating to both extraction and surface processing such as crushing and dressing. Most of the buildings were built in the mid-nineteenth century although the steel headgear and the corrugated iron shed date to the 1950s.

Detailed descriptions of the site are to be found in the listing on the Historic England website and in a guidebook published by the Peak

District Mines Historical Society. Copies are on sale at the Peak Mining Museum (open only at weekends) in Matlock Bath and from the Agent's House or Smithy when Society members are on site.

Haddon Grove area of Lathkill Dale: a geological idyll (SK 183659; 53.1903, -1.7270)

There is very limited parking on the verge **(53.1933, –1.7314)** *along the narrow lane by Mill Farm (with millstone-shaped name sign also inscribed 'Haddon Grove'). Take care not to obstruct other traffic. Cycling is not permitted in Lathkill Dale.*

This is an easily reached and very attractive locality at which to see ramp facies of the Monsal Dale Limestones and a section of beautiful traffic-free Lathkill Dale. This sector of the Derbyshire Platform, between the Cronkston–Bonsall Fault to the south and the Taddington–Bakewell anticline to the north, developed into a ramp (Gutteridge, 1987) perhaps the cause of deeper-water facies at the base of this section.

Walk through Mill Farm on a public track continuing sharp left at the farm, curving right and then turning left through gate **(53.1908, –1.7301)** to enter a small wood at the head of a secluded and idyllic

Figure 11.37 A single bed within the thicker Lathkill Shell Bed in the upper part of the Monsal Dale Limestones at Haddon Grove. The shells are mainly gigantoproductus with the shells showing a mixture of concave-up and convex-up orientations, suggesting that some are in life position whilst others have been disturbed and redeposited, probably by storm events.

pair of small, steep tributary valleys of Lathkill Dale. The track seems to have been carefully engineered from here on down, with a drystone embankment, to maintain an even incline throughout, perhaps because it was used for transporting ores up from Mandale Mine in the main valley. The uppermost part of the path is in woodland and on the side of a deep little ravine. Cuttings (or natural faces?) alongside the path expose some 10 m of continuous gigantoproductid beds, referred to in this area as the Lathkill Shell Bed (Fig. 11.37). Whilst the majority of the brachiopods are convex-down suggesting life-position, a significant proportion are in concave-down position, suggesting intervals of reworking by higher energy, probably storm events that penetrated into a normally relatively deeper, lower-energy setting (Le Yao *et al.*, 2016; Nolan *et al.*, 2017). The brachiopod beds end in a high bluff where the ravine begins to open out and a steep grassy side valley joins from the left. Bearing right round a near-hairpin curve at the foot of this side valley, the path reaches a second set of exposures on the left. These rubbly pale limestones contain two coral beds, 2–3 m apart, vertically. They consist of small scattered *Siphonodendron* colonies and possibly other phaceloid rugosans. These and other fossils can be found in the clitter.

Figure 11.38 Well-bedded dark limestones of the Monsal Dale Limestones at the lowest end of the Haddon Grove section near the floor of Lathkill Dale. Although not within the Intra-Platform basin, the dark facies and parallel bedding suggest greater water depth than for the pale rubbly limestones with the coral beds, and the Lathkill Shell Bed, higher up the section. The facies looks identical to the well-bedded dark limestones of the Intra-Platform basin as seen in the Monsal Dale Viaduct Cutting.

Continuing straight ahead to the foot of this little valley, a small, incised section on the right **(53.1888, –1.7265)** shows 3–4 m of dark, thin, parallel-bedded limestones regularly alternating with thin shaly partings (Fig. 11.38). These are similar to facies in the Monsal Dale Limestones of the Monsal Dale area (Fig. 11.50) and around Wirksworth (e.g. Fig. 11.6). They are event beds, conceivably proximal turbidites deposited down the ramp slope, in deeper water than the coral and brachiopod beds higher up the section.

In summary, the Haddon Grove section shows upwards shallowing in time from deeper-water conditions down the ramp slope to shallower conditions recorded first by coral beds and then brachiopod beds. The latter are largely in life position, suggesting sheltered but shallow conditions.

Return back up the path, or for a longer walk in the main Lathkill valley, join the riverside trail just beyond the last outcrop above and walk in either direction along it. There, further outcrops and various mine/cave entrances can be seen. This part of the valley is wooded but further upstream, towards Monyash and beyond the woodland, it becomes a dry karstic gorge, although in the past the river rose further upstream. An important anthropogenic aspect of Lathkill Dale is the fall of its water table over the last few hundred years. The reduced flow is due, in particular, to the remarkable hand-hewn, often long and deep underground drains (soughs) made to drain the lead mines, as mentioned above for Magpie Mine and on information boards. Banks (2017) provides the wider context and states that Derbyshire soughs were first constructed in the seventeenth century. Although lead mining has ceased, large volumes of water still flow through the Derbyshire soughs, some of which are even used for water supplies.

One of the best mining remains in Lathkill Dale can be seen by proceeding eastwards (downstream) from Haddon Grove for about 2 km to the once important Mandale Mine complex **(53.1921, –1.7075)**, which includes the ruins of its former Cornish engine house. The mine exploited the minerals in the long, rich Mandale Rake. There are excellent explanations on information boards. A further example of regularly bedded darker (?deeper-water) facies of Monsal Dale Limestones is exposed close to the mine.

Continue downstream to reach the lane at Lathkill Lodge (**53.1922, –1.698**). Ascend from the dead-end of a steep lane here to the village of Over Haddon (**53.1949, –1.6959**) with pub, car park and toilets. The slabs of the clapper bridge over the River Lathkill at the foot of the lane and in the nearby path are full of large brachiopods. Return to Mill Farm either via Haddon Grove or along the lanes from Over Haddon to Mill Farm.

Furness Quarry (also known as Horseshoe Quarry), Stoney Middleton: glimpse of an ancient sea floor (SK209761; 53.2816, –1.6886)

This attractive, disused quarry is notable for the relatively rare occurrence of a substantial bedding plane outcrop with corals on the quarry floor. It lies 1 km southwest of Eyam and 2.3 km west-northwest of Stoney Middleton, immediately north of the A623. There is parking for a few smaller vehicles immediately off the A623, and also in small laybys a little further to the west. There is no sign for the car park and the entrance is not apparent until quite close. The road is fast and busy, so care is needed when turning into or leaving the car park.

Vehicle access to the quarry is prevented by large blocks of rock between the car park and the access path but this also hinders wheelchair users. Otherwise, entry into the quarry is along an almost level path which opens out directly on to the quarry floor, which is an irregularly oval area, c.75 m x 55 m (0.32Ha). The high, near-vertical faces form an impressive amphitheatre on three sides. Most quarry levels can only be readily examined by climbers. The faces are in two main tiers, a continuous lower one c.20 m high and a discontinuous more irregular higher set of faces up to 20m high. There is a substantial and accessible ledge separating the two faces, which opens out into a wider upper floor at the western end. This level can be easily accessed by even but steep ramp-like tracks, one rising along the southwestern side of the main part of the quarry and turning back on itself through a hairpin on to the upper level, the other branching off to the right of the main entrance path and ascending around the south-eastern end of the quarry. Further ramps lead to a path around the tops of the main upper faces and this can be followed continuously from one ramp to the other, giving good views over the quarry and the Middleton

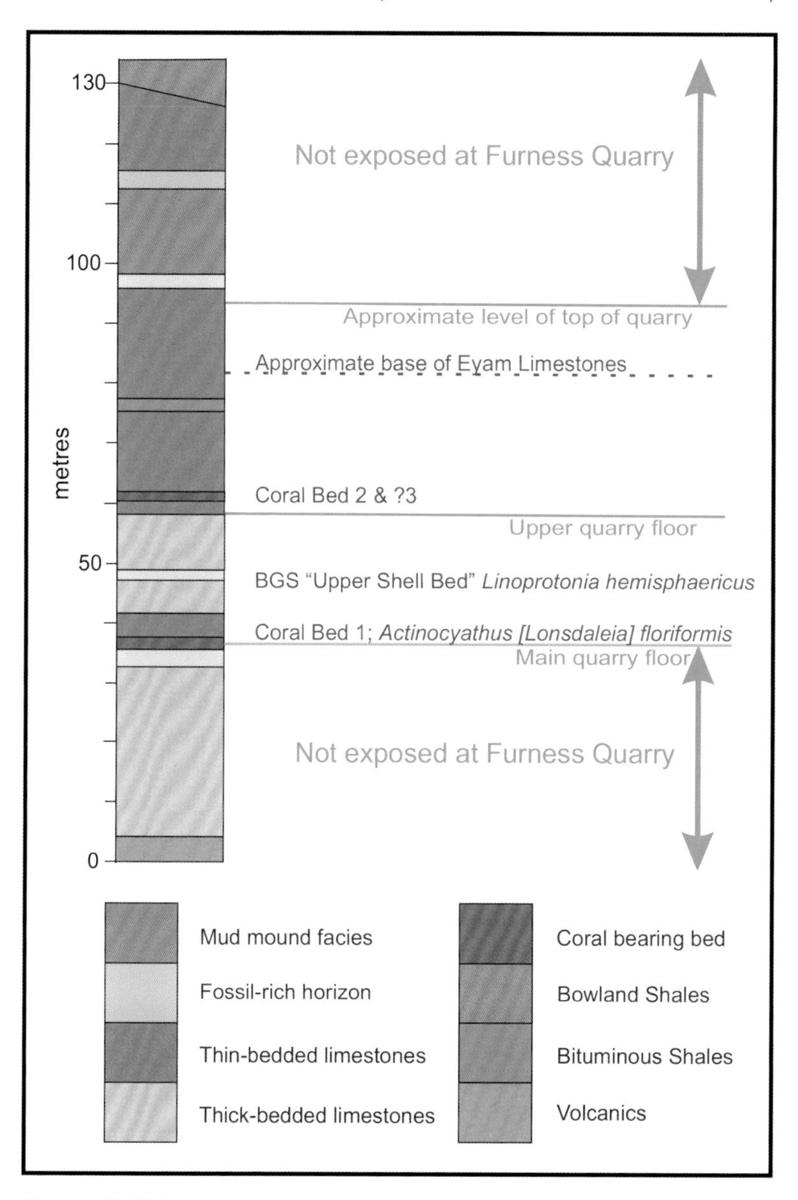

Figure 11.39 A composite log of the area around Furness Quarry with the interval exposed in quarry indicated. The majority of the limestones exposed in the quarry are within the Monsal Dale Limestones, with a small thickness of Eyam Limestones at the top (after Morris, 1929).

valley. The ramps, the main ledges and the upper floor area are bounded by steep slopes or rock faces without railings.

Although the quarry floor is part of an SSSI it is not protected in any way, and it is potentially very easy for the corals exposed on the quarry floor to be damaged by continual footfall, especially by visitors who are not geologically aware. Please do not hammer or try to collect coral samples, however small, without a genuine research reason.

There are limited facilities in Stoney Middleton about 2 km east from the quarry along the A623, and a little further to the east at Calver.

This quarry, popular with climbers, is owned by the British Mountaineering Council (BMC) who have placed a useful information board where the access path starts from the car park. BMC's ownership ensures indefinite open public access. The quarry lies within the Stoney Middleton Dale SSSI. Although the quarry is disused, vegetation overgrowth is relatively restricted and is managed to keep quarry faces clear. At one time, there were lower faces between the road and the path into the quarry, but these are now very overgrown. Therefore, the quarry section visible today is incomplete compared with older published sections (Fig. 11.39).

There appear to be few recent studies of Furness and neighbouring quarries, the older literature concentrating mostly on the local stratigraphy with much emphasis on the palaeontological (taxonomic) identification of 'marker bands' rather than on understanding the palaeoecology and palaeoenvironment of these interesting intervals. This account offers perhaps the first such published observations and interpretation of the coral bed on the floor of the quarry.

The quarry exposes an incomplete section of about 35–40 m of Brigantian limestones in the upper part of the Monsal Dale Limestones, overlain in places by an incomplete section of about 10 m of Eyam Limestones which are mostly set back from the top of the main faces. The quarry succession is in a mid-platform setting. The main accessible features of immediate interest are two, possibly three, coral beds within the Monsal Dale Limestones, the lowest of which occurs on the floor of the quarry as an extensive bedding plane exposure suitable for making basic palaeoecological measurements.

A generalized section is given for this quarry by Stevenson & Gaunt (1971). It is, however, difficult to match their details with the present

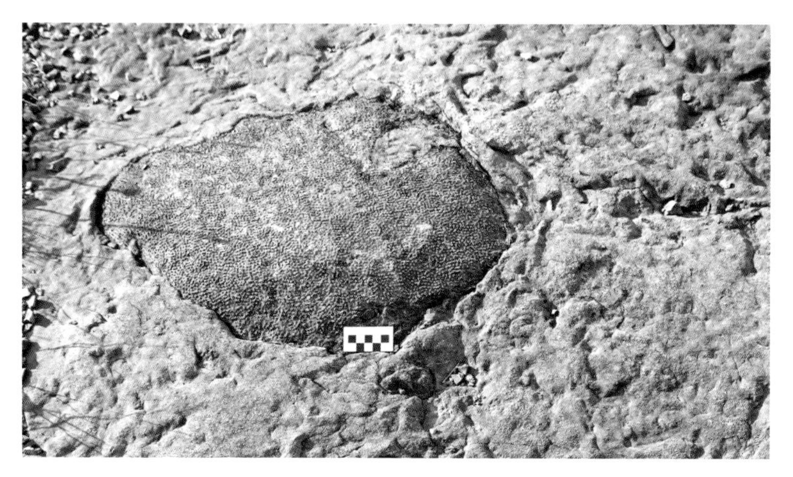

Figure 11.40 The upper surface of a colony of the phaceloid tabulate coral *Syringopora geniculata* (Phillips, 1836) on the floor of Furness Quarry. The surrounding (overlying) peloidal packstone grainstones are highly disturbed by burrowing and, in this example, drapes the margins of the colony.

Figure 11.41 The upper surface of a colony of the cerioid rugose coral *Actinocyathus floriformis* (Martin, 1809), previously called *Lonsdaleia floriformis*. This example is partially silicified and thus stands out clearly in relief.

configuration of the quarry. Morris (1929) gave a generalized section for the Stoney Middleton and Eyam area (Fig. 11.39) and a generalized one specifically for the limestones. Orme (1967) provided further useful details. It is clear from Morris's map that Furness Quarry did not exist at his time of writing. Nevertheless, his thicknesses accord better with our observations than with those of Stevenson & Gaunt.

The main quarry floor consists of a dark carbonate mudstone to wackestone whose thickness cannot be seen. This is overlain by a thin rather paler, sheet of bioturbated peloidal packstone-grainstone which has been partially removed by quarrying and which locally drapes scattered coral colonies (Fig. 11.40). Together, these two beds appear to correspond to literature descriptions of the '*Lonsdaleia* Bed' also known as the '*Lonsdaleia floriformis* Bed', although this coral is now referred to the genus *Actinocyathus*. The corals are at least in part silicified, their corallites projecting a little from the rest of the rock surface (Figs 11.40 & 11.41).

The Monsal Dale Limestones in the lower part of the quarry are in well-bedded, thick cyclical units. Compared with localities further south, around Wirksworth and closer to the palaeoslope of the platform margin, the units are somewhat thinner, with thin breaks between them, and show fewer obvious emergent surfaces. At the top of the lower face, the limestones weather to noticeably paler shades of grey, corresponding to the 'White Bed' marker horizon. Higher up, above the main ledge, the appearance of the beds is less regular, and they look thinner due to frequent cherty layers.

Given that only some of the levels in the quarry can be readily examined, we suggest visitors follow the itinerary provided here to look at selected details arranged as Stops in upward stratigraphical order. The focus is on the coral beds, as this is the best locality described in the Guide to observe easily accessible, well-preserved corals. Leaders of student groups and others with time and a palaeoecological interest might like to consider first the questions below about the coral beds, and then compare their answers against the notes provided at the end of this account.

Stop 1: Main quarry floor

Overview of the whole quarry section and coral palaeocology of the bedding plane on the quarry floor.

> Assuming that the bedding plane represents a snapshot of a Brigantian sea floor, are the corals in life position or nearly so? If so, what is the density of coral cover? Did the corals live mostly above the sediment surface (suprastratally) or mostly within it, just keeping their distal surfaces above the sediment surface (constratally)? Is this a coral reef? And if not, why not?

Stop 2: Large fallen blocks with numerous large brachiopod shells

Walk about halfway up the ramp path on the south side of the quarry to Stop 2. Fallen blocks contain numerous brachiopods. The position of these blocks suggests they represent the '*Hemisphaericus* Bed' with the brachiopods identified in the literature as *Linoprotonia [Productus] hemisphaericus*, apparently distinct from true gigantoproductids reported from Ricklow Quarry (Nolan *et al.*, 2019).

Stop 3: Fallen blocks at foot of upper quarry face

Continue up the rest of the ramp path, passing around the hairpin at the top (with view across Middleton Dale), to reach the upper ledge of the quarry. About 50 m along the path, examine faces and fallen blocks. Corals and brachiopods occur in a cherty horizon <30 cm thick about 3 m above the path level. Note the much higher proportion of chert in these upper faces generally.

> Compare the corals of this Stop and Stop 4 with those on the quarry floor. Which if any of them represents a life assemblage? In what ways do the corals at these two stops differ/resemble each other and the coral bed on the quarry floor (taphonomically and ecologically)?

Stop 4: Coral bed in quarry face

Continue to the end of this upper path where it opens out on to a wider floor within its own smaller amphitheatre. In the furthest southwest corner, a face on the south side contains a thin (<30 cm) coral bed about 3 m above the upper quarry floor. It is not clear whether this is the same coral bed as at Stop 3, as there is insufficient intervening exposure, but it seems to be a little higher stratigraphically. The same question as for Stop 3 can be addressed here too.

Another ramp ascends from between these two upper coral bed exposures and follows back on itself to become a path around the top of the quarry. Further stops can be made along this to see exposures of thinly bedded Eyam Limestones before eventually descending to the entry of the quarry.

Corals and coral palaeoecology

1. The coral bed on the quarry floor

The corals at Stop 1 are almost entirely colonial and mostly phaceloid. Preservation is good enough for provisional field identifications to be made. These include the tabulate coral *Syringopora* (Fig. 11.40), (about one-third of the total number of corals) and the rugosans *Siphonodendron* and *Diphyphyllum* (most of the remaining two-thirds), with a minority occurrence of the cerioid form *Actinocyathus [Lonsdaleia] floriformis* (Fig. 11.41). The latter gives its name to this coral bed. Nearly all the corals are more or less domal (or have weathered out in this way, often rising slightly above the even quarry floor) and visually of similar size (30–50 cm diameter), although they may be wider below the floor level. Some may be overturned, but the outcrop does not allow this to be properly observed. Provisionally, therefore, the corals are assumed to be more or less in life position and all living at around the same time as part of a single coral assemblage.

Work on growth rates in living and fossil corals gives a working figure for linear extensional increase in size of *c*.4 mm or less per year (Scrutton, 1999). As the corals on this single horizon are of a similar size to each other, this suggests that they represent a single larval recruitment event, seeded perhaps from a coral assemblage(s) elsewhere, rather than the result of ongoing longer-term recruitment (although detailed study of the sediment might reveal some much smaller colonies due, say, to some later recruitment, too). Colony sizes (above) in this assemblage are generally a little smaller than, rather than typical of, many Palaeozoic corals (Scrutton, 1999) suggesting that the assemblage survived for *c.*100 years before ambient conditions became unfavourable to corals or before they were overcome by an increase in sedimentation rate.

The dark sediment of the lower, finer-grained mudstone-wackestone part of the '*Lonsdaleia* Bed' suggests quiet conditions below wave-base, and Stevenson & Gaunt (1971) noted thickening southwest-wards of this dark facies into Cressbrook Dale and further towards the Wye Valley into the intra-platform basin of Gutteridge (1989; see the Monsal Trail itinerary below). However, Stevenson & Gaunt also noted that over its outcrop area, it is unclear whether the corals in this bed grew just above its base, or within it, or just above it. If within it at Furness Quarry, they seem to have been exhumed on the sea floor prior to deposition of the draping bioturbated bed. If just above it, then colonisation by the corals was perhaps triggered by more oxygenated conditions associated with the start of a shallowing (regressive) event which then also favoured the bioturbated facies which drapes the corals (Fig. 11.40). This suggests that ideas which Cossey *et al.* (2004) put forward about the relationship of various coral beds to regressive or transgressive facies sequences in their own studied areas, may not apply to the present coral bed.

Unlike most living corals, rugosans and tabulates appear much more commonly to have lived in or on soft substrates in muddy environments than modern scleractinian corals, even though the former commonly commenced growth on shells or other hard fragments (Scrutton,1999) as can sometimes be observed here. However, it is hard to judge whether phaceloid forms, like most of this assemblage, grew suprastratally or constratally. One possible indication of the former, but not readily checked in this field situation, would be the presence of epibionts on the exterior surfaces of the corals, assuming one could also show that these were living at the time of the corals and not after the corals had died. It is possible that the corals lived during a lapse in sedimentation and were later smothered by the conditions represented by the bioturbated bed. Cerioid corals, like the *Actinocyathus* colonies here, can spread their colonies outwards across a soft substrate helping them to avoid sinking into the sediment. Phaceloids would more likely have sunk into soft ongoing sediment as they grew and must, in general, have been either constratal or only slightly suprastratal.

Density of coral cover is a common ecological metric in living coral environments as well as fossil ones. However, it is generally more difficult to establish in fossil ones, as most outcrops do not provide large enough suitable surfaces, whether bedding planes or vertical sections. A simple approach, used here, was to make a series of improvised random line transects a few metres apart in various directions across the quarry floor, by pacing along each line and counting the paced distances between any colonies encountered along the line. Some of the surface is obscured by clitter and low vegetation, which likely hides a proportion of the existing corals, so the amount of this was also estimated along the transects. Converting paces to metres and allowing for the clitter and vegetation, corals have a mean spacing of *c.*9.5 m. Subjectively, there appear to be no obvious differences in cover between the different directions of transects, suggesting that the corals were part of a level bottom community without obvious influences of wave action or currents. Taking colony diameter into account and, assuming that the line transects represent narrow strips of substrate 1 m wide, the areal cover of the corals here is *c.*2.5%.

As the corals are well spaced and without any observed clustering in their distribution, they were obviously not building any kind of framework, and no other obvious framework organisms are present in the bed. Therefore, this is not in any sense a reef nor a 'coral carpet' (*sensu* Riegl & Piller, 2000). Although the colonies might be considered too far apart to qualify for any of Insalaco's (1998) categories of growth fabrics, this assemblage might be described as a uniform sparse pillarstone. Compared with Aretz's (2010) classification of colonial rugose habitats, the present assemblage clearly belongs with his B1 category, described as a 'regular coral meadow' of 'moderate' diversity.

Aretz (2010) also discussed how his scheme of rugosan habitats might provisionally correlate with the taxonomically-based classification of rugosan coral 'associations' (RCAs or more more correctly 'assemblages') by Somerville & Rodríguez (2007). The assemblage in the Furness Quarry floor is closest to their RCA 4 ('*Siphonodendron* – *Diphyphyllum* – *Actinocyathus*').* Aretz did not correlate this assemblage

* Guide users interested in identifying the corals mentioned here and elsewhere should refer to the illustrated field key to Lower Carboniferous genera by Mitchell (2003) and the illustrated descriptions of Scrutton (1985).

with his B1 meadow category as seems to occur here in Furness Quarry, but found instead that it correlated with his B2 habitat ('patchy [i.e. clumped] coral meadow'). Since Aretz says that his habitat types are variable and intergradational and 'not fixed boxes', this apparent discrepancy is probably not very significant. Moreover, he points out that Somerville & Rodríguez indicated strong stratigraphical limitations of their model, leading to Aretz's view that a successful classification of rugosan coral assemblages is only practical for regional models within a narrow stratigraphical range.

Integrating the various observations, some features of the population ecology of this assemblage can be reconstructed. At the point in time when water and substrate conditions favoured coral larval settlement, the assemblage developed from a single larval recruitment event. Either larval supply was relatively low, or there were insufficient amounts of hard substrate fragments for larvae to settle, so colonies were few and cover sparse. There may also have been a high rate of larval mortality. The sparseness of the coral cover was also limited by a balance between some kind of determinate limits of growth size in the corals themselves (seemingly *c.*50cm – see above) and the time interval, *c.*100 years (above) during which suitable growth conditions prevailed. In any case, the corals either died off, all at more or less the same time, perhaps smothered by the later bioturbated sediment event. As this '*Lonsdaleia* Bed' also occurs beyond Furness Quarry, further understanding of the coral assemblage here might benefit from comparisons with its other occurrences.

2. The upper coral beds

The upper coral beds are very different from the '*Lonsdaleia* Bed', even allowing for them being visible only in vertical section rather than on a bedding plane. At Stop 3, the corals are incomplete and broken, rarely more than 20 cm across, and lie in different orientations. They include *Siphonodendron*. They are in a very dense, hard pale matrix along with brachiopod fragments in different orientations and showing normal grading, suggesting a high-energy event (cf. Le Yao *et al.*, 2016). The matrix is cherty, and/or perhaps also with microbialite. In contrast to corals on the quarry floor this is not, in any way, a life assemblage. It must have resulted from reworking in relatively high-energy conditions.

This bed corresponds to the '*Dibunophyllum* cf. *muirheadi* Bed' in older literature and, although a few solitary corals occur here, this particular species was not obvious. At Stop 4, the coral bed seems less broken and more continuous with intermittent silicified phaceloid and solitary corals. These might be in place but disturbed by current action.

Monsal Trail: geobiking through the Mid-Platform Basin (▶)

*The Monsal Trail (see websites below), 13.7 km long, occupies a segment of the trackbase of the former London–Derby–Manchester main railway line built in 1863, more or less following the course of the River Wye, often high above. We promote the idea of 'geo-biking' because the Trail connects a series of easily accessible, though somewhat widely separated, important localities which provide a transect through an intra-platform basin which developed and infilled during the Brigantian. One of us (BRR) 'geo-biked' the Trail in course of preparing this Guide. The Matlock–Buxton line was closed in 1968 for money-saving reasons and the track between Matlock and Topley Pike was completely removed. The designated Trail currently runs between the area (***53.2084, −1.65664***) southeast of Bakewell, near Haddon Hall, and Topley Pike (***53.2505, −1.8341***), c 5 km E of Buxton. About 150 m to the west of the latter, freight lines from nearby limestone quarries that join the remaining stub of the former main line remain in routine operation. The stub at the opposite end, at Matlock, was reconnected to Rowsley South as a preservation line run by Peak Rail, with a Derwent Valley Cycleway alongside most of it. Plans to extend this cycleway to meet the Haddon Hall end of the Monsal Trail have been taking many years to realize.*

The Monsal Trail localities treated here can be visited individually, or in small neighbouring groups. Cyclists in particular can cover them all in a single day. The Trail is often high above the valley floor, and as the latter is often narrow and gorge-like, close access by vehicle and parking, mostly paid, is surprisingly limited. Access points with paid parking are as follows.

- *Topley Pike (Wye Dale Car Park) (**53.2495, −1.8457**) and A6 layby (**53.2493, −1.8334**)*
- *Miller's Dale Station (**53.2563, −1.794**)*
- *Upperdale (**53.2463, −1.7367**) (very small, but free)*

- *Monsal Head (**53.2401, –1.7250**)*
- *Hassop Station (**53.2314, –1.6756**)*

Of these, Topley Pike, Miller's Dale and Hassop give the most direct access. Access from the A6 layby, Upperdale and Monsal Head, is via footpaths that are not ideal for cyclists and disabled people. Many other access points exist for walkers using footpaths, including from Bakewell. The relatively few car parks are very popular in holiday seasons and, in good weather, at weekends all the year round. Note that Hassop Station is not close to the village of Hassop.

For bus route access, Guide users should consult the webpage below.

https://derbysbus.info/maps/county.htm

The Monsal Trail is dedicated for use by walkers, cyclist, horse-riders and disabled people. It is fully surfaced in places like the tunnels (which are lit) but otherwise consists of fine compacted chippings with broad grassy verges.

*Bicycles can be hired at Hassop Station and Blackwell Mill (**53.2511, –1.83301**) near Topley Park, but not at Miller's Dale Station. Cafés and toilet facilities are limited to Miller's Dale Station and Hassop Station, but others are within footpath distance at (e.g.) Monsal Head.*

The length of the Trail, and the distances between key geological points, particularly favour 'geo-biking'. The entire Trail with points of geological interest can easily be covered 'there and back' in a single day. Surfaces are suitable for almost any kind of bicycle and gradients are scarcely noticeable, having been engineered for fast expresses and heavy freight trains. Some of the Trail can be followed remotely using Google Street View although the images were taken when vegetation was prolific.

Somewhat confusingly, the Wye valley has different 'Dale' names along its length, including Monsal Dale, Wye Dale, Miller's Dale and Chee Dale.

Surprisingly, there is no published geological guide that specifically treats the Monsal Trail. Cope's (1999) Geologists' Association Guide follows the Wye Valley downstream from Buxton to Ashford in the Water with only passing mention of the Trail. His route uses roads and footpaths as well as some of the Trail, and information usually has to be mentally transposed to outcrops along the Trail. This is supplemented by the BGS sheet memoir (Aitkenhead *et al.*, 1985). Another useful source is Butcher & Ford (1973) which treats localities along the line as a transect but includes numerous other places in the Wye Valley area. Walkden (1977) includes a transect along some of the line and Cossey *et al.* (2004) directly follow 9 km of the then apparently still undesignated Trail in locality descriptions. They also include side-trips to other localities, notably Cressbrook Dale, White Cliff and Hobs House, but footpaths in some cases are not straightforward.

The geological trend along the route going eastwards is from oldest strata near Topley Pike, which provides a relatively rare sighting (for this Guide) of Holkerian Woo Dale Limestones to youngest Dinantian (Brigantian Monsal Dale and Eyam Limestones etc.) around Bakewell. There are particularly good exposures of volcanic lavas. The Monsal Trail section and immediate area has attracted research effort for at least a century, and this still continues. The present itinerary concentrates on the main features exposed in cuttings along the Trail between Topley Pike and the eastern end of the Monsal Viaduct cutting (**53.2412, −1.7288**).

The Intra-Platform Basin developed during Brigantian times between the syndepositional movements of Longstone Edge Monocline and the Taddington–Bakewell Anticline (Fig. 4.10). Much of it shows deeper-water redeposited facies, sometimes slumped down a southeast-facing palaeoslope (Gutteridge, 1989). The facies make an interesting contrast with the time-equivalent shallow-water facies in the Monsal Dale and Eyam Limestones to the west and south. BGS maps use a darker colour shading for the 'dark lithofacies' of the basinal Monsal Dale Limestones. The Basin also includes some distinctive facies, including resedimented turbidites and slumps as well as the unusual, almost black Ashford Marble (not seen along the Trail) and problematic laminites such as the Rosewood Marble. This occurs at the eastern end of the Monsal Dale Viaduct Cutting and, though apparently basinal, seems not to be a deeper-water deposit. The influence of the syndepositionally growing Cressbrook Anticline can also be inferred from localities along the Trail.

There has been a long tradition of basing stratigraphy of Dinantian limestone successions on 'marker bands', generally designated by an index name of its more abundant or significant fossil(s), e.g. Hobs House Coral Bed, Upper *Girvanella* Band. Some of these evidently have stratigraphical value over relatively local areas, whilst some apparently extend further afield. Emphasis has largely been on their empirical stratigraphical value, and less attention has been paid to the palaeo-ecological implications of their often apparently homogeneous, thin, basin-wide bio-assemblages, apart from the Hobs House Coral Bed (see also Furness Quarry for palaeo-ecological discussion of the *'Lonsdaleia' floriformis* Band). The same applies to many other named bands in Carboniferous limestones throughout the rest of the Pennines. Some authors have referred to some of them as biostromes, but in particular for those bands in which the fossils appear to be in life position, it is difficult to think of modern analogues on a corresponding lateral scale (tens of kilometres). Guide users with palaeontological interests are invited to consider the taphonomic and palaeo-ecological nature of these marker bands where the quality of exposures is sufficiently good. There are several examples along the Trail.

Cossey *et al.* (2004) suggested that the coral beds, in particular, may represent the bases of transgressive sequences overlying earlier regressive units, although unlike living photosymbiotic scleractinian corals, most Palaeozoic corals were apparently not restricted to shallow euphotic waters. Manifold *et al.* (2020, 2021) concluded that emergent surfaces in the Brigantian were not as laterally persistent as workers had previously believed, and perhaps the same applies to many of the supposed marker bands.

Stop 1. Blackwell Mill: Woo Dale Limestones

Starting from the Topley Pike car park (**53.2495, −1.8457**), or from the Blackwell Mill bicycle rental (**53.251, −1.8330**), dark, relatively thinly bedded, somewhat rubbly facies of highest Woo Dale Limestones can be seen along the Trail east of Blackwell Mill before reaching the first of the two viaducts where the Trail crosses the Wye. The limestones are lenticular-bedded wackestone-packstone beds 0.5–1.0 m thick with fenestrae, gastropods, bivalves and brachiopod valves (Fig. 11.42). Sharp, erosively based beds of fine-grained bioclastic grainstone are also present. These limestones are thought to have accumulated peritidally on a shallow carbonate platform surrounded by tidal channels and areas of subtidal deposition.

Figure 11.42 Thinly bedded, rather lenticular limestones in the uppermost part of the Woo Dale Limestones, east of Blackwell Mill.

Stop 2. Chee Tor Rock in Chee Dale: Bee Low Limestones

The previous outcrops are overlain beyond the first viaduct by Bee Low Limestones, which have previously been called Chee Tor Rock and Miller's Dale Limestones. They occur in imposing cliffs, crags and quarry faces and in cuttings as far as Miller's Dale Station (approx. **53.2563, −1.7944**). The Trail does not pass Chee Tor proper (**53.2562, −1.8157**) but this can be seen by following the attractive riverside footpath below the level of the Trail from Blackwell Mill and then either returning to Blackwell Mill to regain the Trail, or continuing until the riverside path reaches a footbridge and the steps beyond it, to ascend on to the Monsal Trail. (If the river is in spate, do not follow this riverside path, especially this last part, which runs along the foot of a cliff on stepping stones. If in doubt, seek local advice.)

The Bee Low Limestones are particularly thickly bedded and massive in this area, but as with the formation over much of the rest of the Derbyshire Platform, they show strongly developed cyclothems with emergent surfaces and palaeosols, at least in part involving volcanic ash (Fig. 11.43). The limestones consist of 'pale bioclastic peloidal

Figure 11.43 Irregular relief on a palaeokarstic emergence surface within thickly bedded, massive limestones of the Bee Low Limestones. A thin layer of clay is present on the surface.

packstone-grainstone with minor wackestone and with calcrete features associated with palaeokarstic surfaces' (Cossey *et al.*, 2004). As in many places, Bee Low Limestones here formed on a shallow platform.

Stop 3. Miller's Dale Station: Monsal Dale Limestones on Bee Low Limestones (53.2563, -1.7952)

The wider regional significance of the exposures at Stops 3, 4 and 6 to 11 relates to the evolution of the Intra-Platform Basin (Walkden, 1977; Gutteridge, 1989 & 1990b; Cossey *et al.*, 2004, particularly pp. 318–319). As its rather complex development is summarized in the introduction to this itinerary and is outlined in Chapter 4, it is not repeated in individual locality descriptions.

The station car park is bounded on its north side by faces of the former Station Quarry, *c.*15 m high, running parallel to the Monsal Trail. Since disuse, these faces have become much overgrown, and it is difficult to see many of the details described by Walkden (1977) and Gutteridge (1990). However, a recent rock-fall (seen by chance by BRR in February 2023 and just visible in Google Streetview, shot in October 2022) combined with seasonal loss of tree leaf and vegetation die-back, has revealed more detail (Fig. 11.44). A few tens of cubic metres at most have fallen, but with numerous saplings and ivy still in place on the rock-fall scar, this new exposure may not be visible for long especially in summer. Much of the face is still loose, and the area has been railed off. The risk of injury is obviously high and access should not be attempted. Accordingly, the following notes are not based on recent close-up examination, but on the description of Gutteridge (1990) (Fig. 11.45).

Early literature relevant to this quarry refers to Station Quarry Beds overlying Miller's Dale Limestones, but current stratigraphical terminology places Monsal Dale Limestones above the Bee Low Limestones. The Chee Tor Rock and Miller's Dale Limestones are now combined as Bee Low Limestones, and the Station Quarry Beds are now placed in the lowest part of the Monsal Dale Limestones below the Upper Miller's Dale Lava.

Gutteridge's (1990) account focuses on the Station Quarry Beds and his measured log (Fig. 11.45) records just the uppermost 1 m of

Figure 11.44 The scar of a recent rock fall at Miller's Dale Station Quarry exposing a lenticular body of dark clay at the boundary between the Bee Low Limestones and the Monsal Dale Limestones. This is probably 'Cope's Pit' interpreted by previous authors as a palaeokarstic dissolution feature related to the emergence of the Bee Low LImestones during a relative low stand of sea level at the Asbian–Brigantian boundary. The first thick bed of the Monsal Dale Limestones passes horizontally over the pit.

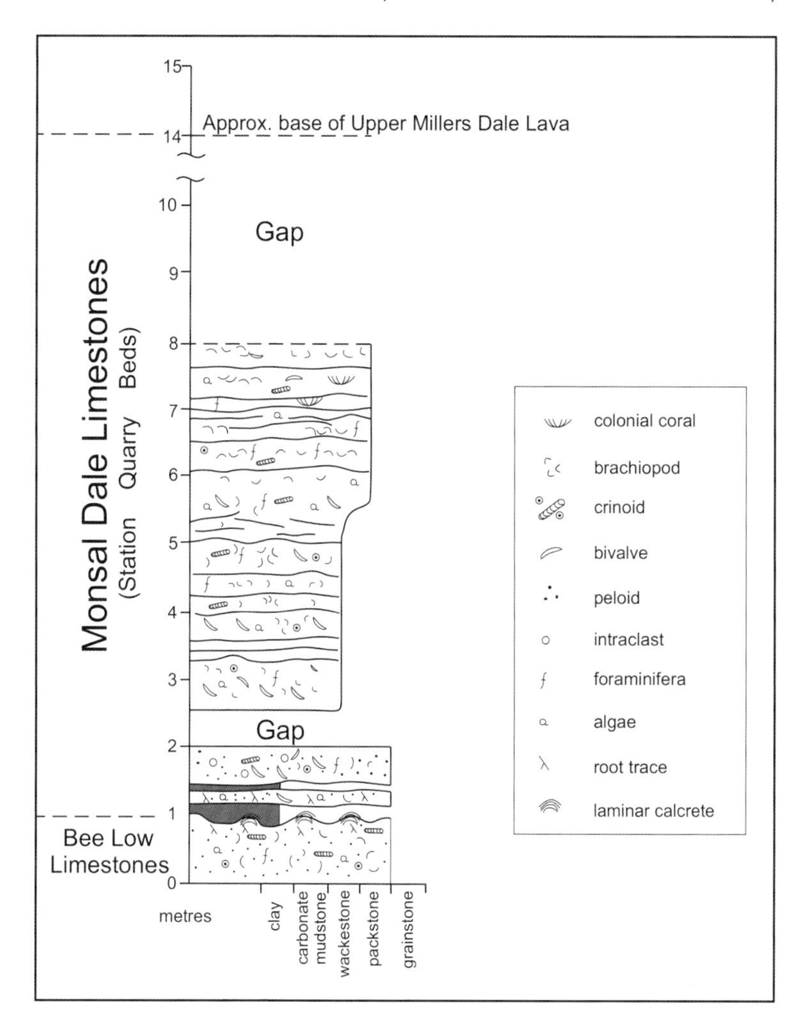

Figure 11.45 Measured sedimentological section through the top of the Bee Low Limestones and the lowest part of the Monsal Dale Limestones at Miller's Dale Station Quarry. There are about 5 m of Bee Low Limestones below the base of the section exposed at the quarry. The clay beds directly above the unconformity probably thicken to fill the pit shown in Figure 11.44. (after Gutteridge, 1990b)

Bee Low Limestones below which the full quarry face actually shows 6–7 m of regularly and thickly bedded limestones. The irregular and mammilated top characterizes the boundary with the Monsal Dale Limestones. These start with *c.*50 cm of poorly exposed, very thinly bedded irregular limestones, above which is *c.*1 m of what appears to be a single bed (Fig. 11.44). The densely vegetated ledge above this bed obscures the rest of the section, although this can be extrapolated from Figure 11.45 and appears to correspond to a further 6 m of Monsal Dale Limestones with a further 6 m exposure gap to the approximate base of the Upper Miller's Dale Lava.

The thinly bedded basal beds of Monsal Dale Limestones (Station Quarry Beds) here have been interpreted as the earliest Brigantian transgression over Bee Low Limestones. This is now regarded as a major sequence boundary that is seen widely across the Derbyshire Platform (e.g. Harboro' Rocks, Redhill Quarry, Baileycroft Quarry, and at Litton Mill Cutting West (Stop 6 of this itinerary). However, Walkden (1977) and Gutteridge (1990) argue that the transgression began earlier in this area because of downwarping of the platform during an early phase in the formation of the Intra-Platform Basin.

The facies of the lowest one metre or so of the Monsal Dale Limestones are described by Gutteridge (1990) as filling a palaeo-karstic surface initially by a non-marine volcanic clay and subsequently by shallow-water, peritidal then restricted subtidal limestones. They were deposited in a sheltered embayment, which later deepened to form the Intra-Platform Basin, whose characteristic dark limestones are exposed at numerous points further along the Monsal Trail.

Walkden (1977) and Gutteridge (1990) both refer to earlier observations of the Asbian–Brigantian boundary and document several substantial depressions on its irregular palaeokarstic surface. One example was described by Cope (1999) with a sketch of a hollow or pit, or possible channel (his Fig. 2). He later redrew it differently for his GA Guide (1999, Fig. 7). The pit was apparently obscured by the time of later authors' observations. It was reported to be *c.*3 m deep and somewhat wider and contained a large 'water-worn boulder' *c.*1.5 m in longest dimension. Walkden suggested that the boulder might have been part of the irregular surface of the pit. Gutteridge regarded the depression as karstic, i.e. a solution hollow, rather than a channel.

It is not clear where exactly along the quarry face 'Cope's Pit' (named herein) occurs. He says it was *c.*150 m 'east of the road' but this cannot be within this quarry area, and it only makes sense if he meant 'west' of the road. If so, its position closely matches that of the recent rock-fall, which seems to have revealed the same depression. Its dark clayey infill likely facilitated the rock-fall. Figure 11.44 clearly shows basal beds of the Monsal Dale Limestones spanning the depression as in Cope's sketches, although some of these have slipped down in front of the rock-fall scar. The depression must therefore have formed on the surface of the Bee Low Limestones prior to the deposition of the basal shallow-water limestones of the Monsal Dale Limestones.

Stop 4. Upper Miller's Dale Lava: Miller's Dale near Ravenstor (53.2556, −1.7787)

At 1.04 km eastwards along the Monsal Trail from Miller's Dale Station area near Stop 3. Both sides of the cutting show 6–8 m of dark basaltic lava overlain by *c*.2 m of thinly and irregularly bedded limestone (Fig. 11.46). The limestones and the contact are too high to see details, but the contact is clearly undulating and irregular. BGS map the limestone as dark facies of the Monsal Dale Limestones and do not indicate the presence of the lava, probably because the cutting is almost vertical and capped by limestones, so the lava cannot be depicted in plan view. BGS puzzlingly indicates Bee Low Limestone along the Trail at this point, but it is not visible. Perhaps it underlies the trail surface and therefore the lava. Judging from the horizontal scale in Cossey *et al.* (2004, fig.7.7), the

Figure 11.46 Massive dark basalts of the Upper Miller's Dale Lava overlain by *c*.2 m of thinly bedded Monsal Dale Limestones; Monsal Trail, near Ravenstor.

lava at this point lies entirely within the Monsal Dale Limestones, and only rests on Bee Low Limestones further east.

The lava is irregularly jointed but in one place shows a rounded 'nose' facing westward within its mass. This suggests that there are different, discrete flow lobes within the two major flows mentioned by Waters (2003). The Upper Millers Dale Lavas have resulted from two major extrusive subaerial flows on an emergent surface of Brigantian strata within the lowermost part of the Monsal Dale Limestones. They flowed broadly eastwards, the upper phase terminating *c*.860 m eastwards along the Trail (**Stop 6**). The lava is highly altered, fine-grained, amygdaloidal, and in places with olivine phenocrysts.

Stop 5. Ravenstor view: Bee Low Limestones (53.2550, –1.7769)
View across Wye Valley from Monsal Trail 140 m east of Stop 4. The viewing point is on the left (N) side of the trail through a break in the trees. It gives a striking view north-eastwards across the Miller's Dale gorge of the River Wye (Fig. 11.47). The view is dominated by a near-vertical cliff of very pale-weathering Bee Low Limestones beneath the summit of Ravenstor. The Lower Miller's Dale Lava outcrops at the foot of the cliff below a marked overhang (Cope, 1999) with cave-like

Figure 11.47 Ravenstor seen across the Wye gorge from the Monsal Trail. The cliff comprises entirely Bee Low Limestones, more massive and thickly bedded in the lower part and with clearer, thinner bedding in the top. Figures for scale.

hollows, but it is not really visible from here. About half the cliff above the overhang (say 15–20 m) consists of massive limestones with almost imperceptible bedding that suggests the presence of at least five units 3–4 m thick. At the top of these massive beds is a marked discontinuity, somewhat irregular, and probably with a wayboard, above which are *c*.15–20 m of more thinly bedded limestones mostly in units of 1–2 m. The massive lower section corresponds visually to what BGS and others have mapped as 'Chee Tor Rock' as seen at Stop 2. The more clearly bedded higher section corresponds visually to 'Miller's Dale Beds', as seen at Stop 3. Both are now included in the Bee Low Limestones.

Stop 6. Litton Mill: Upper Miller's Dale Lava front and Monsal Dale Limestones (53.2537, –1.7666)

Exposures are on both sides of Trail cutting *c*.856 m east of Stop 5.

6a. Cutting on the south side of the Trail

This is one of the geological highlights of the Monsal Trail, being a rare example of an advancing lava front 'frozen in time'. This is an SSSI and there is an information board giving a summary of the main geological features and their importance. Over the years, however, it gradually became badly obscured by vegetation, but in April 2022, a work-party spent a day cleaning it (Waltham, 2022) with excellent results. Unfortunately, finely fragmented volcanic material facilitates quick recolonization and the exposure will need quite frequent maintenance.

The cutting exposes the Upper Miller's Dale Lava and overlying Monsal Dale Limestones. The height of the exposure is 3–4 m but as the lava is dipping, the total thickness exposed is *c*.9 m. As at Stop 4, the general nature of the basalt is highly altered, amygdaloidal, fine-grained, and in places with olivine phenocrysts. The texture is blocky and finely brecciated. However, in contrast with Stop 4, the lava proper here occurs in striking, more or less rounded structures up to 1 metre across separated by a matrix of fine dark granular material (Fig. 11.48).

The lava mass is wedge-shaped both eastwards and westwards. The downslope of its upper surface on the westward side (right when facing the outcrop) dips westwards, though broken into scarp-like steps. This seems to be its present-day erosion surface, but it might have an older origin. The dip of the eastern surface (left) is eastwards as is that of the

Figure 11.48 The Upper Miller's Dale Lava, showing the chaotic fabic made of volcanic pillows within a poorly sorted matrix of highly fragmented volcanic debris.

Figure 11.49 The 'frozen' inclined eastern flow margin of the Upper Miller's Dale Lava near Litton Mill. A wedge of orange clay of probable volcanic origin separates the lava from thinly bedded Monsal Dale Limestones that drape the inclined lava front.

limestones which overlie it, although steeper than most of the limestone beds (Fig. 11.49). Within the lava mass, the dip is also eastwards, but very variable and irregular, often less steep than the front itself. The lowest, most easterly point of the lava is at the level of the Trail, and all accounts say that this is not the true toe of the lava, which actually lies quite close beneath Trail level.

Petrographically, the dark granular material consists of small glassy shards and fragments (hyaloclastics). These are interpreted as resulting from the advancing margins of hot moving lava being suddenly quenched to form glassy (palagonitic) material as it entered water, while at the same time being explosively shattered into hyaloclastic fragments. Lava rubble is also present, but there appears to be no published mention of pumice, commonly associated with underwater eruptions. The bomb-like bodies are not volcanic bombs but represent separate lava masses (pillows) showing internal brecciation and exfoliation. Unlike many pillow lavas, however, the pillows are not contiguous, but are supported by hyaloclastic material, suggesting that they represent individual innermost zones within a more rapidly cooled outer zone of hyaloclastics and palagonite. Much of the hyaloclastics were redeposited.

Earlier authors have described this outcrop as a flow-foot breccia, but the volcanic processes merit further discussion. Waltham (2022) states that after forming, pillows 'rolled down the underwater slope and settled within the mass of debris'. Similarly, Walkden (1977) had proposed a 'hyaloclastic carpet' which accumulated in front of a lava that was shattering as it advanced across the floor of a stretch of water. The disposition of the pillows within the lava mass, and its irregular imbricated dips, suggest that the lava advanced in short-lived eastward pulses over each other, shattering in the process. On the other hand, superficially, the overall fabric is reminiscent of a succession of submarine debris flows, perhaps with phases of reworking by water movements. Indeed, Walkden went so far as to suggest that the imbricate sheets might be 'analogous to the foresets of a sedimentary cross-bedded unit.'

The volcanic interval apparently does not end with the upper surface of its main body. Recent clearance of the face revealed a long-unexposed metre-thick orange-brown layer overlying its eastern end (Fig. 11.49). Waltham (2022) believes this to be either a hyaloclastite or tuff, and offers explanations for its striking colour, which have yet to be resolved by microscopic examination.

Above the orange deposit is a wedge about 2 m at its thickest point, which according to Gutteridge (1989) consists of 'bioclastic grainstone/

packstone facies'. They were transported and deposited in a 'high-energy environment derived from a shallow-water environment over the nearby lava flow, then washed off the lava and deposited in this wedge banked up against the flow front'. These are poorly fossiliferous and more or less concordant with the lava front angle. Walkden (1977) did not think that the dip of this wedge was tectonic, but because they contain horizontal geopetals, he suggested that, like the lava imbrication below, it probably consisted of foresets, in this case of sediments banked against the foot of the lava. Gutteridge & Walkden (1988) thought slumping was a possibility too. As shown by Cope (1999, fig. 11), the bioclastic wedge is overlain with angular unconformity by a thicker set of limestones with a shallower dip. It oversteps the lower set, apparently cutting them out erosively. These higher beds are better seen on the opposite side of the cutting. See further under 6b below.

6b. Outcrop on north side of Trail cutting; opposite 6a

On the north side are excellent exposures of dark thinly and regularly parallel-bedded Monsal Dale Limestone in 20–50 cm units (Fig. 11.50). These limestones are mapped by BGS as dark lithofacies. Waters (2003) gives a thickness of 17.95 m and describes them as dark grey, thinly-bedded,

Figure 11.50 Dark, thinly- and parallel-bedded Monsal Dale Limestones on the Monsal Trail near Litton. The bed of comminuted coral fragments (Fig. 11.51) occurs in the lower part of this interval.

Figure 11.51 A limestone bed, in the lower part of the section in Figure 11.50, containing broken and clearly re-deposited fragments of coral.

very fine-grained with chert nodules, whilst Cossey *et al.* 2004 characterize them as 're-sedimented and periplatform carbonates'. They are very similar lithologically to the upper beds on the south side of this cutting above the wedge of bioclastic grainstone/packstones. Gutteridge (1989, Figs 6 & 7) shows the relationships of the two limestone facies to the lava front, and refers these higher ones to his 'bioclastic wackestone-packestone facies', while Gutteridge & Walkden (1988) state that they are probably turbiditic. At about 2 m above Trail level there are fossiliferous horizons, including corals (e.g. *Siphonodendron*) but they are generally broken and not in place (Fig. 11.51). Other horizons seem to be (auto?-)brecciated. The sedimentary characteristics here are consistent with deposition in the Intra-Platform Basin.

Stop 7. Litton Tunnel West Portal Cutting (53.2530, –1.759)
Exposures are on both sides of the Trail in the portal cutting about 1.03 km east of Stop 6.

About 12 m of rather thin and well-bedded dark, subtidal Monsal Dale Limestones overlie *c.*4 m of thickly bedded platform carbonates of the Bee Low Limestones, with marked palaeokarstic discontinuity at the contact – a consistent feature across the platform (e.g. Harboro' Rocks and Redhill Quarry). However, the cutting is now rather overgrown. Cossey *et al.* (2004) indicated that a 0.5 m thick interval comprises deep brown to bluish-grey clay-filled pits containing pebbles of the Station Quarry Beds (i.e. Monsal Dale Limestones below the Upper Miller's Dale Lava; see Stop 3).

The overlying Monsal Dale Limestones apparently comprise bioclastic grainstones with reworked gigantoproductids, crinoids and solitary corals. These were laid down in the Intra-Platform Basin during highstands when the surrounding carbonate platform was flooded. Coarser bioclastic beds may have been deposited by storm events or were generated by slumping, perhaps, as turbidites (Butcher & Ford, 1973). Gutteridge & Walkden's (1988) reconstruction of the effects of the syndepositional uplift of the Cressbrook Anticline is consistent with the Monsal Dale Limestones here belonging to Gutteridge's bioclastic grainstone/packstone facies. Between Stops 6 and 7, the basinal environment apparently became deeper, and the facies finer-grained before returning to shallow-water coarser facies here close to the axis of the Cressbrook Anticline to the east.

Stop 8. Litton Tunnel East Portal Cutting (53.2510, −1.7523)

Exposures occur on both sides of the Trail in the portal cutting 520 m (entirely in the Litton Tunnel) east of Stop 6. The contact between the Monsal Dale Limestones and Bee Low Limestones at this locality has been widely discussed. In one respect, the relationship is the same as that at Stop 7 and elsewhere, and it invariably incorporates a palaeokarstic surface with palaeosol wayboard(s). However, here, in addition, the Monsal Dale Limestones occupy a substantial karstic depression *c.*7 m deep and 25 m wide (Fig. 11.52). The basal fill consists of buff clay with pebbles and thin shales. Below this contact, the Bee Low Limestones consist of *c.*4 m of thickly bedded platform carbonates. The overlying Monsal Dale Limestones are well bedded dark limestones described by Gutteridge & Walkden (1988) as bioturbated packstones with a diverse bioclast fauna recording deposition above storm wavebase. However, the bedding seems very irregular. Here, and in the slopes above, are well-reworked bioclastic grainstone intervals deposited above fair-weather wavebase consistent with deposition on the axis

Figure 11.52 The unconformable contact between the Bee Low Limestones and the Monsal Dale Limestones at the eastern portal of Litton Tunnel, marked by a large palaeokarstic trough or depression. Its surface is overlain by a thin clay representing a palaeosol and/or volcanic material.

of the syndepositional Cressbrook Anticline. Continuing further east, the upper surface of the Bee Low Limestone rises until the formation occupies much of the height of the cutting.

The excellent views from the north side of the Trail, just a few metres from the cutting between here and the west portal of Cressbrook Tunnel, show almost continuous outcrops of dark facies (though very pale-weathered) of the Monsal Dale Limestones above Water-cum-Jolly Dale.

Stop 9. Cressbrook Tunnel East Portal Cutting (53.2485, −1.7441)
Exposures on both sides of the Trail in the portal cutting, 620 m east of Stop 7, much of this distance in the short (430 m) Cressbrook Tunnel.

In the cutting east of the tunnel there are *c.*8 m of dark, fine-grained cherty limestones in the lower part of Monsal Dale Limestones. Two slumped units of recumbently folded bioclastic packstone-wackestone are visible on the south side of the cutting (Fig. 11.53). The limestones are evenly bedded bituminous-smelling packstone-wackestone (Gutteridge, 1989) deposited in deeper water than the limestones at **Stop 8**. Here they were laid down on the eastern flank of the syndepositional Cressbrook Anticline where they slid down slope and underwent plastic deformation. The recumbent nose (Fig. 11.53) points approximately SE consistent with and anticlinal axis running NE-SW (Gutteridge, 1989, Fig 6).

Figure 11.53 Recumbently folded, thin-bedded Monsal Dale Limestones in the eastern portal cutting of the Cressbrook Tunnel. The folding resulted from plastic deformation during slumping on the flank of the syndepositionally growing Cressbrook Anticline.

Butcher & Ford (1973) and Gutteridge (1989) report the 1 m thick Upper Dale Coral Bed halfway up this section. The corals are mainly phaceloid lithostrotionids, not in growth position and evidently reworked. The bed rests on a scoured surface and is graded. Erosively based, graded coral beds like this, containing reworked colonies, indicate transportation of colonies into deeper parts of the Intra-Platform Basin, probably by debris flow, although not over great distances (Gutteridge, 1989).

Stop 10. Monsal Viaduct Cutting, Monsal Head and Hobs House

Since this is also a stop along the present geobiking itinerary on the Monsal Trail, the geological part of any walking route begins on the Trail just west of the Viaduct cutting at **53.2445, −1.7352**. *This is the nearest Trail point on the walking route suggested below to Stop 9 (eastern portal of Cressbrook*

Tunnel) and lies 750 m along the Trail east of that Stop. However, since the suggested route also applies to walkers visiting the area as a one-off trip, and to accommodate both walkers and cyclists, the present circular walking route begins at the free small Upperdale Car Park **(53.2464, –1.7367).** *Overall, the walk is not suitable for most disabled people. Apart from the Trail itself and some short lengths of road, the walk uses steep unsurfaced paths, sometimes with irregular steps.*

There are a hotel and café at Monsal Head, and toilets adjacent to the Monsal Head car park.

The precise name for the famous viaduct here is 'Headstone Viaduct' although it is now invariably known as the 'Monsal Dale Viaduct'. Although the adjacent cutting is an integral part of any cycling trip along this part of the Trail, the outstanding and famous view from Monsal Head (**53.240784, –1.72499**) over the surrounding area is less easily accessed by bicycle from the Trail, although straightforward for those cycling the roads to Monsal Head. Since this immediate area is one of the most interesting and attractive along the Wye valley, Guide users with less time might want to visit it as a one-off trip using the following circular route of less than 2 km with some alternative starting points.

Butcher & Ford (1973) provide field notes and a section of the cutting. It would be worth making a scan or photo of these illustrations for one's device, or a hard copy, to use whilst exploring the Viaduct cutting. Extensive geological details of the Viaduct cutting were given by Gutteridge & Walkden (1998) but this conference field guide is not widely available. If Guide users can obtain a copy, it is well worth bringing it into the field. But see also Gutteridge (1989). However, since their account, the cutting has become much more overgrown and many of the details are now hard to see. Discreet and judicious use of secateurs might help.

10a Upperdale to Putwell Hill Vein

The free Upperdale car park is served by bus and additional services pass Monsal Head from where it is a short (1.03 km) downhill walk to the starting point. Monsal Head can also be used if the Upperdale car park is full, as there is much more (paid) parking space there.

From the Upperdale car park, take the track to the southwest past the riverside cottages, cross the bridge and ascend the valley side south-south-eastwards to pass under the railway bridge carrying the Monsal Trail.

*Almost immediately after the bridge turn sharp right to climb up to the level of the Trail near the southeast end of the now derelict Monsal Dale Station (***53.2441, −1.7347***). This track was the access route to the station before the line closed. Here, those using the present geobiking route from Stop 9 join the walking route as they head east along the trail. From the Upperdale car park to this point, the walk is about 350 m. At this point, walkers should turn sharp right (south-eastwards), away from the station area and, like the cyclists, continue from here along the Trail for 160 m to reach the Putwell Hill Vein (***53.2437, −1.7330***) at the start of the viaduct cutting, c 50 m W from a footbridge over the Trail.*

The mineralized Putwell Hill Vein is also an important fault running east-southeast– west-northwest, down-throwing to the north-northwest. The vein itself, which has a mine adit entrance, can be seen on the south side of the Trail (Fig. 5.3). The main remaining mineral appears to be calcite in columnar masses, in several parallel zones, with long axes of crystals normal to the steeply sloping vein wall. The adit presumably coincides with where the ores, mostly lead, have been extracted.

10b Monsal Viaduct Cutting: northwest end

A few paces from 10a, and on both sides of the cutting, exposures of dark, relatively thin, and regularly bedded limestones are visually similar to those seen around the Litton and Cressbrook cuttings. They are mapped by BGS as dark lithofacies of the Monsal Dale Limestones. They dip to the east towards the footbridge, exposing some 50 m of beds through the whole cutting. Gutteridge (1989) shows 'thickly bedded, graded bioclastic calcisiltites with erosive bases interpreted as low-density turbidites' at his Locality 9 which is in, or close to, this Viaduct cutting. He assigns this lithology to a 'resedimented' facies described as 'impure limestones 0.1 m–2 m thick comprising fine sand- to silt-sized comminuted bioclasts'. The beds are graded with sharp and sometimes erosional bases. Their petrography indicates derivation from a shallow marine environment, but the lack of reworking and the interbedded shales suggest deposition in a low-energy environment, below wavebase, from waning flows, probably low-density turbidity currents. The setting is the Intra-Platform Basin on the eastern side of the Cressbrook Anticline.

Two particular details should be noted at this western end of the cutting: the Hobs House Coral Bed and notable slumping.

The Hobs House Coral Bed, *c.*20 cm thick here, is seen 2 m above track level on the south side of the cutting. The bed is densely packed with the clisiophyllids *inter alia*, which seem to be mostly broken and eroded. Here they are also affected by mineralization and slickensides, presumably related to the proximity of the Putwell Hill Vein and Fault. Consistent with the turbiditic deposition of adjacent beds, the corals seem to have been transported within one or more turbidity currents. However, at the type locality (Hobs House (10e): about 700 m directly south-southeast from here, the bed is several metres thick, the corals are *in situ*, with the uppermost levels showing the most reworking. This direction to Hobs House suggests that it lay further down the palaeoslope of the Intra-Platform Basin and therefore is unlikely to be the proximal source for the redeposited corals at the same stratigraphic horizon seen here. Bearing in mind that isolated coral colonies occur elsewhere along the Viaduct cutting, it is possible that Hobs House corals were generally living in these slightly greater depths, and instead of being transported wholesale, were disturbed, rolled and broken locally by turbidity currents.

Gutteridge & Walkden (1987) report better outcrops of the Hobs House Coral Bed in the more or less opposite side of the cutting, 40 m from the footbridge. However, these faces, which also show some notable large-scale slumping, are now rather overgrown, and substantial removal of vegetation would be required to reveal the coral bed. Guide users might like to try and locate it, especially in winter, and, if successful to assess the taphonomy of the corals (e.g. whether transported or not). One conspicuous large slump near the footbridge, now very overgrown, shows strata tilted steeply eastwards. Like other slumps along the cutting, it reflects unstable conditions on the southeast-facing palaeoslope of the Intra-Platform Basin.

Gutteridge & Walkden (1987) and Butcher & Ford (1973) described other interesting details including redeposited and *in situ* productid beds.

*10c Monsal Viaduct Cutting: Southeast end: Rosewood Marble (**53.2417, –1.7294**)*

This end of the Viaduct cutting is 350 m from the Putwell Hill Vein (10a).

On the southwest side of the cutting where it ends, there is a gate to several paths that follow the valley sides and river bank downstream towards Brushfield and other places. Exposures of Rosewood Marble can be seen in the rocky parts of the path by the gate and in the nearest part of the cutting proper. **(Please do not hammer or remove specimens from the exposures)**. Quite substantial young trees now make it difficult to reach the cutting sides and, in the deep shade, it is not easy to make it out the Marble horizon. This is not a true marble, of course, but one of Derbyshire's notable limestones that can be polished for ornamental purposes.

Gutteridge & Walkden (1987) and Gutteridge (1989) describe the Marble as a fine laminite consisting of up to 1 m or more of finely interleaved (*c.*0.5–5 mm thick) layers of dolomitized grainstone and calcite mudstone in striking convolutions, small pull-aparts and microfaulting (Fig. 11.54). The laminae show evidence of normal sea-water salinity. This intriguing deposit has proved to be an enigmatic challenge for research. Unsurprisingly, there have been several very different, sometimes contradictory, interpretations, which are also, in some cases, difficult to reconcile with its occurrence within the Intra-Platform Basin juxtaposed with turbidite deposition. More importantly, interpretation also rests on the question of whether the dolomitization is primary or secondary.

Such fine duplex lamination brings deposits like varves to mind, and older interpretations include (1) a deep-lagoonal deposit with alternating deposition of carbonate mud and dolomite due to seasonal climatic variation and (2) an epigenetically dolomitized off-shelf storm laminite. Peter Gutteridge (unpublished), however, found that dolomitization post-dates cementation of the grainstone laminae and concludes that the dolomitization is secondary, which rules out earlier interpretations. He argues that it is a partially intertidal deposit formed in ponds that received occasional storm wash-over from an adjacent marine area, with the laminites deposited in palustrine (marshy) to lacustrine conditions related to a short-lived drop in relative sea level, which led to widespread

Figure 11.54 The Rosewood Marble, a thinly laminated interval of dolomitized grainstones and calcite muds deformed by small-scale fracturing and folding that probably relates to small-scale movements on a slope.

subaerial emergence. During such a break in marine deposition, the floor of the Intra-Platform Basin was possibly just submerged, perhaps giving rise to marsh, wetland or ponds, albeit of normal salinity. It is difficult to picture such a shallow environment located on or even at the shallow margin of a palaeoslope.

The convolutions, overthrusts and pull-aparts within the laminites probably indicate instability of the sediment on the southeast-facing palaeoslope of the Intra-Platform Slope, analogous to the much larger slumps seen elsewhere along the Viaduct cutting, leading Gutteridge to infer that the Rosewood Marble is allochthonous. Perhaps the entire body of finely bedded deposits was still in a relatively plastic, deformable state when relative sea level rose, and shortly afterwards it was triggered into down-slope slumping by a single event such as an earthquake, or by a sequence of such events.

10d Monsal Head view

*Parking is possible either at Monsal Head (paid) (**53.2401, −1.7250**) or in the small free car park at Upperdale (**53.244, −1.7367**). Buses also stop at Monsal Head. For those doing the geobiking itinerary, there is no easy access* *to Monsal Head from the Viaduct unless they onerously carry their bikes or lock them in a safe place near the Viaduct and walk up. Alternatively, they may prefer to omit the view and either continue on the Trail to Hassop Station, via Stop 11, or return to Topley Pike or wherever else their trip began. The total distance from Monsal Dale Station along the Trail to the foot of the path up to Monsal Head is 740 m.*

For walking up to Monsal Head from the Viaduct, there is a steep narrow path which starts from the northeast side of the Trail between the southeast end of the Viaduct and the western portal of the Headstone Tunnel. The path ascends northeast through trees with many steps in places to reach the well-known viewing point over Monsal Dale by a lane and close to the Hotel, café and toilets. From the Rosewood Marble outcrop at 10c, the walk to Monsal Head is 340 m.

The only other obvious cyclists' route from 10c is to return across the Viaduct towards Monsal Dale Station, and to take the track just to the east of it (10a), which descends to Upperdale, where this circular walk begins. This is probably better suited to mountain bike users as it is steep and rough until the valley floor is reached. At Upperdale, turn right onto the steep lane that goes directly up to Monsal Head (1.03 km). Return to the Trail by retracing the route.

The view from Monsal Head is one of the most popular in the Peak District, with the strongly incised 'hairpin' meander of the River Wye below giving two views at once, one straight ahead upstream (northwest) towards Upperdale (at the start of this round walk), and the other to the left, downstream towards Brushfield. The valley sides are very steep, almost gorge-like, and the meander system is strongly incised into the White Peak plateau. The geomorphology and palaeohydrology of the Wye and Lathkill catchment area has been studied by Banks *et al.* (2012). See also Banks & Allen (2017).

The main meander, below the viewpoint, is bridged by the Monsal Dale Viaduct, one of the most famous railway viaducts in the country, not so much for its size as for its location and the way it emphasizes the scale of the landscape. It draws the eye to the focal point of the meander. Hard to believe now, but its construction in 1863 was famously subjected to a bombastic rant by John Ruskin (https://en.wikipedia.org/wiki/Monsal_Dale). Notwithstanding, most people, consciously or otherwise, recognize the aesthetic effect of a well-placed focal detail like this viaduct in a wider landscape, although Ruskin seemed (unusually for him) to be unaware of it.

10e Hobs House Coral Bed at Hobs House

It is possible to do a side-trip walk from the Monsal Head viewpoint to the Hobs House Coral Bed (c.950 m each way), by taking the footpath southwest from the viewing point. Hobs House (**53.2384, −1.7382**)*, is a series of large land-slipped, castle-like blocks. The path initially contours along the valley and later descends toward the footbridge over the Wye. However, turn off left before that, after c.700 m from Monsal Head, to walk in a more westerly direction from around* **53.2385, −1.7348** *and to go up through the woods. However, the former paths are very overgrown and very difficult to locate and follow, and Hobs House does not come into view until further up.*

Figure 11.55 The uppermost layer of the Hobs House Coral Bed at Hobs House. The bed is in the Monsal Dale Limestones *c.*40 m below the Rosewood Marble. In this coral rudstone the corals are partially or entirely silicified and with large solitary rugosans belonging to the family Aulophyllidae. They are mostly lying on their side and appear to be somewhat abraded and broken, probably transported only a short distance from their life positions suggesting a low-energy environment, below fair-weather wave-base, but above storm-base. Blue-black cherty nodules also occur within the coral bed.

Furthermore, the approach to Hobs House crosses almost totally overgrown loose, blocky scree that requires great care. Hobs House lies about 250 m from the main path to the footbridge and the coral bed is exposed on the faces of the main landslipped blocks (Banks & Allen, 2017).

The Hobs House Coral Bed (see also 10b) occurs high in the Monsal Dale Limestones, *c.*40 m below the Rosewood Marble. The corals are partially or entirely silicified. They typically include large solitary rugosans belonging to the family Aulophyllidae, which mostly lie on their sides, abraded and broken (Fig. 11.55), and were probably transported only a relatively short distance from their life positions. The lower part of the Coral Bed is dominated by phaceloid rugosans like *Siphonodendron*. This kind of solitary coral assemblage suggests a low-energy, quiet environment in moderately shallow water. The top of the bed is marked by an undulating discontinuity, thought to be palaeokarstic with a calcrete (Cossey *et al.*, 2004). Blue-black cherty nodules occur within the coral bed here and continuous chert overlies the discontinuity.

From here, return to Monsal Head, to complete the circular walk back to Upperdale.

Return to Upperdale car park.The lane by the viewing point (not the main road, B6465) descends steeply north-westwards all the way to the Upperdale car park. The distance is 1.03 km. The entire circular walk is less than 2 km excluding the detour to Hobs House.

Stop 11. Headstone Tunnel East Portal Cutting (53.2393, −1.7193)

From Monsal Viaduct, cyclists can continue their geobike ride ESE-wards from the Viaduct to this final stop along the Monsal Trail by passing through the Headstone Tunnel (425 m long) towards Hassop Station.

The cutting begins directly one emerges from the tunnel at its eastern portal. Gutteridge (2023) has recently re-evaluated his own work and others' previous work on the origin and nature of the laminite seen here at the boundary between the Monsal Dale Limestones and the Eyam Limestones. In essence, this represents shallow freshwater carbonate deposition in the Intra-Platform Basin when the surrounding Derbyshire carbonate platform was emergent, and differs in some respects from the Rosewood Marble laminite (Stop 10c above) which, amongst other things shows small-scale slumped contortions. As the cutting section is too complex to summarise briefly, Guide users are recommended to refer to Gutteridge's paper for details, and to allow plenty of time to gain a good understanding of them. Examples of Headstone Laminite are also abundant in the walls of buildings in Ashford in the Water and can easily be examined there too.

The total length of the geobiking itinerary from Topley Pike to the Headstone Cutting East Portal is 9.3 km, although the Trail continues to Hassop and beyond Bakewell.

Chapter 12

Staffordshire Basin and its Eastern Margin

This area extends from the western edge of the Derbyshire Massif across large areas of the Staffordshire Moorlands to the west (Fig. 12.1). The western margin of the massif coincides roughly with the Derbyshire–Staffordshire border and, in places, has a spectacular topographic expression. The southwest corner of the limestone platform is complex with a belt of 'Waulsortian' mud mounds interbedded with limestone turbidites that have been quite intensively folded by Varican compression.

Across the moorlands, Namurian sediments occur in broad folds with axes trending north–south. The sediments mainly range in age from Pendleian to Yeadonian but, in the centres of some synclines, strata extend into the early Westphalian (Langsettian). Most of the succession, up to mid-Marsdenian, comprises mudstone with marine bands, with thin turbidite sandstones derived from the south. The main basin fill, which culminates in the Roaches Grit, is of mid-Marsdenian (R_{2b}) age and is overlain by a series of deltaic cyclothems.

The folding of the Roaches Grit at the southern end of the Goyt Syncline gives rise to the spectacular scenery of the Roaches, Hen Cloud and Ramshaw Rocks and to other edges and crags further north (Fig. 2.3).

Brown End Quarry (SK 292230; 53.0486, –1.8667)

This locality is not included in Fig. 12.1 and lies some 8 km to the south of Ape's Tor. The disused quarry on the north side of the A523 lies about 1 km east of Waterhouses village. It is accessed through a narrow bridge over the River Hamps, immediately west of a roadside cottage. There is a parking place for quarry visitors just over the bridge, with space for around 12 cars. The lane to the right is only for visitors to Brown End Farm. The quarry, which ceased operations in the 1960s, is now an SSSI and is owned and managed

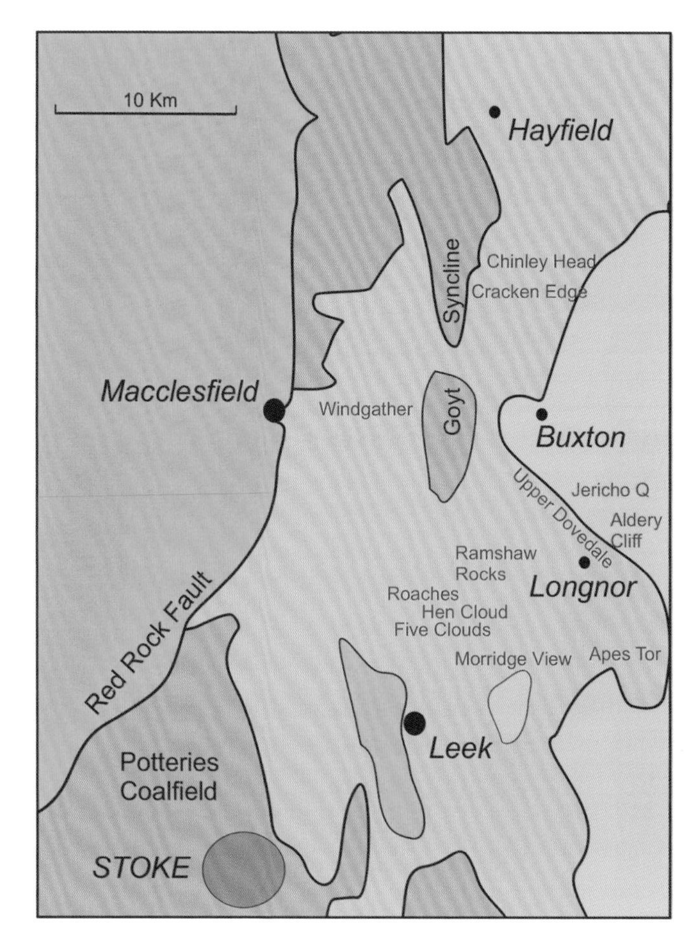

Figure 12.1 Map of the western part of the Derbyshire Massif and the Staffordshire Basin, showing positions of localities described in this chapter.

by the Staffordshire Wildlife Trust. It is managed both for geology and for its biological interest with several useful and graphic information boards around the quarry. Unlike many disused quarries, the faces are kept relatively clear of vegetation, providing good views of the geology. The faces, which are up to around 12 m high, are fenced off at a safe distance and the fence should not be crossed nor the faces climbed. There is a good footpath, which can be rough and muddy in places, along the length of the quarry, and there are higher viewing points along the SW side. Wheelchair access is possible, although the footpath just above the car park is quite steep. Anyone seeking close access to the faces for serious research should contact the Staffordshire Wildlife Trust.

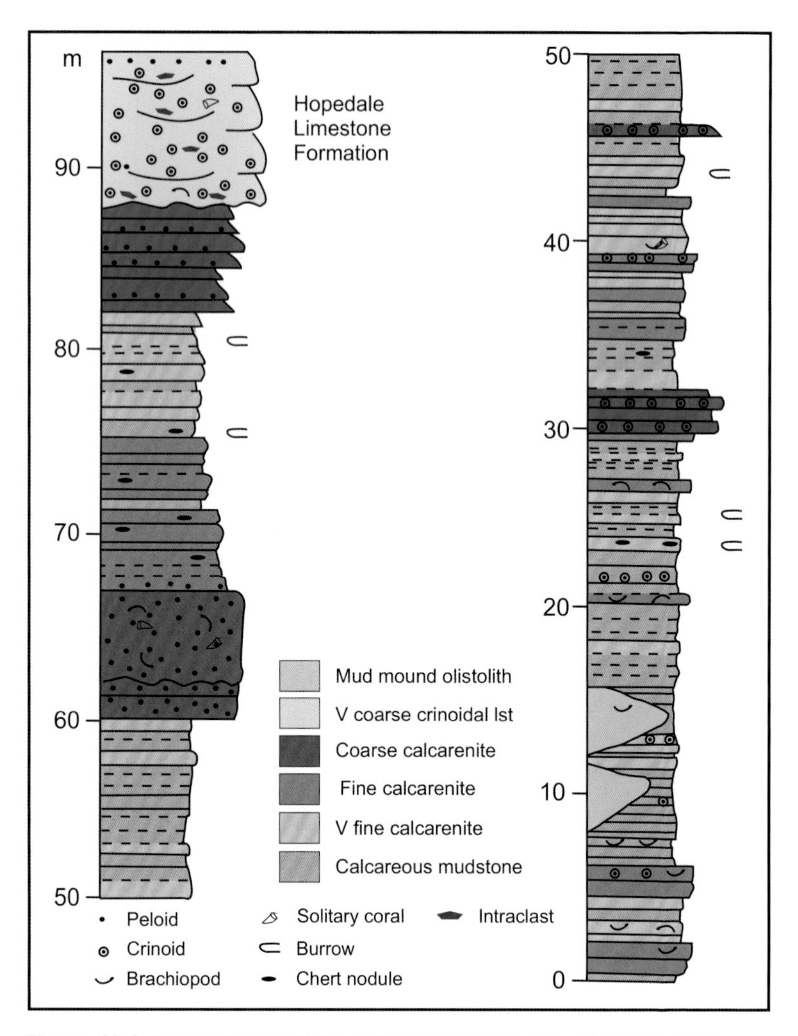

Figure 12.2 A measured section through the succession exposed in Brown End Quarry, near Waterstones. The lower part of the sequence is within the Milldale Limestones, whilst the top 4 m are in the base of the Hopedale Limestones (after Cossey *et al.*, 2004).

These sediments were deposited in the south-eastern corner of the Staffordshire Basin, an off-platform or basinal setting. It lies close to the northern margin of the Staffordshire Shelf, which is an extension of the Midland Landmass. The area immediately to the north is a belt of deep-water ('Waulsortian') mud mounds and associated sediments that lies south of the main Derbyshire platform. The locality provides a good example of off-platform facies, contrasting with the platform and platform-margin settings described elsewhere.

The beds dip vertically, striking NNW–SSE and young to the WSW. The steep dip is a result of Variscan deformation with rather ductile, basinal sediments being compressed between the more rigid massifs to north and south. Such deformation clearly involved significant tectonic shortening, and this makes it difficult to estimate horizontal distances at the time of deposition.

The quarry exposes around 95 m of sediments, the lowest 87 m of which are within the Milldale Limestones and the uppermost 8 m are assigned to the Hopedale Limestones (Fig. 12.2). The two units have contrasting facies and are separated by an erosion surface but with no angular discordance. Hopedale Limestones are confined to the NW end of the quarry and the bulk of the exposure is within Milldale Limestones.

The ages of both formations have proved contentious with conflicting interpretations of their included fossils and microfossils. The Milldale Limestones have yielded both Courceyan and Chadian microfossils. A Chadian age is now accepted, as Courceyan microfossils could have been reworked from older sediments, and the presence of olistoliths of mud-mound material equates with Chadian deep-water mud mounds (Waulsortian) present in Dovedale and around the southwestern margin of the massif, as far distant as Ecton. These limestones are, therefore, amongst the oldest exposed Carboniferous sediments in the southern Pennines.

The Hopedale Limestones, here, have yielded microfossils typical of Chadian and Arundian ages and reworking of older microfossils can again be invoked. The Milldale and Hopedale limestones appear conformable, despite their erosional contact, and an age around the Chadian/Arundian boundary for these basal Hopedale Limestones seems reasonable. The Hopedale Limestones extend to the north and their upper part is equivalent to the Ecton Limestones, which are of Asbian age and are exposed in the Manifold Valley, some 8 km to the north.

The Milldale Limestones here comprise interbedded limestones and calcareous mudstones with limestones beds ranging from very fine to very coarse bioclastic calcarenites. The clasts comprise mainly peloids, crinoid fragments and brachiopods, but colonial and solitary corals also occur and, exceptionally, trilobites have been found. Bed thicknesses are typically centimetres to tens of centimetres, although there are some thicker beds, most noticeable the 4.6 m unit of very coarse calcarenite above the 62 m point in Figure 12.2. This unit has a basal erosion surface that shows minor discordance with the beds below. The package of very coarse calcarenites at the top of the Milldale Limestone interval comprises a series of sharp-based graded beds, some of which are reported to have flute marks, but no palaeocurrent directions have been recorded. Within some finer-gained intervals, horizontal burrows (e.g. *Zoophycus, Planolites*) are present on bedding surfaces, showing that the sea floor was oxygenated. In the lowest 20 m seen high on the quarry face, are two rounded bodies, a few metres in diameter, made up of rather massive micrite (Fig. 12.3). Their lithology is typical of that seen in mud mounds generally, particularly the local Chadian deep-water examples seen in Dovedale. These mud-mound bodies have scattered fossils and stromatactis cavity fills. Infills of cavities within one body shows it to be inverted; it must have been detached and re-sedimented from an *in-situ* mud mound.

The facies of the Milldale Limestones point towards deposition in relatively deep water. The lack of evidence for high basin energy such as ripples or cross-lamination suggests a setting below storm wave base. Calcareous mudstones record the quiet background sedimentation whilst the coarser beds, from calcareous siltstone to very coarse calcarenites, record resedimentation of material generated in shallower water by density underflows of one sort or another. The thin calcisiltite beds were probably deposited from dilute turbidity currents, possibly generated by storms in shallower water. Thicker, coarser beds are likely the deposits of more powerful turbidity currents that may have been generated by slumping at a steep basin margin. The thick interval of very coarse calcarenite may record a major debris flow or a series of closely spaced flows, perhaps generated by slumping of a shallow platform margin. The lack of palaeocurrent indicators makes the source hard to specify. The Staffordshire Shelf is a possible source, but crinoid debris commonly produced thick flanking beds to mud mounds when mound growth ceased, and these could have become unstable and been resedimented. The bodies of mud-mound material are olistoliths that were transported to the area by sliding or rolling, from nearby, up-slope mud mounds, probably located to the north. The

Figure 12.3 Milldale Limestones in lower part of the exposed section at Brown End Quarry, showing the generally parallel bedding of the limestone beds and a large block of mud mound which is a displaced olistolith. The block is around 3 m wide.

gradient for such displacement need not have been high. We have not included specific localities to illustrate the *in-situ* Waulsortian mounds of this area, but accessible examples can be found around Wetton and Thor's Cave in the Manifold Valley, and along the gorge stretch of Dovedale south of Milldale.

The 8 m interval of Hopedale Limestones at the top of the quarry section comprises very coarse bioclastic limestones with a high proportion of coarse fossil fragments and some intraclasts. The interval lacks any finer-grained partings but is divided by internal erosion surfaces above which are graded intervals. This unit overlies the Milldale Limestones without any significant break, suggesting a continuation of deep-water conditions. The sediments probably record a succession of sediment gravity flows, probably triggered by sediment instability at a shallow-water shelf edge, probably the Staffordshire Shelf. It is possible that phases of instability resulted from simple sediment accretion or were triggered by seismic shocking during a period of active tectonic extension.

Apes Tor, Ecton (SK 098586; 53.1251, -1.8535) (▶)

Park at the roadside about 400 m north of the junction with the lane to Warslow. Walk along the road to the small quarries immediately above the road. This is a National Trust property and there are marker posts for a small geological trail. It is also a UKRIGS and should be treated with respect, including not hammering. Good disabled access along the road.

These old quarries are in the Ecton Limestones (Asbian; probably Bee Low Limestones equivalent) and show limestones and interbedded mudstones deformed in sharp folds. The rather dark limestones are broadly parallel bedded and are separated by thin, dark muddy partings. The limestones are bioclastic and have sharp bases and some erode into one another. Some have larger intraclasts of early cemented limestone. The beds lack clear internal lamination and are interpreted as proximal turbidites, derived from the adjacent limestone platform. It is highly likely that they thin out into the basin in more distal settings.

The folds have north–south axes, rather flat, steeply dipping limbs and sharp axial zones, suggesting significant bedding-plane slip during folding (Fig. 12.4). They formed during Variscan deformation when rather ductile interbedded and rather muddy sediments were squeezed against more rigid sediments, between the Manifold Valley and Dovedale. This rigidity may have resulted from large, deep-water mud mounds, such as those that characterize early Dinantian succession to the south and east.

Figure 12.4 Tight folding of limestone turbidites of the Ecton Limestones (Asbian) at Ape's Tor, Manifold Valley.

About 500 m down the road, just south of the road junction to Warslow, is the site of the Ecton Copper Mine, high on the eastern side of the valley. Some of the old workings and spoil heaps are now rather obscured by vegetation but an impression of the layout can be gained from the lane to Warslow that goes steeply up the valley side. This lane is quite narrow and twisting so that stopping a vehicle to view the opposite hillside is tricky. Walking up the lane from the crossing at the River Manifold is a safer way to get a view. Whilst Ecton Limestones are at outcrop here, the copper mineralization probably extended down into the older Milldale Limestones.

Mining was carried out here from the Bronze Age until the end of the nineteenth century. Production was at its peak in the eighteenth and nineteenth centuries when it was one of the main sources of copper in Britain. Working extended to around 300 m below the level of the River Manifold and the workings involved some of the earliest winding and pumping systems. The mine and some its surface buildings, including the Engine House, high on the hillside, are now preserved and the site is managed as an educational trust.

The Ecton copper deposits, hosted in limestone, are something of an anomaly. Most copper mineralization in the North Midlands occurs in Triassic sandstones, as in the Cheshire/Shropshire area (e.g. Alderley

Edge and Clive). Apart from Ecton, all mineralization in the Derbyshire limestones involves lead, zinc and various subsidiary minerals.

Upper Dovedale, NW of Longnor (SK 083657; 53.1874, –1.8758) (▶)

From the crossroads in the centre of Longnor, take the road to the north, towards Buxton. After about 800 m, park in the pull-in area on the left-hand side just before the road junction. Cross the road with care to view the valley below and the hills on the eastern side of Dovedale. It is possible to walk a little way down the track below to avoid vegetation that can obscure the view. Disabled access from car park and across the road but the track is steep.

This roadside stop provides a view of the western margin of the Derbyshire Massif where it passes abruptly into the Staffordshire Basin. The road runs along a ridge formed by the Longnor Sandstone (Namurian, Kinderscoutian, $R_{1c(v)}$), a turbidite unit equivalent to the fluvial Upper Kinderscout Grit of the main Pennine Basin some 10 km to the north. These turbidite sandstones represent the first arrivals in the Staffordshire Basin of feldspathic sands derived from the northern source area. These sands dominated the fill of the Pennine Basin to the north and arrived in the Staffordshire Basin from the latest Kinderscoutian times onwards. The turbidity currents that deposited the Longnor Sandstone are thought to have spilled from a barrier located near the northwest corner of the Derbyshire Massif.

The view to the northeast, across Upper Dovedale, shows the steep margin of the limestone platform, extending to the south for several kilometres. Behind the upper break of slope are flat-lying bedded limestones of the carbonate platform top. The limestones at the break of slope are predominantly rather massive lime mudstones that developed as a fringing belt of mud mound build-up, sometimes termed a 'fringing reef'. At the foot of the slope, and extending across the floor of the valley, are mudstones of Namurian age (Morridge Formation) which drape the lower part of the limestone topography, but which initially covered the whole limestone massif with a thinned succession. The north-eastern side of Dovedale is therefore an exhumed end-Viséan topography, complicated by faulting. This morphology is preserved through having

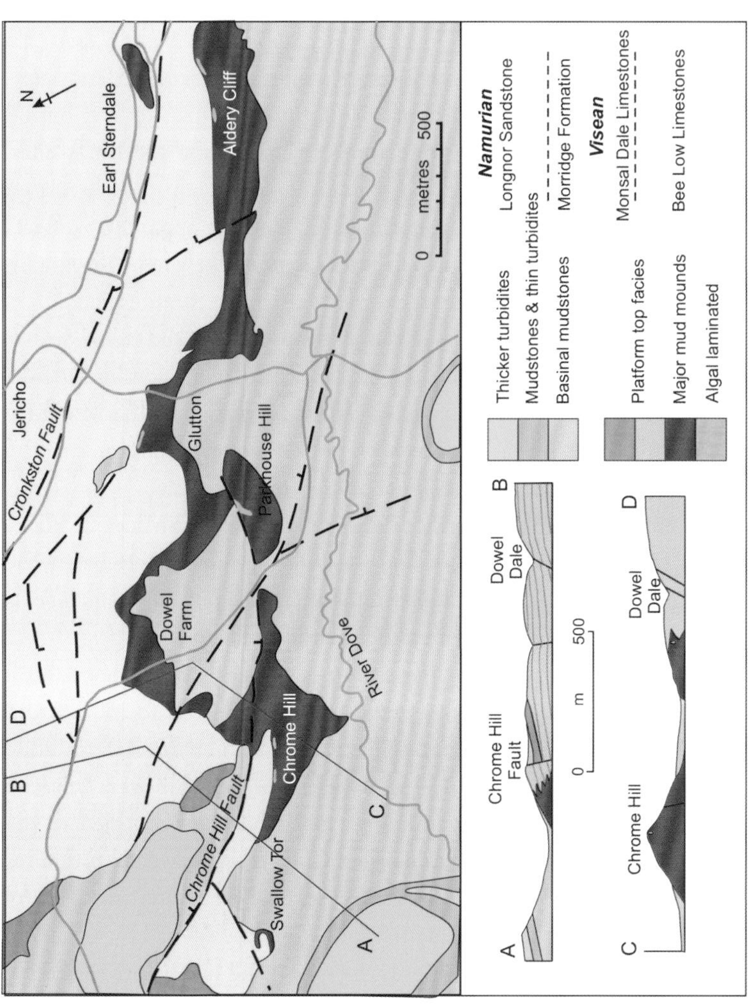

Figure 12.5 Map of the margin of the Derbyshire Massif in Upper Dovedale showing how the spectacular topography results from the exhumation of Asbian platform-edge mud mounds displaced by later pre-Namurian faulting (based on Aitkenhead *et al.*, 1985).

been buried by Namurian mudstones before it was exhumed through more recent erosion (Figs 12.5; 12.6).

The limestones involved mainly fall within the Bee Low Limestones (Asbian). Here, there are no equivalents of the Eyam Limestones (late Brigantian), and the Monsal Dale Limestones (early Brigantian) are represented on the platform by only small erosional remnants. This means that the surface of the limestone massif in this area is largely the product of Asbian deposition with some possible modification by later Brigantian erosion. Some carbonates may have spilled over the platform edge in Brigantian times, even though few Brigantian limestones are found on the platform top in this area.

The situation in Upper Dovedale is complicated by the fact that Chrome and Parkhouse Hills, which are major features of the landscape, are separated from the main limestone margin by faults. The distribution of the Morridge Formation mudstones, which drape the limestones, suggests that this faulting was pre-Namurian and is most likely related to an intra-Brigantian tectonic episode. The faulting apparently reflects an extensional stress regime, which makes it more likely to be Brigantian than Variscan, when stresses here were dominantly compressional. It is possible that detached masses of the platform margin slid basinward as it was uplifted. However, the faults on the north-facing sides of the hills suggest some additional fault activity after deposition of the Morridge Formation. There is clearly more work to be done on reconstructing the tectonic, depositional and erosional history in this area.

The mudstones of the Morridge Formation (Bowland Shale equivalent) occupy the present valley floor and the hillside immediately below the viewing point. They are mainly unfossiliferous, but the sequence is punctuated by thin, goniatite-bearing marine bands indicating ages from Pendleian to Kinderscoutian. These crop out extensively in the banks of the River Dove, and permission would have to be sought to collect from these. In addition, the mudstone sequence has thin quartz-rich turbidite sandstones (sometimes called protoquartzites), which were derived from the south (Midlands Landmass), and which are confined to the Staffordshire Basin and the Widmerpool Gulf. These quartzitic turbidites are more thickly developed in the central parts of the Staffordshire Basin.

Further views of this spectacular landscape can be had by following the lane to Hollinsclough that joins the main road just before the traffic

Figure 12.6 The western margin of the Derbyshire Platform in Upper Dovedale. Chrome and Parkhouse Hills are faulted fragments of the marginal mud mound complex (Asbian; Bee Low Limestones), sometimes referred to as a 'fringing' or 'apron reef'. The fragments became detached from the main platform prior to the deposition of Namurian Bowland Shales that wrap around their bases. Morridge Formation (Bowland Shales), ranging in age from earliest Namurian up to Kinderscoutian, occupy the valley floor.

lights. The lane drops down onto the valley floor, giving an impressive view of the platform-margin relief. This route shows the highly embayed nature of the margin as mudstones drape the flanks of the hills and extend back into the embayment valleys. Turning right in Hollinsclough village takes you up the Longnor Sandstone escarpment, the lower part of which is occupied by the Morridge Formation (Bowland Shale).

Aldery Cliff (SK 098664; 53.1941, -1.8547) (▶)

This locality can be approached either from Crowdecote or from Earl Sterndale. The lane is quite narrow and, near Earl Sterndale, rather twisting with blind bends. There is space to park several cars on the side of the road at the quarry, which is entered through a small gateway. Disabled access to the quarry floor is uncertain but much can be seen from road. The quarry is a climbing face run by the British Mountaineering Council. The face is fairly stable, but there is always a danger of falling rocks, especially if climbers are active. Helmets are a good idea. It is easy to climb around on the limestone steps and platforms on the lower part of the cliff. A hand lens is helpful as the rocks are rather homogeneous.

This abandoned limestone quarry is cut into the flank of one of the large mud mounds that fringe the western margin of the limestone platform. The face is around 30 m high and shows a mass of rather featureless carbonate mudstone within the Bee Low Limestones (Asbian). The lime mudstones have rather dispersed small fragments of crinoids and brachiopods but there is no evidence for any binding organisms that might have stabilized the lime mud during growth of the mound. It appears that the generation and binding of the carbonate mud were both part of the same complex, mainly bacterial, process which is still not fully understood. Bedding is very poorly defined but the steps in the exposure at the foot to the face suggest bedding surfaces with a low dip. The main face of the quarry is cut by a series of near-vertical joints with other joints at a high angle to the face together forming a conjugate set which is probably a response to Variscan tectonic compression.

Across the road, small natural exposures in the hillside show poorly defined bedding dipping towards the basin to the south.

Morridge view (SK 028595; 53.1337, –1.9586) (▶)

This roadside viewpoint is about 1.5 km northeast of the hamlet of Thorncliffe and about 1 km from the former Mermaid public house. There is a pull-in for several cars and there is no need to cross the road. This stop is pointless if visibility is poor.

The magnificent view to the west shows very clearly the geology of the southern end of the Goyt Syncline (Fig. 2.3). The hillsides directly below the road are underlain by early Namurian (up to early Marsdenian) sediments of the Morridge Formation, mainly mudstones, with interbedded thin protoquartzitic turbidite sandstones that were derived from the south. These beds dip to the west and are exposed in stream sections, but they are not recommended except for specialist studies. More distant to the west is the round form of Gun Hill, which is underlain by the same Morridge Formation succession, folded in a broad anticline. An exploration well was drilled on this structure in the 1940s and penetrated a Viséan basinal succession of mudstones and limestones.

To the northwest, the main feature in the landscape is the ridge formed by the Roaches Grit (Marsdenian, R_{2b}) which wraps around the nose of the northward-plunging Goyt Syncline. The eastern limb of the fold is picked out by the prominent feature of Ramshaw Rocks, just above the A53 road. The western limb is picked out by the prominent ridge that constitutes The Roaches. At the fold axis, the Roaches Grit feature is cut by small faults so that the sandstone stands proud as a separate crag, Hen Cloud ('Cloud' is a local term for a prominent crag). To the north, the two limbs of the Roaches Grit diverge either side of the syncline axis, and further ridges, formed by stratigraphically higher sandstones in the Upper Namurian and the Lower Westphalian, wrap around the axial zone. These include the Chatsworth Grit (Marsdenian, R_{2c}), the Rough Rock (Yeadonian G_1) and the Woodhead Hill Rock (Langsettian). Further north in the core of the syncline, thin coal seams were once worked on a small scale.

The westward-dipping strata (not exposed) around the viewpoint extend to the east, where they form the western flank of the Mixon Anticline, a domal structure that brings latest Viséan shales to outcrop, with slightly older limestone at the core. This structure occurs in an area of poor exposure but is interesting in that it exposes a silty and calcareous sandstone, the Onecote Sandstone, at the top of the Viséan succession. This sandstone is petrographically different to the protoquartzitic sandstones derived from the south, which characterize the lower part of the Namurian succession. Its provenance remains enigmatic.

In the further distance to the west, beyond Gun Hill and The Roaches, is the Cheshire Plain, a flat expanse underlain by Triassic sediments and largely blanketed by Quaternary deposits. It is separated from the folded Carboniferous succession by the Red Rock Fault that was a major extensional structure in Permian and Triassic times, down-throwing to the west. Whilst Triassic sediments are largely outside the scope of this guidebook, it is interesting that, in road cuttings within and around the town of Leek, there are exposures of red, pebbly sandstones which correlate with the very extensive Sherwood Sandstone of Early Triassic age. These fluvial sediments rest unconformably on Carboniferous rocks and show conspicuous cross-bedding, indicating palaeoflow to the north. They occur in areas that are topographically lower than the surrounding countryside, which is made up of older Carboniferous rocks, demonstrating large-scale relief on the Variscan unconformity. The Triassic sediments lie within a palaeovalley, probably a wadi-like feature, that fed sand and gravel from the south into the Cheshire Basin.

Five Clouds Sandstone (SJ 993630; 53.1624, −2.0152) (▶)

This stop is for the view only. It is worth doing as it sets the scene for The Roaches locality in a clearer way than going direct to The Roaches. Turn off the A53 at the Three Horseshoes at Blackshaw Moor and follow the road past the Tittesworth Reservoir visitor centre to Meerbrook. Take a turn to the right and follow the lane for about 1 km before turning sharply right. Follow this lane up the hill and stop where you have the best view up towards The Roaches, with the Five Clouds Sandstone features clear in the slope below. The lane is narrow and so park carefully.

At this locality, the five discrete sandstone features in the hillside are exposures of the Five Clouds Sandstone, which lies stratigraphically around 100 m below the Roaches Grit (Fig. 12.7). The features occur at the same stratigraphic level and, on close examination (see The Roaches locality), comprise rather massive, structureless sandstones, thought to be thick and probably amalgamated unit of turbidites, possibly part

Figure 12.7 Five Clouds Sandstone (Marsdenian, R_{2b}) stands out as prominent features in the hillside, with the Roaches Grit on the skyline behind. These sandstone features probably result from differential fracturing, cementation and/or weathering of a laterally extensive unit of massive, turbidite sandstones.

of a channel fill. Without seeing exposed contacts with neighbouring sediments, it is not possible to confirm this. The fragmentation into five distinct features probably results from small-scale faults or joints within a laterally continuous sandstone. Zones of increased joint intensity seem to have enhanced weathering. There is no evidence of any relative displacement between the features.

The Roaches Grit succession (Marsdenian, R_{2b}), of which this locality shows the lower part, represents the progradation of a major turbidite-fronted delta into the Staffordshire Basin, supplied with feldspathic sand derived from the northern source area. However, the direction of progradation was from the southeast, suggesting that the course of the river that fed the system was influenced by earlier structure, palaeo-bathymetry and on-going differential subsidence. The river system appears to have avoided the Derbyshire Massif by following a route along the Widmerpool Gulf.

The Roaches (ca SK 005625; 53.1563, -1.9954)

This locality can be approached either via the Five Clouds Sandstone viewpoint or, more directly, from Upper Hulme. From the Five Clouds view, continue up the lane to the junction and turn sharply right. Follow the road to the parking spaces below The Roaches. Via Upper Hulme, leave the A53 and follow lane through the village and up the hill below Hen Cloud. Continue to the parking spaces near the main entrance to The Roaches. At weekends, Bank Holidays and in school holidays it can be very busy as this is one of the prime climbing edges in the Pennines. As with other popular localities, it is important to store valuables in vehicles as securely as possible. Disabled access beyond the road is limited in scope and challenging.

As well as being important for climbing, this locality is an important conservation area, managed by the Staffordshire Wildlife Trust. Large areas were devastated by fire in 2018. Take the greatest care especially when conditions are dry. The area is criss-crossed by paths and there are many ways of finding your way around. Use existing paths and do not pioneer new routes through the heather. Trees can obscure views close to the face.

This long scarp exposure on the western limb of the Goyt Syncline shows an extensive section in the Roaches Grit (Marsdenian, R_{2b}). The sandstone forms the top of the escarpment, in the lower part of which are less well-exposed components of the local basin-filling succession. To appreciate the nature and scale of this succession, it is best to take a somewhat circuitous route through the locality.

Immediately on entering through the main gate, turn left and follow a small footpath that rises gradually to the features formed by the Five Clouds Sandstone. Here there are exposures of rather massive and structureless sandstone, similar to other thickly bedded, commonly channelized turbidites seen across the Pennine Basin (e.g. Shale Grit; Pendle Grit). The Five Clouds Sandstone forms a distinct, discontinuous feature along the hillside (see previous locality description). Whilst no palaeocurrents can be derived directly from the Five Clouds Sandstone, those from thin-bedded turbidites in the underlying section, seen in nearby stream sections, show palaeoflows to the northwest.

Standing on top of the feature made by the Five Clouds Sandstone and looking towards the Roaches Grit at the top of the escarpment allows an appreciation of the thickness (around 100 m) of the dominantly fine-grained sediments that separate these two sandstone intervals. This interval is not exposed here but it is poorly exposed in rather inaccessible stream sections elsewhere. It includes, in its upper part, siltstones, ripple cross-laminated sandstone, thin sharp-based sandstones and small channel sandstones along with abundant bioturbation by *Lockeia* (*Pelecypodichnus*). Together these features suggest a delta-front or mouth-bar setting, dominated by flood events and frequent underflows. The unseen lower part of the interval, immediately above the Five Clouds Sandstone, probably comprises mudstones and siltstone and the whole interval is thought to record a major deltaic advance. This caused shallow-water conditions to develop in the basin for the first time since its formation. Prior to the arrival of these sands, the Staffordshire Basin had remained deep, probably from earliest Dinantian times with a Namurian basinal succession (Morridge Formation) dominated by mudstones and thin quartzitic turbidites derived from the south. The sandstones in the succession that culminates in the Roaches Grit are feldspathic and, therefore, of northern derivation, like the sandstones that dominate the Namurian fill of the basins to the north of the Derbyshire Massif. Prior to deposition to the Roaches Grit system, the only northerly-derived sand that entered the Staffordshire Basin was the turbidite Longnor Sandstone (Late Kinderscoutian, $R_{1c(v)}$) in the north-eastern part of the basin.

The top of the Five Clouds Sandstone feature gives the best general view of the Roaches Grit, which is dominated by giant cross-bedding that characterizes the lower part of the exposure. The base of the Roaches Grit is not seen but, from comparison with other similar examples such as the Lower Kinderscout Grit, it is probably an erosional channel base. There are probably several metres of unexposed sandstone between the base of the exposure and such an erosion surface. The giant cross-bedding extends laterally along the face for several hundreds of metres in sets up to around 20 m thick. Above the giant foresets are cosets of medium-scale cross-bedding with mainly tabular sets up to 2 m thick.

Detailed mapping within the Roaches Grit shows that the interval is cut by a succession of major erosion surfaces demonstrating that sub-channels within the complex were progressively offset to the northeast. Mapping both at The Roaches and at Ramshaw Rocks suggests that the individual channels were of the order of 1 km wide. This pattern probably developed under a regime of progressively rising relative sea level that may have resulted from a combination of eustatic sea-level rise and subsidence due to tectonic subsidence and the compaction of the thick underlying unit of fine-grained sediment. The pattern of giant cross-bedding being overlain by a coset of medium-scale cross-bedding is a feature of each of the channel components. In some examples, individual medium-scale sets within the coset expand down dip to evolve into giant foresets. Such lateral changes, along with large-scale reactivation surfaces within the giant sets, give rise to quite complex internal geometries (Fig. 12.8).

The accessible three-dimensional exposure along the top of The Roaches allows the smaller scale internal organization of the cross-bedding to be investigated in some detail. Some medium-scale sets have

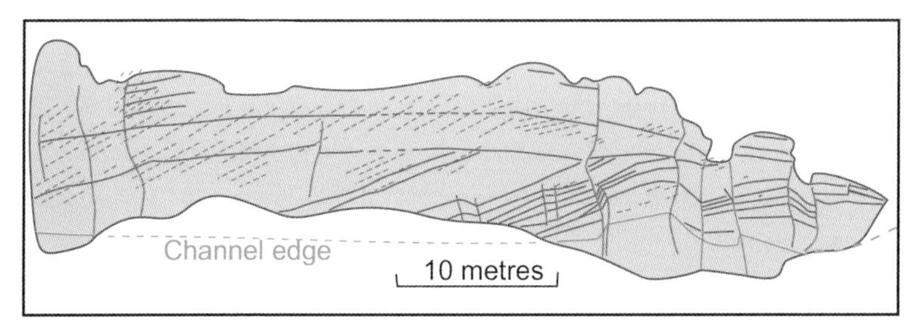

Figure 12.8 The Roaches Grit at The Roaches, showing the giant cross-bedding and the various erosion surfaces that cut through it (after Jones & McCabe, 1980).

toesets of ripple cross-lamination with a counter-current orientation, whilst others show overturned foresets and more complex deformation related to liquefaction of the sand soon after deposition (Fig. 12.9).

The giant cross-bedding has been variously interpreted as large bar forms within channels and as in-channel deltas related to relative base-level changes. This recurring issue is discussed fully in Section 4.3, 'Turbidite-fronted deltas'.

Figure 12.9 Soft sediment deformation in the Roaches Grit at The Roaches. The structures suggest quicksand conditions and water escape soon after deposition, which in turn, suggest rapid deposition on the bed of the river channel.

Hen Cloud (SK 008615; 53.1504, −1.9925) (►)

This crag is a southern extension of The Roaches and is best viewed from the road below to see the large-scale organization of the giant cross-bedding. The co-ordinates above indicate the best viewing point on the road, but a car may have to be parked further up the road. Disabled access is good along the road. To access the crag on foot, use the same parking as for The Roaches, follow the main path from the gate for around 300 m and then take the path over the field to the south and then follow the well-defined path up the hill.

Viewed from the road, a thick interval of the Roaches Grit is seen in a section orientated virtually parallel with the palaeoflow direction. The lowest 40–50 m is poorly exposed in the hillside and comprises

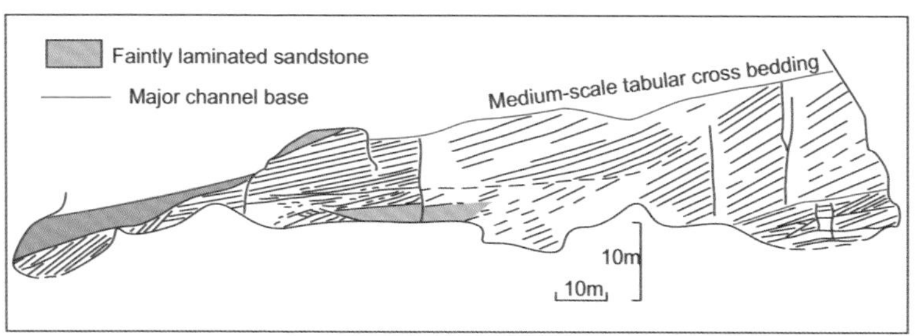

Figure 12.10 The Roaches Grit at Hen Cloud showing the scale and complexity of the giant cross-bedding and associated erosion surfaces (after Jones & McCabe, 1980).

structureless sandstone. This seems to be the oldest and most deeply incised of a series of channels that characterize the Roaches Grit. It may record the fill of an incised valley fill at an early stage in a eustatic sea-level rise, or it may result from rapid deposition following deep scour by a major flood event. The main face of the crag is made up of giant cross-bedding including a 16 m set with internal erosion surfaces overlain by a 21 m thick coset of medium-scale cross-bedding that is probably multi-storey in character (Fig. 12.10).

Ramshaw Rocks (SK 018610; 53.1556, -1.9753) (▶)

These crags rise steeply above the A53 and are best approached from the small lane, accessed at the southern end, in front of the cottage. Follow the lane around the bend to the right and park at the side of the lane. Follow the path up the dip slope of the sandstone to the crest of the ridge. It is possible to follow a rough path along the ridge and generally get around with some gentle scrambling. The east-facing scarp face is sometimes used for climbing. It is possible to walk along the bottom of the face with care, but the steep dip of the rocks, the steep rock face and the steep hillside below make detailed observation awkward.

This locality exposes the Roaches Grit dipping at around 40° on the eastern limb of the Goyt Syncline and provides a section with a different orientation to that seen at The Roaches. Despite the high dip and the steep topography, it is possible to recognize concave-upwards erosion surfaces cutting into quite large-scale cross-bedding in the main face. There are at least three major erosion surfaces, some with steep margins overlain by massive, coarse and pebbly sandstone. This suggests that there was little time between erosion and deposition. The orientation of the exposure means that the face at Ramshaw Rocks is close to being normal to the dip of the giant foresets. Medium-scale cross-bedding, indicating the palaeoflow direction, can be seen in the higher part of the sandstone, in the dipping beds at the crest of the ridge.

Further discussion of the Roaches Grit is set out in the description of The Roaches locality and in the general description of 'Turbidite-fronted deltas' in Section 4.3.

Chinley Head (SK 049845; 53.3592, –1.9274)

This stop, which is for the view alone, is on the A624, about 10 km south of Glossop and 3.5 km north from Chapel-en-le-Frith. Park in the large lay-by on the eastern side of the road. The best view is obtained by crossing the road, but take great care of traffic.

The view to the west, across the valley, shows a hillside capped by Cracken Edge and its quarries in the Rough Rock (Yeadonian, G_1), which are described separately. On the hillside below the quarries, the gradient is broken by a step feature that dies out abruptly to the north (right) (Fig. 12.11). This step is formed by the top of the Chatsworth Grit (Namurian, Marsdenian, R_{2c}), with the *Ca. cancellatum* Marine Band present closely above the sandstone. The sharp northern termination of the sandstone feature has been interpreted as the northern margin of the Chatsworth Grit palaeovalley whose fill of coarse sandstone is seen at Stanage Edge and Windgather Rocks. The coarse pebbly sandstones

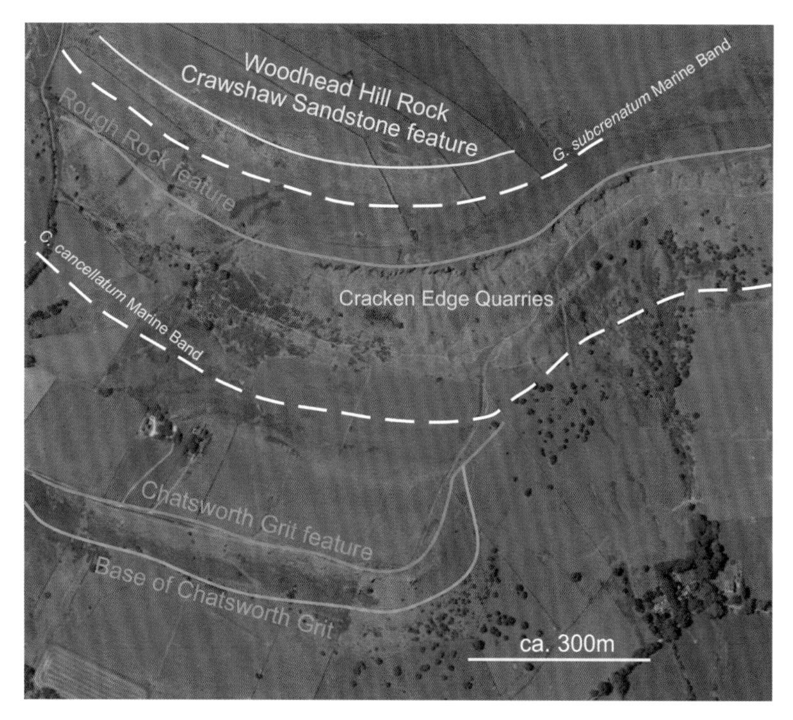

Figure 12.11 The hillside at Chinley Head. The main sandstones are picked out by features in the landscape. The feature made by the Chatsworth Grit dies out quite sharply to the right (north) and is thought to be the northern edge of the Chatsworth Grit palaeovalley that trends east–west across the area. The Rough Rock feature is Cracken Edge with its extensive quarries. Google Earth image.

of the palaeovalley fill are more resistant to erosion than the finer-grained deltaic sediments into which the palaeovalley is cut, and hence the expression in the hillside morphology. In the east, the palaeovalley can be inferred to be around 25 km wide (Figs 4.14; 10.19), but in the western outcrop, the southern margin is unconstrained.

Cracken Edge Quarries, near Chinley (SK 037835; 53.3456, -1.9565)

This locality is high on the hill north of Chinley. Travelling from Chinley, follow the main street (Green Lane) to the T junction just beyond the railway bridge and turn left (west) into Stubbins Lane. Follow this lane to the very top of the hill and park in spaces at the roadside. Walk back down the lane for about 400 m to a stile up the bank on the left-hand side. From here take a well-defined path that follows the route of an old trackway from the quarries. After about 400 m of steady up-hill walking, the quarry exposures are seen higher on the left. Leave the main path and follow smaller paths up to the south end of the exposure. There are paths along the base of the quarry face, which extends for around 400 m to the north. It is not necessary to explore the full length to see the main features of the succession.

These disused quarries are mainly in the Lower Rough Rock (Yeadonian, G_{1b}) and trend roughly normal to the palaeoflow direction, which is to the west, into the face. The quarries show the nature of the channelling within and at the base of the sandstone, and some of the lateral facies variability within the channel complex. The hillside immediately below is the coarsening-upward sequence above the *Ca. cancellatum* Marine Band. The step in the slope, lower down the hillside, is formed by the top of the Chatsworth Grit, as described at the Chinley Head locality. The Rough Rock is closely overlain by the *G. subcrenatum* Marine Band, which marks the base of the Westphalian. The Upper Rough Rock appears to be absent here. The hillside above the quarry covers a further coarsening-upward cyclothem (Langsettian), capped by a widespread fluvial channel sandstone, locally called the Woodhead Hill Rock and equivalent to the Crawshaw Sandstone seen at Birchen Edge on the eastern side of the Derbyshire Massif.

The sandstones at Cracken Edge comprise a lower interval of essentially parallel-bedded sandstones with some finer interbeds, and an upper, sharp-based unit of more massive, coarser sandstone that shows good examples of medium-scale cross-bedding. The lower unit is the Rough Rock Flags. It contains various trace fossils and is thought to record a mouth bar at the top of a progradational cyclothem (Fig. 12.12). The cross-bedded sandstone above is the product of an extensive river channel complex, probably braided in character, which extended across virtually all the Pennine Basin from Airedale to North Staffordshire. Whilst the upper and lower sandstones are both derived from the same northern source area, it seems that they are not necessarily the product of one continuous delta progradation. The Lower Rough Rock may record a period of sea-level low stand, when the high-energy braided river system was forced rapidly basinward.

Figure 12.12 Thinly bedded fine-grained sandstones of the Rough Rock Flags (Yeadonian, G_1) at Cracken Edge Quarries. These sandstones are thought to be mouth-bar deposits eroded by the channel base of the Upper Rough Rock.

The quarries are also interesting from a structural standpoint. The workings have exploited the main joint sets in the sandstone, giving a saw-tooth profile to the quarry faces. This conjugate joint set can be seen particularly well in satellite images and is related to the east–west compression associated with the Variscan folding. The westerly structural dip here is the eastern limb of the Goyt Syncline, seen to the south at The Roaches and Ramshaw Rocks. It is one of several north–south trending folds in the area.

The view to the east from these quarries is extensive. The western edge of Kinder Scout forms the skyline to the north with Kinder Downfall clearly visible if conditions are good. To the south, successive Namurian and Lower Westphalian (Langsettian) cyclothems give excellent examples of stepped topography, whereby the tops of the steps are the dip slopes of the upper surfaces of sandstones and the steeper scarps are underlain by the coarsening-upward cyclothem, usually with a marine band at its base. Understanding such relationships between the topographic features and the underlying geology is key to geological mapping, especially where exposures are scarce.

Windgather Rocks, near Kettleshulme (SJ 997784; 53.3008, -2.0010) (▶)

From the B5470 at Kettleshulme take a lane to the south at the southern end of the village. Follow this lane for about 1.5 km and park on the roadside near the stile/gate that leads to the footpath across a field to the crag. It is a short walk up to the main face, which is a popular rock-climbing locality. If busy, try parking at the southern end, near the stile into the small quarry. Take care to hide any valuables in your vehicle. We have first-hand experience that this is a popular spot for thieving. The stiles may make disabled access difficult.

This exposure is in the Chatsworth Grit (Marsdenian, R_{2c}). It is in a coarse pebbly sandstone facies that is thought to be within the main palaeovalley fill of the Chatsworth Grit as seen at Stanage Edge, some 25 km to the east (Figs 10.15; 10.16). There, the palaeovalley is demonstrably at least 25 km wide and palaeocurrents at both localities are broadly towards the west. Here, on the western side of the Pennine Anticline, the northern margin of the palaeovalley is thought to be below Cracken Edge as seen at Chinley Head (Fig. 12.11). If that is the case, Windgather Rocks lie about 4 km from the northern margin.

The natural gritstone edge of Windgather Rocks provides accessible exposure to examine cross-bedding styles and the various internal erosion surfaces with their associated pebble lags. It is a good example for geologists and reservoir engineers to discuss geological influences on the internal heterogeneity of similar sandstone reservoirs, particularly the relationships between pebble lags and possible high-permeability streaks. The small quarry at the southern end of the exposure shows the three-dimensional geometry of the cross-bedding very clearly.

Glossary

Words in ***bold italic*** are explained further in the glossary.

A

Accommodation space: Space that allows sediment to accumulate on a geological time scale. It can be created by ***tectonic*** subsidence, by compaction and by a rise in water level or a combination of these.

Aggradation: The accumulation of a sediment unit by essentially vertical growth. It contrasts with ***progradation*** where a body of sediment grows horizontally by building out into a basin.

Allocyclic: Vertical repetition of a pattern of sedimentation (***facies*** sequence) where the cause of the repetition (***cyclicity***) is externally imposed by, for example, changes of sea level.

Amalgamation: The creation of thick beds of sandstone through a succession of episodic flows that were erosive prior to becoming depositional.

Anthropogenic: A term used to characterize features of a landscape that have been significantly influenced by human activity.

Aragonite: A mineral form of calcium carbonate ($CaCO_3$) that is precipitated from sea water by many organisms, as shells or as fine needles by some plants.

Arkose: A sandstone with a high content of ***feldspar***.

Autocyclic: Vertical repetition of a pattern of sedimentation (***facies*** sequence) where the cause of the repetition (***cyclicity***) is an inherent feature of the depositional setting.

Avulsion: The rapid switching of a river or distributary channel, cutting off sediment supply in one area and initiating deposition elsewhere.

B

Bedform: A structure formed on the surface of sand as a result of wind or water currents, or from waves. Small bedforms are ***ripples*** and larger ones are ***dunes*** or sand waves.

Benthos: Communities of animals that live most of their lives on the seabed or lakebed, or below the sediment surface in burrows.

Bentonite: A clay of volcanic origin, usually having a high content of potassium and a high gamma-ray count, making them useful for correlation between boreholes.

Bioclast: A sedimentary particle from an organic source, typically a shell fragment.

Bioherm: A general term for organic build-ups. In this book, almost all recognized build-ups are termed ***mud mounds***.

Biostratigraphy: The subdivision and correlation of successions of strata using fossils.

Bioturbation: The disturbance of sediment by the activity of animals or plants shortly after deposition, producing ***trace fossils***, typically burrows, surface trails, borings or root traces.

Blue John: A distinctive purple banded form of ***fluorite*** found only near Castleton in Derbyshire where it has been mined as a decorative stone.

Bouma sequence: A succession of types of lamination within sandstone beds deposited by short-lived, decelerating currents, typically flood surges or ***turbidity currents***.

Boundstone: a limestone with evidence that the original components were organically bound at the time of deposition, although the mode of binding is not identifiable.

Brachiopods: Marine bivalves (not to be confused with Molluscan bivalves) with calcareous shells that commonly occur as fossils, particularly in limestones.

Bryozoa: Colonial invertebrate organisms, mainly marine, that typically form calcareous encrustations on a sediment surface or on a rocky seabed, or as leaf-like laminae.

Bullion: A calcite-cemented ***concretion*** in mudstone.

C

Calcarenite: A limestone made up predominantly of sand-size particles that are composed of ***calcite*** or ***aragonite***. The particles may be fragments of shells, **ooids** or faecal pellets.

Calcite: The stable and most common mineral form of calcium carbonate.

Calcrete: A fossil soil (***palaeosol***) characterized by calcite concretions that in some cases form a distinct profile.

Caledonian: A phase of long-lasting and complex ***tectonic*** activity that affected large areas of northern Europe, Greenland and North America. It culminated in late Silurian times.

Cerioid: A coral colony where cloned individuals remain attached to each other throughout their growth, with no intervening tissue, giving colonies a honeycomb appearance.

Charnian: A sequence of Precambrian rocks that form the basement rocks in the East Midlands.

Chert: A form of micro-crystalline silica that commonly occurs as ***concretions*** in limestones.

Chronostratigraphy: The organization and correlation of strata in relation to geological time. It leads to successions being assigned to ***stages*** and sub-stages.

Clast: Any fragment of rock or unconsolidated sediment, regardless of size. Clastic sediments are commonly more tightly defined by composition; e.g. silici-clastic, bioclastic.

Clay wayboards: Thin beds of mudstone (***bentonite***) of volcanic origin, typically draping an irregular surface between thicker beds of limestone.

Colony: (with reference to corals and ***bryozoa***) Groups of individuals cloned to and remaining attached to each other.

Corallite: An individual in a coral colony; the skeletal part of a polyp.

Coset: (pronounced co-set) A unit made up of multiple ***sets*** of ***cross-bedding*** or ***cross-lamination***.

Crinoid: A fossil echinoderm characterized by a stem made up of ***calcite*** discs (ossicles). They commonly occur as small lengths of stem (columnals) or as individual ossicles.

Cross-bedding: Inclined lamination (***foresets***) commonly occurring in sandstones or ***calcarenites***. Foresets are commonly arranged in ***sets*** of similarly inclined laminae, which are usually separated by erosion surfaces. The dip direction of foresets indicates the direction of flow of the current responsible (***palaeocurrent***).

Cross-lamination: Similar to ***cross-bedding*** but at a smaller scale (cm) and commonly involving finer-grained sediment. It results from sediment aggradation during the migration of ripples.

Cyclothem: An interval of sedimentary rock, typically metres to tens of metres thick, that shows a particular vertical succession of rock types or ***facies***. Cyclothems typically occur in cyclical or repetitive stacks of broadly similar intervals.

D

Debris flow: A cohesive gravity-driven flow of water-saturated poorly sorted sediment which moves downslope through deformation of the fine-grained, commonly clay-rich, matrix. The deposit is typically poorly sorted, with larger clasts supported by the matrix.

Delta front: The area in front of a river mouth where it enters the sea or a lake and where most fine-grained, suspended sediment, is deposited.

Delta plain: The area on top of a delta that is variably sub-aerially exposed.

Diagenesis: The processes that alter sediment after deposition and during burial.

Dinantian: A commonly used European ***chronostratigraphic*** term for the early part of the Carboniferous period. It has been superseded by ***Tournaisian*** and ***Viséan***, which together now make up the former Dinantian.

Dolomite: A mineral form of calcium magnesium carbonate [$CaMg(CO_3)_2$], which is an alteration product of limestones that have reacted with magnesium-rich pore fluids. The term is also used for rocks that have been altered in this way.

Dune: A large, usually repetitive ***bedform*** that develops in sand moved by rapid and deep-water currents. Aeolian dunes form through the action of wind in coastal and desert settings.

E

Elaterite: A type of bitumen that has been highly degraded, mainly by bacterial activity.

Eustacy: Worldwide sea level whose changes through time exert widespread controls on sedimentation and erosion, particularly close to shorelines.

Euphotic (Zone): Uppermost part of a water column into which sufficient light penetrates to permit growth of photosynthesizing organisms (including those that live symbiotically in host animals, like reef corals).

F

Facies: A term used in the description and characterization of sedimentary rocks. It is a way of grouping sediments into classes with similar features that may be descriptive or interpretive.

Floatstone: A coarse-grained limestone in which larger particles are supported by the matrix.

Flooding surface: A well-defined surface within a sedimentary succession that is inferred to result from a relative rise in base level (sea or lake).

Fluorite: (or fluorspar) A mineral form of calcium fluoride (CaF_2) that occurs in veins. It has cubic crystals and occurs in a range of colours, the most famous being purple ***Blue John***.

Flute marks: Protruding marks on the bases of some sandstone beds where interbedded with mudstone (as in ***turbidites***). They are used to determine the ***palaeocurrent direction***.

Footwall: The side of an inclined ***fault*** that occurs below the fault plane.

Forced regression: The basinward shifting of a shoreline that is accelerated by an externally induced loss of accommodation space such as fall of sea level.

Foresets: Inclined laminae that make up ***cross-bedding*** and ***cross-lamination***. They form by deposition on the lee side of a ***bedform*** or on a steep delta slope.

G

Galena: The crystalline form of lead sulphide (PbS), the lead ore that was mined extensively in the Pennines.

Ganister: A highly quartz-enriched sandstone that occurs as a very mature, leached ***palaeosol***, commonly as a ***seat earth*** to a coal seam.

Geopetal infill: A cavity in limestone, commonly the interior of a shell, partially filled with sediment prior to burial and lithification. The space above the sediment is filled with sparry (crystalline) ***calcite***, precipitated later.

Gondwanaland: A supercontinent dating back to late Precambrian times and persisting into the Mesozoic. It broke up to give present-day fragments that include Africa, South America, Antarctica, India and Australia.

Goniatite: Free-swimming coiled marine ammonoid molluscs that thrived during the ***Devonian*** and ***Carboniferous***. Goniatites evolved extremely rapidly and are the main fossils for establishing the ***biostratigraphic*** framework of ***Namurian*** and ***Westphalian*** strata.

Graben: An elongate, ***rifted*** basin bounded on either side by normal ***faults***.

Grainstone: A limestone dominated by sand-size particles (***calcarenite***) which might be ***bioclasts***, ***ooids*** or ***pellets***. The spaces between grains are dominantly filled by ***sparry calcite cement***.

Growth fault: A fault (usually a normal fault) that is active during sediment deposition, leading to differences in thickness and ***facies*** across the fault.

H

Hanging wall: The side of an inclined ***fault*** that occurs above the fault surface.

Hybrid event beds: Beds resulting from sub-aqueous gravity-driven flows that are transitional between ***turbidity currents*** and debris flows.

Hyperpycnal flow: A sub-aqueous gravity-driven flow generated when the inflow of a river is denser than the basin water.

I, K

Intraclast: A fragment of sediment that has been eroded and re-deposited within a slightly younger sediment.

Inversion: The reversal of displacement on a fault or a switch from subsidence to uplift resulting from a change in tectonic regime.

Karst: A general term for the suite of features that result from the dissolution of limestones and dolomites at outcrop by freshwater precipitation and run-off at the present day and in the recent past.

L

Lag: A concentration of coarser ***clasts***, often above an erosion surface, formed through winnowing by currents. The clasts may be ***intraclasts*** or of exotic origin.

Lateral accretion: The growth of a unit of sediment through deposition on its flank, typically by flow along the flank. The most common example is deposition on the ***point bar*** of a meandering river.

Laurussia: A supercontinent that was assembled on Silurian and Devonian times by the collision of continental components that presently comprise North America, Greenland and northern Europe.

Listric: A curved surface with a concave-upwards form. The term is used to characterize faults that flatten with depth.

Lithostratigraphy: A scheme of stratigraphic subdivision based on the correlation of strata of similar character.

M

Marine band: A unit of sediment, usually mudstone, containing marine fossils, that can be the basis for ***chronostratigraphic*** correlation.

Micrite: Very finely crystalline ***calcite*** that makes up a large part (typically the matrix) of many limestones. It is mainly the result of recrystallization of depositional lime mud. The term is also used for limestones made up predominantly of micrite.

Microbialite: A body of carbonate mud that was deposited and preserved in situ by the mediation of microbes.

Microfacies: A scheme of sediment classification that depends largely on microscopic examination of thin sections. It applies particularly to limestones where identification of the constituent grains is important.

Miocene: A period of geological time between 23 and 5.3 Mya.

Mouth bar: The area in front of a river mouth where most of the sandy sediment load is deposited.

Mud mound: The term used in this book for all for all localized build-ups of micritic limestone that resulted from the precipitation and stabilizing of carbonate mud by biogenic processes. These features have been variously called ***bioherms***, ***reefs***, ***reef knolls***, ***apron reef*** and ***Waulsortian reefs*** depending on their shape, size and setting.

N, O

Namurian: A regional stage within the Carboniferous period, between the ***Viséan*** and the ***Westphalian*** stages. It lasted from 330 to 318 Mya and is subdivided into seven sub-***stages*** and some seventy sub-***zones***.

Ooid (oolite): Ooids are sand-size spherical particles of ***calcite*** that accrete in shallow-marine tropical settings through agitation by waves and currents.

P

Packstone: A sand-grade limestone (***calcarenite***) with a grain-supported framework but with a dominant fine-grained matrix (***micrite***) between the grains.

Palaeocurrent: The inferred direction of flow of the current that deposited a unit of sediment.

Palaeokarst: Surfaces or features within a sedimentary succession that indicate that the sediment surface was subaerially exposed and subjected to ***karstic*** processes.

Palaeosol: A unit of sediment that has been subjected to soil-forming processes (pedogenesis).

Palaeovalley: A deep and often wide channel eroded by a major river into a ***delta top*** or a continental shelf due to a fall in sea level.

Palynology: The study of spores and pollen as a means of stratigraphic correlation or of environmental interpretation.

Pellet (peloidal): A sand-size particle, usually of lime mud, produced as faecal material. They were sufficiently cohesive to withstand reworking.

Phaceloid: Coral colony in which individuals have grown clonally by branching with free space between them.

Plate tectonics: The global model for the tectonic behaviour of the Earth's crust and mantle, whereby large slabs move relative to one another. Collisions, separations and lateral contacts of plates ultimately cause most sedimentary basins, oceans, mountain chains, volcanoes, and most major earthquakes.

Platform: An area of sea floor significantly shallower than surrounding areas. The sea bed in such settings may be reworked by high-energy processes such as waves and tides.

Point bar: The inner bank of a bend in a channel, most commonly a meandering river. This is the main site of sedimentation through ***lateral accretion***.

Pro-delta: The area of deeper water in front of delta, beyond the ***delta front*** (slope).

Progradation: Sedimentation leading to building out into a basin of a delta or other shoreline. It is one process by which ***regression*** occurs.

Provenance: In a geological sense, the source area from which detrital sediment was derived.

Pyrite: A crystalline form of iron sulphide (FeS_2) which forms in sediment, usually mudstones, under reducing conditions.

Q

Quaternary: A period of geological time that spans the most recent 2 My and includes the Pleistocene and the Holocene.

R

Ramp: An area of basin (sea) floor across which depth gradually increases.

Reactivation surface: An inclined erosion surface, within a set of cross-bedding, that truncates underlying ***foresets*** and above which normal foresets resume.

Reef: In nautical terms, a hazardous zone of near-shore shallow water caused by emergent or near-emergent rocks. Present-day carbonate reefs are dominated by coral/algal complexes that form rigid frameworks of skeletal material. In the Pennines, within the ***Dinantian*** succession, the term has been used for some of the ***mud mounds***, especially those close to the margins of carbonate platforms.

Reef knoll: Large biogenic ***mud mounds*** which, when exhumed through erosion, create hills in the present-day landscape.

Regression: A basinward shift in the position of a shoreline, often through the ***progradation*** of a ***delta***. Regression also results from a fall of water level in a basin, when the regression is described as 'forced'.

Rift: A fault-bounded basin caused by extensional tectonic forces. Rifts are associated with crustal thinning and subsidence.

S

Seat earth: A fossil soil (***palaeosol***) in coal-bearing successions, usually occurring directly below a coal seam.

Siderite: A mineral form of iron carbonate ($FeCO_3$) that commonly occurs as ***concretions***.

Sole marks: Structures found on the bases of some sandstone beds, especially where sandstones and mudstones are interbedded. The marks are casts of erosional features cut into the underlying mudstone before sand was deposited.

Solifluction: The movement of loose material on a steep slope, typically promoted by freeze-thaw processes. Its results are the 'head' of many geological maps.

Sparry cement: Crystalline calcite cement in limestones that were deposited in high-energy environments where mud (***micrite***) was winnowed away.

Stage: A period of geological time that makes up the ***chronostratigraphic*** framework at a coarse level. Carboniferous examples are the ***Viséan*** and the ***Namurian***.

Stratotype: The location at which a particular stratigraphic unit is defined. The unit is commonly a ***stage*** or a sub-stage and may be defined by ***biostratigraphic*** markers.

Stromatactis: A structure of uncertain origin occurring in ***micritic*** limestones that occurs as a flat-bottomed cavity infilled by ***sparry*** calcite.

T

Talus: Loose fragmented material that has accumulated after falling under gravity down a steep slope, as in scree deposits.

Taphonomy: A wide suite of processes by which an organism, after death, becomes a fossil. Applicable to individuals or to whole assemblages which can be *autochthonous* (in life position), *parautochthonous* (locally moved from life position or *allochthonous* (transported).

Tool marks: A type of ***sole mark*** produced by the dragging or bouncing of large clasts over a muddy surface prior to burial by sand. A common feature of ***turbidite*** sandstones.

Tournaisian: The earliest stage of ***Carboniferous*** time spanning 359–346 Mya. It equates to the lowest part of the Mississippian. It coincides with the Courceyan sub-stage.

Trace fossils: Tracks, trails and burrows that disturb sediment soon after deposition. The intensity of disturbance by trace fossils (***bioturbation***) gives a relative measure of the rate of sedimentation.

Transgression: A landward shift in the position of a shoreline, often through a rise in base level, as in a ***eustatic*** sea-level rise. Transgressions are often associated with ***flooding surfaces***.

Tuff: Fine-grained volcanic ash that has settled through the air following an explosive eruption. In the rock record, tuff layers are commonly thin but very widespread and can give a basis for correlation.

Turbidite: A bed of sandstone or sand-grade limestone, interbedded with mudstone, that is inferred to have been deposited by a ***turbidity current***.

Turbidity current: A current that flows down a slope, driven by gravity acting on the excess density created by its load of suspended sediment. Turbidity currents are the main process whereby coarser sediment is carried into deeper water where they decelerate and deposit their sediment loads.

V, W

Variscan: A complex phase of orogeny that spanned Carboniferous and early Permian time. Its effects were related to closure of an ocean between the ***Laurussia*** and ***Gondwana*** tectonic plates.

Viséan: The stage of Carboniferous time spanning 346–330 Mya. It falls within the middle part of the Mississippian sub-period. Along with the ***Tournaisian***, it was formerly included in the ***Dinantian***.

Wackestone: A limestone with a matrix-supported texture, dominated by fine-grained material (***micrite***).

Waulsortian (reef): Large ***mud mounds*** that are thought to have developed in deeper water. They are named from examples in Belgium.

Westphalian: A regional stage within the Carboniferous period. It lasted from 318 to 310 Mya and is subdivided into four sub-stages**.** In the British Isles, it coincides closely with the Coal Measures.

Yoredale: A pattern of sedimentation (‘Yoredale facies’) comprising a cyclic pattern of upwards-coarsening units (***cyclothems***) with limestones at their bases.

Further support and background information on sedimentology topics may be found in:

Collinson J. & Mountney N. (2019) (4th Edn) *Sedimentary Structures.* Dunedin Academic Press, Edinburgh. 340pp.

Jones S.J. (2023) (2nd Edn) *Introducing Sedimentology.* Dunedin Academic Press, Edinburgh. 128pp.

Bibliography/References

Adams A.E. (1980) Calcrete profiles in the Eyam Limestone (Carboniferous) of Derbyshire: petrology and regional significance. *Sedimentology* **27**, 651–660.

Adams A.E. & Cossey P.J. (1978) Geological history and significance of a laminated and slumped unit in the Carboniferous Limestone of the Monsal Dale region, Derbyshire. *Geological Journal* **13**, 47–60.

Aitkenhead N., Chisholm J.I. & Stevenson I.P. (1985) *Geology of the country around Buxton, Leek and Bakewell.* Memoir of the British Geological Survey, Sheet 111. 168pp.

Aitkenhead N. *et al.* (2002) (4th Edn) *British Regional Geology: The Pennines and adjacent areas.* British Geological Survey.

Allen J.R.L. (1960) The Mam Tor Sandstones: A 'turbidite' facies of the Namurian deltas of Derbyshire, England. *Journal of Sedimentary Petrology* **30**, 193–208.

Aretz M. (2010) Habitats of colonial rugose corals: the Mississippian of western Europe as example for a general classification. *Lethaia* **43**, 558–572.

Arthurton R.S., Johnson E.W. & Mundy D.J.C. (1988) *Geology of the country around Settle.* Memoir of the British Geological Survey, Sheet 60. 147pp.

Baines J.G. (1977) *The stratigraphy and sedimentology of the Skipton Moor Grits (Namurian E_1c) and their lateral equivalents.* PhD thesis, Keele University.

Banks V. (2017) Hydrogeology of the Peak District and its river basin management planning. *Mercian Geologist* **19**, 94–101.

Banks V. & Allen M. (2017) Excursion guide: Monsal Dale, Derbyshire. *Mercian Geologist* **19**, 122.

Banks V.J., Gunn J. & Lowe D.J. (2009) Stratigraphical influences in the limestone hydrology of the Wye catchment, Derbyshire. *Quarterly Journal of Engineering Geology & Hydrogeology* **42**, 211–225.

Banks V.J., Jones P.F., Lowe D.J., Lee J.R., Rushton J. & Ellis M.A. (2012) Review of tufa deposition and palaeohydrological conditions in the White Peak, Derbyshire, UK: implications for Quaternary landscape evolution. *Proceedings of the Geological Association* **123**, 117–129.

Bijkerk J.F. (2014) *External controls on sedimentary sequences: a field and analogue modelling-based study.* PhD thesis, University of Leeds.

Bisat W.S. (1924) The Carboniferous goniatites of the north of England and their zones. *Proceedings of the Yorkshire Geological Society* **20**, 40–124.

Bond G. (1949) The Lower Carboniferous reef limestones of Cracoe, Yorkshire. *Quarterly Journal of the Geological Society of London* **105**, 157–188.

Brettle M.J., McIlroy D., Elliott T., Davies S.J. & Waters C.N. (2002) Identifying cryptic tidal influences within deltaic successions: an example from the Marsdenian (Namurian) interval of the Pennine Basin, UK. *Journal of the Geological Society* **159**, 379–391.

Brettle M.J, Waters C.N. & Davies S.J. (2023) An integrated sequence stratigraphic analysis of the early Marsdenian sub-stage of the Millstone Grit Group, Central Pennines, UK. *Proceedings of the Yorkshire Geological Society* **64**, 149–189.

Bridges P.H., Gutteridge P. & Pickard N.A.H. (1995) The environmental setting of Early Carboniferous mud-mounds. In: *Carbonate mudmounds; their origin and evolution.* (Ed. C.V.L. Monty, D.W.J. Boscence, P.H. Bridges & B.R. Pratt) *Special Publication of the International Association of Sedimentologists* **23**, 171–190.

Bristow C.S. (1987) *Sedimentology of large braided rivers ancient and modern.* PhD thesis, University of Leeds.

Bristow C.S. (1988) Controls on the sedimentation of the Rough Rock Group (Namurian) from the Pennine Basin of northern England. In: *Sedimentation in a synorogenic basin complex* (Ed. B.M. Besly & G. Kelling). Blackie, Glasgow. 114–131.

Bristow C.S. (1993) Sedimentology of the Rough Rock: a Carboniferous braided river sheet sandstone in northern England. In: *Braided rivers.* (Ed. J.L. Best & C.S. Bristow) *Special Publication of the Geological Society* **75**, 291–304.

Broadhurst F.M. & Simpson I.M. (1967) Sedimentary infillings of fossils and cavities in limestone at Treak Cliff, Derbyshire. *Geological Magazine* **104**, 443–448.

Bromehead C.E.N., Edwards W., Wray D.A. & Stephens J.V. (1933) *The geology of the country around Holmfirth and Glossop.* Memoir of the Geological Survey, Sheet 86.

Buckland W. (1824) *Reliquiae Diluvianae.* John Murray, London. 305pp.

Butcher N.J.D. & Ford T.D. (1973) The Carboniferous Limestone of Monsal Dale, Derbyshire. *Mercian Geologist* **4**, 179–195.

Buxton D. & Charlton C. (2013) *Cromford revisited.* The Derwent Valley Mills World Heritage Site Educational Trust and University of Derby, Matlock. 192pp.

Cameron G.I.F, Collinson J.D., Rider M.H. & Xu L. (1992) Analogue dipmeter logs through a prograding deltaic sandbody. In: *Advances in reservoir geology.* (Ed. M. Ashton) *Special Publication of the Geological Society* **69**, 195–217.

Carniti A.P., Della Porta G., Banks V.J., Stephenson M.H. & Angiolini L. (2022) Brachiopod fauna form the uppermost Viséan (Mississippian) mud mounds in Derbyshire, UK. *Acta Palaeontologica Polonica* **67**, 865–915.

Charlton C. & Buxton D. (2019) *Matlock Bath; a perfectly romantic place.* The Derwent Valley Mills World Heritage Site Educational Trust and University of Derby, Matlock. ix + 246pp.

Chisholm J.I. (1977) Growth faulting and sandstone deposition in the Namurian of the Stanton Syncline, Derbyshire. *Proceedings of the Yorkshire Geological Society* **41**, 305–323.

Chisholm J.I. & Waters C.N. (2012) Syn-sedimentary deformation of the Ashover Grit (Pennsylvanian, Namurian, Marsdenian Substage) deltaic succession around Wirksworth, Derbyshire, UK. *Proceedings of the Yorkshire Geological Society* **59**, 25–36.

Collinson J.D. (1968) Deltaic sedimentation units in the Upper Carboniferous of northern England. *Sedimentology* **10**, 233–254.

Collinson J.D. (1969) The sedimentology of the Grindslow Shales and the Kinderscout Grit: a deltaic complex in the Namurian of northern England. *Journal of Sedimentary Petrology* **39**, 194–221.

Collinson J.D. (1970) Deep channels, massive beds and turbidity current genesis in the central Pennine basin. *Proceedings of the Yorkshire Geological Society* **37**, 495–519.

Collinson J.D. (1988) Controls on Namurian sedimentation in the Central Province Basins of Northern England. In: *Sedimentation in a Synorogenic Basin Complex* (Ed. B.M. Besly & G. Kelling). Blackie, Glasgow. 85–101.

Collinson J.D. & Banks N.L. (1975) The Haslingden Flags (Namurian G_1) of South East Lancashire: bar finger sands in the Pennine Basin. *Proceedings of the Yorkshire Geological Society* **40**, 431–458.

Collinson J.D., Jones C.M. & Wilson A.A. (1977) The Marsdenian (Namurian R_2) succession west of Blackburn: implications for the evolution of Pennine delta systems. *Geological Journal* **12**, 59–76.

Cope F.W. (1999) (3rd Edn) The Peak District. *Geologists' Association Guide* **26.** iv +78pp.

Cossey P.J., Buckman J.O. & Steward D.I. (1995) The geology and conservation of Brown End Quarry, Waterstones, Staffordshire. *Proceedings of the Geologists Association* **106**, 11–25.

Cossey P.J., Gutteridge P., Purnell M.A., Adams A.E. & Walkden G.M. (2004) Derbyshire Platform, North Staffordshire Basin and Hathern Shelf. In: *British Lower Carboniferous Stratigraphy.* (Ed. P.J. Cossey, A.E. Adams, M.A. Purnell, M.J. Whiteley, M.A Whyte, & V.P. Wright) *Geological Conservation Review Series,* No **29**. Joint Nature Conservation Committee, Peterborough. Chapter 6, 304–364.

Cossey P.J., Riley N.J., Adams A.E. & Miller J. (2004) Craven Basin. In: *British Lower Carboniferous Stratigraphy.* (Ed. P.J. Cossey, A.E. Adams, M.A. Purnell, M.J. Whiteley, M.A Whyte & V.P. Wright) *Geological Conservation Review Series,* No **29**. Joint Nature Conservation Committee, Peterborough. Chapter 6, 258–302.

Cotton C. (1699) (4th Edn) *The wonders of the Peake*. Charles Brome, London. 86pp.

Cox F.C. & Harrison D.J. (1980) The limestone and dolomite resources of the country around Wirksworth, Derbyshire. Descriptions of parts of sheets SK25 and 35. *Mineral Assessment Report Institute of Geological Sciences*, No **47**.

Dalton R., Fox H. & Jones P. (1999) *Classic landforms of the White Peak*. The Geographical Association. British Geomorphological Research Group, Sheffield. 51pp.

Darwin C.R. (1842) *The structure and distribution of coral reefs. Being the first part of the geology of the voyage of the Beagle, under the command of Capt. Fitzroy, R.N. during the years 1832 to 1836*. Smith Elder and Co., London. 214pp.

Davies S.J. & McLean D. (1996) Spectral gamma and palynological characterization of Kinderscoutian marine bands in the Namurian of the Pennine Basin. *Proceedings of the Yorkshire Geological Society* **51**, 102–114.

Debout L. & Denayer J. (2018) Palaeoecology of the Upper Tournaisian (Mississippian) crinoidal limestones from South Belgium. *Geologica Belgica* **21**, 111–127.

Defoe D. (1724–1726). Rogers P. (Ed.) *A tour through the whole island of Great Britain*. Penguin, London. 736pp.

Drewery S., Cliff R.A. & Leeder M.R. (1987) Provenance of Carboniferous sandstones from U-Pb dating of detrital zircons. *Nature* **325**, 50–53.

Eagar R.M.C., Baines J.G., Collinson J.D., Hardy P.G., Okolo S.A. & Pollard J.E. (1985) Trace fossil assemblages and their occurrence in Silesian (Mid-Carboniferous) deltaic sediments of the Central Pennine Basin, England. In: *Biogenic structures: their use in interpreting depositional systems*. (Ed. H.A. Curran) *Special Publication of the Society of Economic Paleontologists and Mineralogists* **35**, 99–149.

Earp J.R., Magraw D., Poole E.G., Land D.H. & Whiteman A.J. (1961) *Geology of the country around Clitheroe and Nelson*. Memoir of the Geological Survey, Sheet 68.

Eden R.A., Orme G.R., Mitchell M. & Shirley J. (1964) A study of part of the margin of the Carboniferous Limestone 'Massif' in the Pin Dale area of Derbyshire. *Bulletin of the Geological Survey of Great Britain* **21**, 73–118.

Eden R.A., Stevenson I.P. & Edwards W. (1957) *Geology of the country around Sheffield*. Memoir of the Geological Survey, Sheet 100.

Evans D.J., Walker A.S.D. & Chadwick R.A. (2002) The Pennine Anticline, northern England – a continuing enigma. *Proceedings of the Yorkshire Geological Society* **54**, 17–34.

Farey J. (1815) (2nd Edn) *General view of the agriculture and minerals of Derbyshire*. Sherwood, Neely & Jones, London.

Ford T.D. (1969) The Blue John fluorspar deposits of Treak Cliff, Derbyshire, in relation to the boulder bed. *Proceedings of the Yorkshire Geological Society* **37**, 153–157.

Ford T.D. (1977) Carboniferous Limestone. In: *Limestones and Caves of the Peak District* (Ed. T.D. Ford). Geo Abstracts, University of East Anglia, Norwich. xix, 469.

Ford T.D. (1999) The growth of geological knowledge in the Peak District. *Mercian Geologist* **14**, 161–190.

Ford T.D. (2000) The Castleton Area, Derbyshire. *Geologists Association Guide* No **56**. 94pp.

Ford T.D. (2002) *Rocks and Scenery of the Peak District.* Landmark Publishing, Ashbourne.

Ford T.D. (2005) The geology of the Wirksworth mines: a review. *Bulletin of the Peak District Mines Historical Society Ltd.* **16**, 1–42.

Ford T.D. (2019) *Derbyshire Blue John* (Ed. T. Waltham & N. Worley). *East Midlands Geological Society.* 80pp.

Ford T.D. & Jones J.A. (2007) The geological setting of the mineral deposits at Brassington and Carsington, Derbyshire. *Bulletin of the Peak Mining Historical Society* **16**, 1–23.

Ford T.D. & Rieuwerts, J.H. (Eds) (2000) (4th Edn) *Lead mining in the Peak District.* Landmark Publishing, Ashbourne. 208pp.

Ford T.D. & Torrens H.S. (2001) A Farey Story: the pioneer geologist John Farey (1766–1826). *Geology Today* **17**, 59–68.

Frazer M., Whitaker F. & Hollis C. (2014) Fluid expulsion from overpressured basins: Implications for Pb-Zn mineralization and dolomitization of the East Midlands Platform, Northern England. *Marine and Petroleum Geology* **20**, 1–19.

Frost D.V & Smart J.G.O. (1979) *Geology of the country north of Derby*. Memoir of the Geological Survey, Sheet 125.

Gawthorpe R.L. (1986) Sedimentation during carbonate ramp-to-slope evolution in a tectonically active area: Bowland Basin (Dinantian) N. England. *Sedimentology* **33**, 185–206.

Gawthorpe R.L. (1987) Tectono-sedimentary evolution of the Bowland Basin, north England, during the Dinantian. *Journal of The Geological Society of London* **144**, 59–71.

Gawthorpe R.L., Gutteridge P. & Leeder M.R. (1989) Late Devonian and Dinantian basin evolution in northern England and North Wales. In: *The Role of Tectonics in Devonian and Carboniferous Sedimentation in the British Isles.* (Ed. R.S. Arthurton, P. Gutteridge & S.C. Nolan) *Yorkshire Geological Society Occasional Publication* **6**, 1–23.

Gawthorpe R.L. & Gutteridge P. (1990) Geometry and evolution of platform-margin bioclastic shoals, late Dinantian (Mississippian), Derbyshire, UK. In: *Carbonate Platforms: Facies, Sequences and Evolution.* (Ed. M.E. Tucker, J.L. Wilson, P.D. Crevello, J.R. Sarg & J.F. Read) *Publication of the International Association of Sedimentologists* **9**, 39–54.

Gilligan A. (1920) The petrography of the Millstone Grit of Yorkshire. *Quarterly Journal of the Geological Society of London* **75**, 251–296.

Glennie K.W. (2005) Regional tectonics in relation to Permo-Carboniferous hydrocarbon potential, Southern North Sea. In: *Carboniferous hydrocarbon geology: the Southern North Sea and surrounding onshore areas.* (Ed. J.D. Collinson, D.J. Evans, D.W. Holliday & N.S. Jones) *Yorkshire Geological Society Occasional Publication* **7**, 1–12.

Gutteridge P. (1987) Dinantian sedimentation and basement structure of the Derbyshire Dome. *Geological Journal* **22**, 25–41.

Gutteridge P. (1989) Controls on sedimentation in a Brigantian intrashelf basin (Derbyshire). In: *The role of tectonics in Devonian and Carboniferous sedimentation in the British Isles.* (Ed. R.S. Arthurton, P. Gutteridge & S.C. Nolan) *Yorkshire Geological Society Occasional Publication* **6**, 171–187.

Gutteridge P. (1990a) The origin and significance of the distribution of shelly macrofauna in late Dinantian carbonate mud mounds of Derbyshire. *Proceedings of the Yorkshire Geological Society* **48**, 23–32.

Gutteridge P. (1990b) Depositional environments and palaeogeography of the Station Quarry Beds (Brigantian), Derbyshire, England. *Proceedings of the Yorkshire Geological Society* **48**, 189–196.

Gutteridge P. (1991a) Aspects of Dinantian sedimentation in the Edale Basin, North Derbyshire. *Geological Journal* **26**, 33–59.

Gutteridge P. (1991b) Revision of the Monsal Dale/Eyam Limestones boundary (Dinantian) in Derbyshire. *Mercian Geologist* **12**, 71–78.

Gutteridge P. (1995) Late Dinantian (Brigantian) Carbonate Mud-Mounds of the Derbyshire Carbonate Platform. In: *Carbonate mudmounds; their origin and evolution.* (Ed. C.V.L. Monty, D.W.J. Boscence, P.H. Bridges & B.R. Pratt) *Special Publication of the International Association of Sedimentologists* **23**, 289–307.

Gutteridge P. (2003) Landmark of geology in the East Midlands. The reef at High Tor. *Mercian Geologist* **15**, 235–237.

Gutteridge P. (2003) A record of the Brigantian limestone succession in the partly infilled Dale Quarry, Wirksworth. *Mercian Geologist* **15**, 219–224.

Gutteridge P. (2023) Lacustrine and palustrine carbonates in a Brigantian (late Dinantian) intrashelf basin in the Derbyshire carbonate platform. *Geological Journal*, 1–14.

Gutteridge P. & Walkden G.M. (1987) Tectonic influences on Dinantian carbonate sedimentation on the Derbyshire Dome. Field guide associated with conference: *The role of tectonics in Devonian and Carboniferous sedimentation in the British Isles.* Yorkshire Geological Society. 40pp.

Hallsworth C.R. & Chisholm J.I. (2008) Provenance of late Carboniferous sandstones in the Pennine Basin (UK) from combined heavy mineral, garnet geochemistry and palaeocurrent studies. *Sedimentary Geology* **203**, 196–212.

Hallsworth C.R. & Chisholm J.I. (2017) Interplay of mid-Carboniferous sediment sources on the northern margin of the Wales–Brabant Massif. *Proceedings of the Yorkshire Geological Society* **61**, 285–309.

Hampson G.J. (1997) A sequence stratigraphic model for deposition of the Lower Kinderscout Grit delta. *Proceedings of the Yorkshire Geological Society* **51**, 273–296.

Hampson G.J., Elliot T. & Flint S.S. (1996) Critical application of high-resolution sequence stratigraphy to upper Carboniferous fluvio-deltaic strata of the Rough Rock Group (Upper Carboniferous) of northern England. In: *High resolution sequence stratigraphy*. (Ed. J.A. Howell & J.F. Aitken) *Special Publication of the Geological Society of London*, 221–246.

Holdsworth B.K. (1963) Prefluvial autogeosynclinal sedimentation in the Namurian of the southern Central Province. *Nature* **199**, 133–135.

Holdsworth B.K. & Collinson J.D. (1988) Millstone Grit cyclicity revisited. In: *Sedimentation in a Synorogenic Basin Complex* (Ed. B.M. Besly & G. Kelling). Blackie, Glasgow. 132–152.

Hollis C. (2022) Obituary. Gordon Walkden (1944–2022). *Proceedings of the Geologists' Association* **133**, 282–284.

Hunter J. (2016) Recent attempts to reveal a palaeokarst hollow in the station car park at Miller's Dale, Peak District. *Mercian Geologist* **19**, 47–51.

Insalaco E. (1998) The descriptive nomenclature and classification of growth fabrics in fossil scleractinian reefs. *Sedimentary Geology* **118**, 159–186.

Jones C.M. (1980) Deltaic sedimentation in the Roaches Grit and associated sediments (Namurian R_{2b}) in the south-west Pennines. *Proceedings of the Yorkshire Geological Society* **43**, 39–67.

Jones C.M. (2014) Controls on deltaic sedimentation in glacio-eustatic cycles of late Marsdenian (Namurian R_{2b4} to R_{2c1}, Pennsylvanian) age in the UK Central Pennine Basin. *Proceedings of the Yorkshire Geological Society* **60**, 63–84.

Jones C.M. (2022) Climatic influences on Upper Carboniferous (Serpukhovian to Mid-Bashkirian) sedimentary sequences in the UK Pennine and other European basins. *International Journal of Geosciences* **13**, 715–778.

Jones C.M. & McCabe P.J. (1980) Erosion surfaces within giant fluvial cross-beds of the Carboniferous of northern England. *Journal of Sedimentary Petrology* **50**, 613–620.

Jones C.M. & Chisholm J.I. (1997) The Roaches and Ashover Grits: sequence stratigraphic interpretation of a 'turbidite-fronted delta' system. *Geological Journal* **32**, 45–68.

Jones P. & Banks V. (2014) Generating new geo-data. *Geoscientist* **24**(8), 12–17.

Kane I.A., Catterall V., McCaffrey W.D. & Martinsen O.J. (2010) Submarine channel response to intrabasinal tectonics: the influence of lateral tilt. *Bulletin of the American Association of Petroleum Geologists* **94**, 189–219.

Kirby G.A. *et al.* (11 authors) (2000) *The structure and evolution of the Craven Basin and adjacent areas*. British Geological Survey Subsurface Memoir. HMSO.

Lee A.G. (1988) Carboniferous basin configuration of central and northern England modelled using gravity data. In: *Sedimentation in a Synorogenic Basin Complex* (Ed. B.M. Besly & G. Kelling). Blackie, Glasgow. 69–84.

Lees A. & Miller J. (1995) Waulsortian Banks. In: *Carbonate mudmounds; their origin and evolution.* (Ed. C.V.L. Monty, D.W.J. Boscence, P.H. Bridges & B.R. Pratt) *Special Publication of the International Association of Sedimentologists* **23**, 191–271.

Le Yao, Xiangdong Wang, Wei Lin, Yue Li, Kershaw S. & Wenkun Qie (2016) Middle Viséan (Mississippian) coral biostrome in central Guizhou, southwestern China and its palaeoclimatological implications. *Palaeogeography, Palaeoclimatology, Palaeoecology* **448**, 179–194.

Lister, A. (2014) *Mammoths: Ice Age giants.* The Natural History Museum, London. 128pp.

Manifold L., Strother P. del, Gold D.P., Burgess P. & Hollis C. (2021) Unravelling evidence for global climate change in Mississippian carbonate strata from the Derbyshire and North Wales Platforms, UK. *Journal of the Geological Society* **178**, https://doi.org/10.1144/jgs2020-106.

Martinsen O.J. (1990b) Fluvial, inertia-dominated deltaic deposition in the Namurian of northern England. *Sedimentology* **37**, 1099–1113.

Martinsen O.J. (1993) Namurian (late Carboniferous) depositional systems of the Craven-Askrigg area, northern England: implications for sequence stratigraphy. In: *Sequence Stratigraphy and Facies Associations.* (Ed. H.W. Posamentier, C.P. Summerhayes, B.U. Haq & G.P. Allen) *Special Publication of the International Association of Sedimentologists* **18**, 247–281.

Martinsen O.J., Collinson J.D. & Holdsworth B.K. (1995) Millstone Grit cyclicity revisited, ll: sequence stratigraphy and sedimentary responses to changes of relative sea-level. In: *Sedimentary facies analysis.* (Ed. A.G. Plint) *Special Publication of the International Association of Sedimentologists* **22**, 305–327.

McCabe P.J. (1975) Deep distributary channels and giant bedforms in the Upper Carboniferous of the Central Pennines, northern England. *Sedimentology* **24**, 271–290.

McFarlane D.A., Lundberg J., Rentergem G. v., Howlett E. & Stimpson C. (2016) A new radiometric date and assessment of the Last Glacial megafauna of Dream Cave, Derbyshire, UK. *Cave and Karst Science: Transactions of the British Cave Research Association* **43**, 109–116.

Miller J. & Grayson R.F. (1972) Origin and structure of the Lower Viséan 'reef' limestones near Clitheroe, Lancashire. *Proceedings of the Yorkshire Geological Society* **38**, 607–638.

Miller J. & Grayson R.F. (1982) The regional context of Waulsortian facies in northern England. In: *Symposium on the paleoenvironmental setting and distribution of Waulsortian facies* (Ed. K. Bolton, H.R. Lane & D.V. Lemone). El Paso Geological Society/University of Texas, El Paso. 17–33.

Mitchell, M. (2003) (2nd Edn) *A lateral key for the identification of the commoner Lower Carboniferous coral genera.* Westmorland Geological Society and Manchester Geological Association. 13pp.

Morris T.O. (1929) Geology of Stoney Middleton. *The Natural History of the Sheffield District. The Proceedings of the Sorby Scientific Society* **1**, 37–67.

Morton A.C. & Whitham A.G. (2002) The Millstone Grit of northern England: a response to tectonic evolution of a northern sourceland. *Proceedings of the Yorkshire Geological Society* **54**, 47–56.

Myers K.J. & Bristow C.S. (1989) Detailed sedimentology and gamma-ray characteristics of a Namurian deltaic succession II. Gamma-ray logging. In: *Deltas: sites and traps for fossil fuels.* (Ed. M.K.G. Whateley & K.T. Pickering) *Special Publication of the Geological Society* **41**, 81–88.

Neves R. & Downie C. (Eds) (1967) *Geological excursions in the Sheffield region.* University of Sheffield. 163pp.

Newport S.M., Jerrett R.M., Taylor K.G., Hough E. & Worden R.H. (2018) Sedimentology and microfacies of a mud-rich slope succession: in the Carboniferous Bowland Basin, NW England (UK). *Journal of the Geological Society of London* **175**, 247–262.

Nolan L.S.P. *et al.* (7 authors) (2017) Sedimentary context and palaeoecology of *Gigantoproductus* shell beds in the Mississippian Eyam Limestone Formation, Derbyshire carbonate platform, central England. *Proceedings of the Yorkshire Geological Society* **61**, 239–257.

Okolo S.A. (1983) Fluvial distributary channels in the Fletcher Bank Grit (Namurian R_{2b}), at Ramsbottom, Lancashire, England. In: *Modern and ancient fluvial systems.* (Ed. J.D. Collinson & J. Lewin) *Special Publication of the International Association of Sedimentologists* **6**, 421–433.

Orme G.R. (1967) The Carboniferous Limestone of the Stoney Middleton area. In: *Geological Excursions in the Sheffield region* (Ed. R. Neves & C. Downie). University of Sheffield. 23–31.

Parkinson D. (1957) Lower Carboniferous reefs of northern England. *Bulletin of the American Association of Petroleum Geologists* **41**, 511–537.

Pentecost, A. (1999) The origin and development of the travertines and associated thermal waters at Matlock Bath, Derbyshire. *Proceedings of the Geologists Association* **110**, 217–232.

Phillips J. (1836) *Illustrations of the Geology of Yorkshire; Part 2. The Mountain Limestone.* John Murray, London. 252pp.

Price D., Wright W.B., Jones R.C.B., Tonks L.H. & Whitehead T.H. (1963) *Geology of the country around Preston.* Memoir of the Geological Survey, Sheet 75. 140pp.

Purdy E.G. (1974) Reef configurations: cause and effect. In: *Reefs in time and space.* (Ed. L.F. Laporte) *Special Publication of the Society of Economic Paleontologists and Mineralogists* **18**, 9–76.

Ramsbottom W.H.C. (1966) A pictorial diagram of the Namurian rocks of the Pennines. *Transactions of the Leeds Geological Association* 7, 181–184.

Ramsbottom W.H.C. (1972) Transgressions and regressions in the Dinantian: new synthesis of British Dinantian stratigraphy. *Proceedings of the Yorkshire Geological Society* **39**, 567–607.

Ramsbottom W.H.C. (1977) Major cycles of transgression and regression (mesothems) in the Namurian. *Proceedings of the Yorkshire Geological Society* **41**, 261–291.

Ramsbottom W.H.C., Rhys G.H. & Smith E.G. (1962) Boreholes in the Carboniferous rocks of the Ashover district, Derbyshire. *Bulletin of the Geological Survey of Great Britain* **19**, 75–168.

Reading H.G. (1964) A review of the factors affecting the sedimentation of the Millstone Grit (Namurian) in the Central Pennines. In: *Deltaic and Shallow Marine Deposits* (Ed. L.M.J.V. van Straaten). Elsevier, Amsterdam. 340–346.

Riegl B. & Piller W.E. (2000) Reefs and coral carpets in the northern Red Sea as models for organism–environment feedback in coral communities and its reflection in growth fabrics. In: *Carbonate production systems: components and interactions.* (Ed. E. Insalaco, P.W. Skelton & T.J Palmer) *Special Publication of the Geological Society of London* **178**, 71–88.

Riley N.J. (1990) Stratigraphy of the Worston Shale Group, Dinantian, Craven Basin, north-west England. *Proceedings of the Yorkshire Geological Society* **48**, 163–187.

Riley N.J. (1993) Dinantian (Lower Carboniferous) biostratigraphy and chronostratigraphy in the British Isles. *Journal of the Geological Society of London* **150**, 427–446.

Scrutton C.T. (1985) Cnidaria. Rugosa. Heterocorallia. Tabulata. In: *Atlas of invertebrate macrofossils* (Ed. J.W. Murray). The Palaeontological Association and Longman, Harlow. 13–36.

Scrutton C. (1999) Palaeozoic corals: their evolution and palaeoecology. *Geology Today* (September–October), 184–193.

Shirley J. (1959) The Carboniferous Limestone of the Monyash–Wirksworth area, Derbyshire. *Quarterly Journal of the Geological Society of London* **64**, 411–429.

Simpson I.M. & Broadhurst F.M. (1969) A boulder bed at Treak Cliff, north Derbyshire. *Proceedings of the Yorkshire Geological Society* **37**, 141–151.

Smith E.G., Rhys, G.H., Eden R.A., Calver M.A., Duff P.Mc.D., Harrison R.K. & Ramsbottom W.H.C. (1967) *Geology of the country around Chesterfield, Matlock and Mansfield.* Memoir of the Geological Survey, Sheet 112.

Smith K., Smith N.J.P. & Holliday D.W. (1985) The deep structure of Derbyshire. *Geological Journal* **3**, 109–118.

Somerville I.D. & Rodríguez S. (2007) Rugose coral associations from the late Viséan (Carboniferous) of Ireland, Britain and Spain. In: *Fossil Corals and Sponges.* (Ed. B. Hubmann & W.E. Piller) *Proceedings of the 9th International Symposium on Fossil Cnidaria and Poriifera. Symposium. Österreichische Akademie der Wissenschaften Schriftenreihe der Erdwissenschaftlichen Kommissionen* **17**, 329–351.

Sorby H.C. (1859) On the structure and origin of the Millstone Grit in South Yorkshire. *Proceedings of the Yorkshire Geological and Polytechnic Society* **3**, 669–675.

Sorby H.C. (1908) On the application of quantitative methods to the study of the structure and history of rocks. *Quarterly Journal of The Geological Society of London* **64**, 171–233.

Southern S.J., Mountney N.P and Pringle J.K. (2014) The Carboniferous Southern Pennine Basin, UK. *Geology Today* **30**, 71–78.

Stephens J.V., Mitchell G.H. & Edwards W. (1953) *Geology of the country between Bradford and Skipton.* Memoir of the Geological Survey, Sheet 69. 180pp.

Stevenson I.P. & Gaunt G.D. (1971) *The geology of the country around Chapel-en-le-Frith.* Memoir of the Geological Survey. Sheet 99. 444pp.

Tallis J.H. & Switsur V.R. (1983) Forest and moorland in the South Pennine uplands in the mid-Flandrian period. *Journal of Ecology* **71**, 585–600.

Thomas I.A. (2008) Hopton Wood Stone – England's premier decorative stone. In: *England's Heritage in Stone* (Ed. P. Doyle, T.G. Hughes & I.A. Thomas). English Stone Forum. 90–105.

Thomas I.A. (co-ord.) *Delving along the Derwent: A history of 200 quarries and the people who worked them.* DerwentWISE, the Lower Derwent Valley Landscape partnership. 191pp.

Tiddeman R.H. (1901) On the formation of reef knolls. *Geological Magazine* **8**, 20–23.

Trewin N.H. & Holdsworth B.K. (1972) Sedimentation in the Lower Namurian Rocks of the North Staffordshire Basin. *Proceedings of the Yorkshire Geological Society* **39**, 371–408.

Walkden G.M. (1970) *Environmental studies in the Carboniferous Limestone of the Derbyshire Dome.* Unpublished PhD thesis, University of Manchester.

Walkden G.M. (1972) The mineralogy and origin of interbedded clay wayboards in the Lower Carboniferous of the Derbyshire Dome. *Geological Journal* **8**, 143–159.

Walkden G.M. (1974) Palaeokarstic surfaces in the Upper Viséan (Carboniferous) Limestones of the Derbyshire Block. *Journal of Sedimentary Petrology* **44**, 1232–1247.

Walkden G.M. (1977) Volcanic and erosive events on the Upper Viséan platform, north Derbyshire. *Proceedings of the Yorkshire Geological Society* **41**, 347–366.

Walkden G.M. (1987) Sedimentary and diagenetic styles in late Dinantian carbonates of Britain. In: *European Dinantian Environments* (Ed. J. Miller, A.E Adams & V.P. Wright). John Wiley & Sons, Chichester. 131–155.

Walkden G.M. & Oakman C. (1982) *Field Guide to the Lower Carboniferous of the south-east margin of the Derbyshire Block: Wirksworth to Grangemill.* Department of Geology and Mineralogy, University of Aberdeen. 12pp.

Walker R.G. (1966) Shale Grit and Grindslow Shales: transition from turbidite to shallow water sediments in the Upper Carboniferous of northern England. *Journal of Sedimentary Petrology* **36**, 90–114.

Walsh P.T, Banks V.J., Jones P.F., Pound M.J. & Riding J.B. (2018) A reassessment of the Brassington Formation (Miocene) of Derbyshire, UK and a review of related hypogene karst suffusion processes. *Journal of the Geological Society of London* **175**, 443–463.

Waltham T. (2018) *The Yorkshire Dales: Landscape and Geology.* Crowood Press, Marlborough. 224pp.

Waltham T. (2021) *The Peak District: Landscape and Geology.* Crowood Press, Marlborough. 160pp.

Waltham T. (2022) Litton Mill lava exposure. *Mercian Geologist* **20**, 211.

Waters C.N. (2000) *Geology of the Bradford district.* Sheet description of the British Geological Survey, 1:50 000 Series, Sheet 69 (England and Wales).

Waters C.N., Barclay, W.J, Wright V.P., Cossey P.J. & Bevins R.E. (2003) Carboniferous and Permian igneous rocks of Central England and the Welsh Borderland. In: *Carboniferous and Permian igneous rocks of Great Britain north of the Variscan Front.* (Ed. S.C. Loughlin, D. Millward, C.N. Waters & I.T. Williamson) *Geological Conservation Review Series* **27**, 280–316.

Waters C.N., Chisholm J.I., Benfield A.C. & O'Beirne A.M. (2008) Regional evolution of a fluviodeltaic cyclic succession in the Marsdenian (late Namurian Stage, Pennsylvanian) of the Central Pennine Basin, UK. *Proceedings of the Yorkshire Geological Society* **57**, 1–28.

Waters C.N., Jones N.S., Collinson J.D. & Besly B.M. (2011) Peak District and north Staffordshire. In: *A revised correlation of Carboniferous rocks in the British Isles. Geological Society Special Report* **26**, 66–73.

Waters C.N., Jones N.S., Collinson J.D. & Cleal C.J. (2011) Craven Basin and southern Pennines. In: *A revised correlation of Carboniferous rocks in the British Isles. Geological Society of London Special Report* **26**, 74–81.

Waters C.N. & Condon D.J. (2012) Nature and timing of Late Mississippian to Mid-Pennsylvanian glacio-eustatic sea-level changes of the Pennine Basin, UK. *Journal of the Geological Society, London* **169**, 37–51.

Waters C.N. *et al.* (6 authors) (2017) Mississippian reef development in the Cracoe Limestone Formation of the southern Askrigg Block, North Yorkshire, UK. *Proceedings of the Yorkshire Geological Society* **61**, 179–196.

Waters C.N., Cózar P., Somerville I.D. & Haslam R.B. (2017) Lithostratigraphy and biostratigraphy of the Lower Carboniferous (Mississippian) carbonates of the southern Askrigg Block, North Yorkshire, UK. *Geological Magazine* **154**, 305–333.

Waters C.N., Vane C.H., Kemp S.J., Haslam H.B., Hough E. & Moss-Hayes V.L. (2018) Lithological and chemostratigraphic discrimination of facies within the Bowland Shale Formation within the Craven and Edale basins, UK. *Petroleum Geoscience* **26**, 325–345.

Wells M.R., Allison P.T., Piggott M.D., Pain C.C., Hampson G.J. & De Oliviera C.R.E. (2005) Large sea, small tides: the Late Carboniferous seaway of NW Europe. *Journal of the Geological Society of London* **162**, 417–420.

Wignall P.B. & Maynard J.R. (1996) High resolution sequence stratigraphy in the early Marsdenian (Namurian, Carboniferous) of the Central Pennines and adjacent areas. *Proceedings of the Yorkshire Geological Society* **51**, 127–140.

Wolfenden E.B. (1958) Palaeoecology of the Carboniferous reef complex and shelf limestones of north-west Derbyshire England. *Bulletin of the Geological Society of America* **69**, 871–898.

Wray D.A., Stephens J.V., Edwards W.N., & Bromehead C.E.N. (1930) *The geology of the country around Huddersfield and Halifax*. Memoir of the Geological Survey, Sheet 77. 221pp.

Wright V.P. & Vanstone, S.D. (2001) Onset of Late Palaeozoic glacio-eustasy and the evolving climates of low latitude areas: a synthesis of current understanding. *Journal of the Geological Society of London* **158**, 579–82.

Wright W.B., Sherlock R.L., Wray D.A., Lloyd W. & Tonks L.H. (1927) *The geology of the Rossendale anticline*. Memoir of the Geological Survey, Sheet 76. 182pp.

Zeigler P.A. (1987) *Evolution of the Arctic–North Atlantic and the western Tethys*. Memoir **43**, American Association of Petroleum Geologists, Tulsa.

Index

Occurrences of terms shown in figures or in their captions are shown in italics.